Asheville-Buncombe
Technical Community College
Learning Resources Center
340 Victoria Road
Asheville, NC 28801

DISCARDED

JUL 2 5 2025

**Other Related Fire Service Books
Available from FIRE ENGINEERING**

The Fire Chief's Handbook, Casey
Fire Fighting Principles & Practices, Clark
High Rise Fire & Life Safety, O'Hagan
Fire Service Hydraulics, Casey
Fire Service Hydraulics, A Study Guide, Questions & Answers, Sylvia
Flammable Hazardous Materials, Meidl
Hazardous Materials, Isman & Carlson
Fire Fighting Apparatus & Procedures, Erven
Introduction to Fire Prevention, Robertson
Fire Suppression & Detection Systems, Bryan
Fire Dept. Management: Scope & Method, Gratz
Arson Investigation, Carter
Introduction to Fire Science, Bush & McLaughlin
Fire Dept. Operations with Modern Elevators, McRae
Strategic Concepts in Fire Fighting, McAniff
Investigating the Fireground, Phillipps & McFadden
Modern Suburban Fire Fighting, Sylvia
Fire Service Pump Operator's Handbook, Isman
Managing Fire Services, Bryan & Picard
Fire Dept. Operations in Garden Apartments, Gaines
Hazardous Materials Emergencies, Cashman
Emergency Rescue, Ervin
Practical Approaches to Firefighting
Winning the Fire Service Leadership Game, Caulfield
Vehicle Rescue, Grant
The Common Sense Approach to Hazardous Materials, Fire

**Introduction to
Fire Apparatus
and Equipment**

About the Author

Gene Mahoney is an Associate Professor and Coordinator of Fire Science at Rio Hondo Community College in Whittier, CA. He also was instrumental in the development of a fire science program at Los Angeles Harbor College and a fire administration program at California State College, Long Beach, and taught at both institutions.

Mr. Mahoney has 28 years of active duty in the fire service where he served in many different capacities. He was fire chief of Arcadia, CA, and Garden Grove, CA. Prior to these positions, he served with the Los Angeles Fire Department where he rose to the rank of battalion chief. While at the LAFD, he was also head of the Company Training Unit of the Training Section, and was a task force commander during the famous Watts riots.

In addition to careers in firefighting and fire service training and education, Gene Mahoney is a prolific author. He has written a text, *Fire Department Hydraulics*, as well the first edition of this book. Mr. Mahoney is also the author of many fire training manuals, slide presentations, and films, including ones for command officers, junior officers, supervision and management, firefighter promotion examinations, etc. He has the rare distinction of having had a novel published, entitled *An Anatomy of an Arsonist*.

Gene Mahoney holds both Bachelor of Science and Master of Science degrees from the University of Southern California. He served 22 years in the Navy Reserve as a pilot and retired with the rank of Lieutenant Commander.

> Asheville-Buncombe
> Technical Community College
> Learning Resources Center
> 340 Victoria Road
> Asheville, NC 28801

Introduction to Fire Apparatus and Equipment

SECOND EDITION

Gene Mahoney

a publication of Technical Publishing,
a company of the Dun & Bradstreet Corporation New York

Copyright © 1986, Fire Engineering, 875 Third Ave., New York, NY 10022.
All rights reserved. No part of this book may be reprinted by any process whatsoever without permission in writing from the publisher.

Printed in the United States of America

Library of Congress Cataloging-in-Publication Data

Mahoney, Gene, 1923–
  Introduction to fire apparatus and equipment.

  Includes index.
  1. Fire-engines. 2. Fire extinction—Equipment and supplies. I. Title.
TH9371.M33   1986   628.9'25   85-82103
ISBN 9-912212-12-8

2 3 4 5 6 05 04 03 02 01

To my wife Ethel,
whose patience and understanding
contributed so much

Contents

Preface xi

Acknowledgments xiii

1 **The Engine**
Gasoline Engines 2   Diesel Engines 8   Engine Construction 11   Engine Performance Standards 31   Review Questions 38

2 **Engine Systems**
Fuel Systems 43   Ignition Systems 58   Lubrication Systems 73   Cooling Systems 75   Electrical Systems 81   Review Questions 90

3 **The Chassis and Component Parts**
The Chassis 95   The Power Train 98   Braking Systems 123   Steering 135   Review Questions 141

4 **Engine and Systems Troubleshooting**
Diesel Engines 145   Carburetion Maintenance 148   Carburetor Problems 149   Clutches 151   Drive Train 152   Manual Steering 153   Hydraulic Power Steering 154   Air Brakes 157   Automatic Transmission (Allison) 160   Review Questions 162

5 **Apparatus Testing**
Pumper Testing 168   Aerial Ladder and Elevating Platform Testing 195   Safety Testing and Evaluation 196   Review Questions 199

6 Driving Procedures
Driver Characteristics 205   Driving Techniques 210
Driving Conditions 236   Emergency Response 246
Review Questions 252

7 Fire Pumps
Theory of Positive Displacement Pumps 258   Types of
Positive Displacement Pumps 262   Centrifugal
Pumps 266   Main Pumps 292   Review Questions 312

8 Pump Accessories
The Pump Operator's Panel 318   Priming Devices 320
Pressure Control Devices 327   Pumping Devices 344
Monitoring Devices 351   Automation 358   Review
Questions 365

9 Pumper Operations
Tank Operations 370   Hydrant Operations 371   Drafting
Operations 386   Relay Operations 396   Pumping
Techniques 409   Supplying Master Streams 412
Review Questions 416

10 Aerial Ladder Operations
Types of Aerial Ladder Apparatus 422   The Duties and
Responsibilities of an Aerial Ladder Operator 424   Aerial
Ladder Construction 425   Operating Controls and
Systems 428   Operational Procedures 449   Review
Questions 462

11 Elevated Platforms
Types of Elevated Platforms 469   Elevated Platform
Features 476   The Mack Aerialscope 482   Review
Questions 496

Index 499

Preface

The objective of this book is to provide fire department apparatus operators, and those desiring to become apparatus operators, with enough basic information about fire department apparatus that they can perform their duties effectively. In addition to discussing operational techniques, the book includes information on the construction and operational theory of the essential parts of fire department apparatus. While most fire department apparatus operators are capable of driving an apparatus to an emergency and operating the apparatus effectively under routine conditions, there are many who find it difficult to maintain this efficiency when things go wrong. Only the operator who thoroughly understands what goes on inside the apparatus is capable of making a chaotic situation appear to bystanders to be a routine operation. It is not the intent of the author to provide detailed technical information regarding the various components of a fire apparatus; the goal is to provide a sufficient amount of technical information so a reader may acquire a basic understanding of what goes on inside the apparatus.

Questions have been included at the end of each chapter. These questions were designed to give readers an opportunity to test and evaluate their understanding of the material presented. A reader who has mastered the chapter should be able to answer all the questions.

Some who read this book are technically qualified to criticize certain parts or perhaps all of it and undoubtedly will do so. There are others who will question the wisdom of including certain information in a book for fire department apparatus operators. The author welcomes these criticisms and encourages suggestions, fully realizing that those who submit them are also interested in improving and upgrading the performance of apparatus operators.

Systems and products of various manufacturers are used throughout this text for illustrative purposes. It should not be inferred that the use of these systems and products is endorsed by either the author or the publisher. The illustrations have been used solely to explain the construction and principles of operation of such systems and products and for no other reason.

The text includes a number of recommendations and guidelines regarding the driving and operation of fire department apparatus. Readers should keep in mind that the procedures outlined here are not carved in stone; there is normally more than one way to achieve an objective. Whenever guidelines in the text conflict with departmental procedures, manufacturers' recommendations, or the suggestions of such organizations as the National Safety Council, apparatus operators should adhere to the procedures of these organizations.

Acknowledgments

I was in the fire service for nearly thirty years. To a rookie this seems like a long time. Those who served with me know how quickly it passes. It seems like only yesterday that I was a neophyte fireman, thirsty for knowledge and eager to learn. Unfortunately, I had to dig into many books, gathering a slice here and a slice there, in order to put together the information I needed to compete on my first promotional examination—that of an apparatus operator. How nice it would be, I thought at the time, if all the required information regarding apparatus were gathered in one book. Little did I know that I'd have to wait over thirty years to see such a book in print, and then only because I put it together myself.

Of course, a textbook of this magnitude could not possibly be completed without a tremendous amount of help from many people and many organizations. I'm very appreciative of those who so willingly contributed. My thanks are extended to the following organizations:

    AC Spark Plugs, Division of General Motors
    Aerial Testing Corporation
    American Fire Pump Company
    Bendix Corporation
    Calavar Corporation
    California State Fire Marshal
    Champion Spark Plug Company
    Colt Industries, Holley Carburetor Division
    Dana Corporation, Spicer Clutch Division
    W. S. Darley & Company
    Detroit Diesel Allison, Division of General Motors
    Eaton Corporation, Axle Division
    Emergency One, Incorporated
    Fireman's Inspection & Testing
    Fire Research Corporation
    FMC Corporation, Fire Apparatus Operation
    Fram Corporation, Autolite Spark Plugs
    FWD Corporation
    Grumman Corporation
    Hale Fire Pump Company
    Insurance Services Office
    International Association of Fire Chiefs
    Los Angeles County Fire Department
    Mack Trucks, Inc.
    Muskegon Piston Ring Company
    National Fire Hose Corporation
    National Fire Protection Association
    Nolan Company, Duplex Truck Division
    Peter Pirsch & Sons, Company
    Pierce Manufacturing Incorporated
    Riverside California Fire Department
    Rockwell International
    Seagrave Fire Apparatus, Inc.
    Sealed Power Corporation
    Sunnen Products Company
    TRW
    Underwriters Laboratories, Inc.

Acknowledgments

Ward LaFrance Truck Corporation
Waterous Company

It goes without saying, however, that it is the people within an organization whose assistance is so vitally needed in gathering the information required for inclusion in a text. In addition to those who contributed so much to the original edition, my appreciation is extended to the following individuals whose willing contributions proved so valuable in this revision: Mr. Dennis J. Berry, National Fire Protection Association; Mr. John T. Manion, Hale Fire Pumps; Mr. John W. Stone, Jr., Allyn and Bacon; Engineer Jose Tovar, Santa Fe Springs Fire Department; and Mr. Michael P. Waldoch, Waterous Company.

I extend particular thanks to the following people, who graciously took time to review the revised material and whose constructive criticism contributed so much to the final product: Dick Sylvia, who did the technical editing, and Rose Jacobowitz, who edited the copy and designed the text and cover.

1

The Engine

Apparatus operators cannot expect to become completely competent in the operation of their apparatus until they are familiar with how it functions. The mere fact that they have been trained to drive the apparatus and to follow the proper sequence of operations in the supplying of hose lines does not mean that they are efficient operators. They must know the operational principles of every part of their apparatus, and they must know what happens inside when they move levers and turn valves.

The engine is the heart of the apparatus. It provides the operators with the power to move the heavy piece of equipment safely through the streets and the power to operate the pump, the aerial ladder, or the aerial platform once they have arrived at their destination. Failure of an engine could result in failure to extinguish a fire successfully with a minimum of loss of life and property.

The information given in this chapter, and in Chapters 2 and

3, is not intended to approach that which must be known by a skilled mechanic. The objective is to provide apparatus operators with a sufficient amount of information so that they understand engine terminology and what goes on inside their apparatus.

Engines used in fire apparatus are of the internal combustion, reciprocating type. An internal combustion engine burns fuel within the cylinders, converting the expanding force of the burning fuel into a rotary mechanical force which is used to propel the apparatus. A reciprocating type of engine is one in which the pistons move up and down, causing a shaft to rotate. The rotating motion of the shaft is transmitted to the wheels of the apparatus through a power train.

Fire apparatus use two general types of internal combustion engines. One type mixes air with a fuel before the mixture enters the combustion chamber. The combustible mixture is then ignited by a spark. Gasoline, liquefied petroleum gas, and natural gas have been successfully used in this type of engine, which will be referred to here as a gasoline engine.

In the other type of engine, the fuel is injected into the combustion chamber, where it is mixed with air. The mixture is ignited by the heat of compression. This type of engine is referred to as a diesel engine.

GASOLINE ENGINES

Gasoline engines used in fire apparatus generally have six or more cylinders. Since all cylinders operate in the same manner, it is best to consider the operation of a single cylinder in order to understand the principles involved.

An internal combustion engine uses the expanding force of a burning air-fuel mixture to create a mechanical force. The entire operation required to create the mechanical force is called the power cycle. A number of events must take place in a gasoline engine in order to complete this cycle. The events occur within a given sequence, repeating themselves on every power cycle. The events in proper sequence are as follows:

The Engine

1. The cylinder is filled with a burnable air-fuel mixture.
2. The mixture is compressed into a smaller space.
3. The mixture is ignited.
4. The burning mixture expands, producing power.
5. The burned gases are removed from the cylinder.

This sequence takes place in two strokes in a two-stroke cycle engine. It takes four in a four-stroke-cycle engine. Most fire apparatus engines are of the four-stroke-cycle type. A *stroke* is the movement of the piston from top dead center (TDC) to bottom dead center (BDC) or from BDC to TDC. TDC is the upper limit of piston movement; BDC is the bottom limit (see Figure 1-1).

The air-fuel mixture enters the cylinder from the intake manifold through a port. The movement of the mixture through the

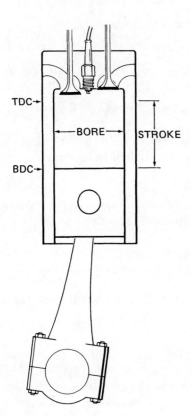

Figure 1-1 A Cylinder of a Four-Stroke Cycle Engine

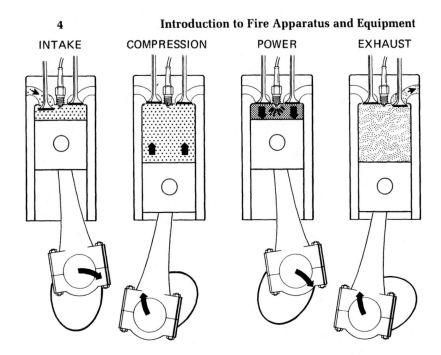

Figure 1-2 The Four Strokes of a Four-Stroke Cycle Engine

port is controlled by an intake valve. The opening and closing of the intake valve is controlled by a cam, which is an integral part of the camshaft.

The burned gases from the mixture are discharged from the cylinder through a port into the exhaust manifold. As with the intake of fuel, control of the discharge of the exhaust gases is through a valve and cam.

The four strokes of the four-stroke-cycle engine are intake, compression, power, and exhaust. For simplicity in explaining the four strokes, it will be considered that the intake and exhaust valves start to open or close at TDC or BDC, although this does not happen in actual practice.

Figure 1-2 illustrates the four strokes.

1. The intake valve is open on the intake stroke. A partial vacuum is created within the cylinder as the piston moves downward. The difference in pressure between the atmosphere and the air within the cylinder causes a movement of air through the carburetor, where a charge of gasoline

The Engine

vapor is picked up. The air-fuel mixture then passes through the intake manifold into the cylinder.

2. Both the intake and exhaust valves are closed on the compression stroke. The upward movement of the piston compresses the mixture, resulting in an increased pressure and temperature within the cylinder.

3. The mixture is ignited by a spark from the spark plug at TDC of the power stroke. At this time both valves are closed. The rapid burning of the mixture results in the creation of a high pressure within the cylinder. The increasing pressure forces the piston downward.

4. The exhaust valve is open and the intake valve is closed on the exhaust stroke. The upward movement of the piston forces the burned gases from the cylinder.

The valves do not start to open and close exactly at TDC and BDC during actual operation. Such a practice would result in poor engine performance. A different timing of the valves is designed into the engine to provide for maximum efficiency and smoother operation.

As mentioned earlier, valves are opened and closed by cam action. The design of the cam does not provide for instantaneous valve opening and closing. As the camshaft turns, the valve (or valve lifter) rides on the lob, starting the valve opening. The valve is fully open at the top of the lob (see Figure 1-3) and starts closing as the valve (or lifter) starts down the opposite side.

Proper Combustion

Efficient power conversion of fuel to piston movement in a gasoline engine occurs when:

1. The engine has proper compression.

2. The ignition system delivers the required voltage to effect an efficient arcing in the gap space of the spark plug during the compression stroke at precisely the right time.

3. The fuel is of the proper octane rating and is mixed with air in the proper proportion.

Normal combustion occurs when all of these factors are

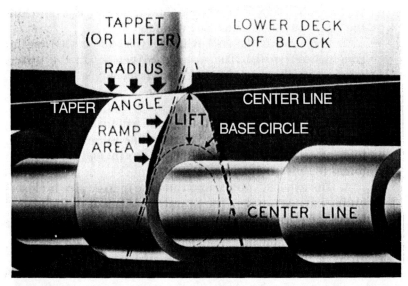

Figure 1-3 The Action of a Valve Cam. *Courtesy of Sealed Power Corporation*

SPARK OCCURS ... COMBUSTION BEGINS ... CONTINUES RAPIDLY ... AND IS COMPLETED

Power developed in an internal combustion engine results from the expanding gases during the burning of the air-fuel charge. If timing is proper and the anti-knock (octane rating) quality of the fuel meets the engine requirements, the burning process should progress in a steady, even flame front.

Figure 1-4 Normal Combustion. *Courtesy of Champion Spark Plug Company*

right. In normal combustion the expanding gases burn evenly across the piston surface, forcing it downward (see Figure 1-4).

Abnormal combustion or misfiring can occur if one or more of these conditions is defective. An example of abnormal combustion occurs when two flame fronts collide in the combustion chamber. This can be caused by either *detonation* or *preignition*,

The Engine

two phenomena that are quite different in origin but very similar in effect. One can cause the other, and either is capable of causing severe damage within the combustion chamber.

Detonation is the ignition of a second flame after the timed spark has taken place (see Figure 1-5). It is caused by the use of the wrong octane fuel, a portion of which starts to burn spontaneously from increased heat and pressure. It can also be caused by such factors as excessive temperature, advanced timing, engine lugging, increased compression rating from deposit build-up, and induction leaks.

Preignition is the ignition of a fuel charge before the spark ignites. It is caused by hot spots within the cylinder, such as overheated plugs, glowing deposits, improperly seated valves, or rough metal edges within the chamber (see Figure 1-6).

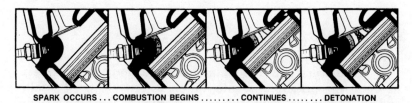

SPARK OCCURS ... COMBUSTION BEGINS .... CONTINUES .... DETONATION

Detonation occurs when the anti-knock quality of the fuel does not meet the engine requirements. A portion of the fuel charge begins to burn spontaneously from increased heat and pressures shortly after ignition. The two flame fronts meet and the resulting "explosion" applies extreme hammering pressures upon the piston and other vital engine parts. The relationship between ignition timing and piston position when burning of the mixture is completed, has been changed by the reduced burning time resulting from two flame fronts. Additional heat is generated along with the abnormal stresses.

Figure 1-5   Detonation. *Courtesy of Champion Spark Plug Company*

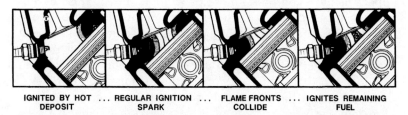

IGNITED BY HOT ... REGULAR IGNITION ... FLAME FRONTS ... IGNITES REMAINING
DEPOSIT   SPARK   COLLIDE   FUEL

Preignition is just what the term implies—ignition of the air-fuel charge prior to the timed spark. Any hot spot within the combustion chamber, such as glowing carbon deposits, rough metallic edges, improperly seated valves, or overheated spark plugs, can be capable of initiating combustion.

Figure 1-6   Preignition. *Courtesy of Champion Spark Plug Company*

DIESEL ENGINES

The diesel engine is fast becoming the primary source of power for fire apparatus. Some of the advantages claimed by proponents of diesel engines are the following:

1. Quicker starting
2. Better dependability
3. Fewer maintenance problems
4. Less "down" time
5. More economy of operation
6. No radio interference
7. Longer life
8. Lower operating temperatures

The Operating Principle

The diesel engine is an internal combustion engine; however, its operating principle is quite different from that of a gasoline engine. The air-fuel mixture is taken into the combustion chamber and compressed in a gasoline engine. Air alone is compressed during the compression stroke in a diesel engine. Diesel fuel is injected or sprayed into the cylinder near the end of the compression stroke.

The compression ratios of diesel engines are much higher than those of gasoline engines. While gasoline engines used in fire apparatus have compression ratios of about 8 to 1, those in diesel engines approach 21 to 1. These high compression ratios provide pressures of approximately 500 pounds per square inch at the end of the compression stroke, raising the temperature of the compressed air to approximately 1000° F. This temperature is sufficient to ignite the diesel fuel without any additional ignition source, thereby eliminating completely the ignition system used in gasoline engines.

A wide variety of fuels can be burned in diesel engines, varying from those approaching volatile jet fuels to those having the characteristics of heavier furnace oil. The lower grade fuels develop more power than the higher grade fuels but also contain more impurities, which contribute to the formation of harmful engine deposits and corrosion. The quality of fuel used for high-speed diesel engine operation is a very important factor in obtaining satisfactory engine performance, long engine life, and acceptable exhaust. Although they develop slightly less power, only high-grade fuels should be used in fire apparatus.

The combustion of the diesel fuel can be controlled by the speed with which the fuel is introduced into the cylinder. In addition, the problem in gasoline engines of equal distribution of mixture to each cylinder is eliminated. The ability to control the speed with which fuel is introduced into the cylinder provides slower burning of the fuel, thereby resulting in an even build-up of pressure.

Both two-stroke cycle and four-stroke cycle diesel engines are used in fire apparatus. The four strokes of the four-stroke cycle diesel engine are similar to those of a gasoline engine. The intake valve is open on the intake stroke. As the piston moves down, it pulls air into the cylinder. Both valves are closed on the compression stroke. The upward movement of the piston compresses the air. Instead of a spark occurring at the end of the compression stroke, as in the gasoline engine, diesel fuel is sprayed or injected into the cylinder. As the fuel burns it creates a high pressure, forcing the piston down on the power stroke. The exhaust valve opens at the end of the power stroke, causing the burned gases to flow from the cylinder as the piston moves upward on the exhaust stroke.

The operation of the two-stroke cycle engine is quite different. There are two revolutions of the crankshaft for every power cycle in a four-stroke cycle engine, while there is a power cycle on each revolution of the crankshaft with the two-stroke cycle engine.

In the two-stroke cycle engine, intake and exhaust take place during part of the compression and power strokes, respectively, as shown in Figure 1-7. A blower is provided to force air into the cylinders for expelling the exhaust gases and to supply the cylinders with fresh air for combustion. The cylinder wall contains a row of ports which are above the piston when it is at the bottom of its stroke. These ports admit the air from the blower into the cylinder as soon as the piston uncovers the ports, as shown in Figure 1-7 (scavenging).

The undirectional flow of air toward the exhaust valves pro-

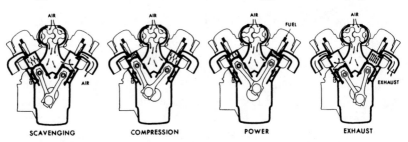

Figure 1-7 The Two-Stroke Cycle Diesel Engine. *Courtesy of Detroit Diesel Allison, Division of General Motors Corporation*

duces a scavenging effect, leaving the cylinders full of clean air when the piston again covers the inlet ports.

As the piston continues on the upward stroke, the exhaust valves close and the charge of fresh air is subjected to compression, as shown in Figure 1-7 (compression).

Shortly before the piston reaches its highest position, the required amount of fuel is sprayed into the combustion chamber by the fuel injector (Figure 1-7, power). The intense heat generated during the high compression of the air ignites the fine fuel spray immediately. The combustion continues until the injected fuel has been burned.

The resulting pressure forces the piston downward on its power stroke. The exhaust valves are again opened when the piston is about halfway down, allowing the burned gases to escape into the exhaust manifold (Figure 1-7, exhaust). Shortly thereafter, the downward-moving piston uncovers the inlet ports and the cylinder is again swept with clean scavenging air. This entire combustion cycle is completed in each cylinder for each revolution of the crankshaft, or, in other words, in two strokes; hence, it is a two-stroke cycle engine.

Fuel Injection

Because of the injection of the diesel fuel into the cylinder and the subsequent burning as a result of the temperatures that have been developed, there is no need for an ignition system. Forcing the fuel directly into the cylinder also eliminates the need for a carburetor. However, the lack of need for these two systems does not eliminate the problems inherent in them. It is still important that the fuel be introduced into the cylinder at exactly the right second and that it be properly vaporized. These functions are the responsibility of the fuel injection system.

ENGINE CONSTRUCTION

The construction of a diesel engine is very similar to that of a gasoline engine. It uses similar pistons, valves, crankshafts, and camshafts. Perhaps the one difference is that diesel engines are generally heavier in structure, in order to withstand the higher pressures resulting from the higher compression ratios. Figure 1-8 is a cutaway view of a diesel engine.

Figure 1-8 A Three-Quarter Cutaway View of a Diesel Engine. *Courtesy of Detroit Diesel Allison, Division of General Motors Corporation*

The size of an engine is referred to by its cubic inch displacement, which is determined by the bore, the stroke, and the number of cylinders. The *bore* is the diameter of the cylinder (see Figure 1-1). The *stroke* is the distance that the piston travels from TDC to BDC.

The displacement of each cylinder is found by multiplying the area of the cylinder bore by the stroke. The engine displacement is determined by multiplying the displacement of one cylinder by the number of cylinders in the engine. For example, the engine shown in Figure 1-8 has a 4.25-inch bore, a 5-inch stroke, and eight cylinders. The area of a cylinder bore is determined by $.7854D^2$ or $(.7854)(4.25)(4.25) = 14.19$ square inches. Multiplying this by the stroke gives 70.95 [(14.19)(5)], which is the piston displacement of one cylinder. The total displacement is determined by multiplying this displacement by the number of cylinders: $(70.95)(8) = 567.6$, or 568 cubic inches.

Cylinders

Cylinders are round holes drilled into the block to receive the pistons, and are also referred to as "bores" or "barrels." It is important that the cylinders be round and true, with dimensions held to a fraction of a thousandth of an inch.

Cylinder walls will experience wear over a long operating period regardless of the material of which they are made. The wear is caused by such factors as pressure of the rings against the walls, kind of lubrication received, type of fuel used, and temperature of operation. The wear takes place primarily in the ring travel area, the area covering the distance from the upper movement of the top ring to the lower movement of the bottom ring.

Some engines use liners or sleeves within the cylinders. The sleeves are inserted between the piston and the cylinder wall. Most sleeves are removable and may be replaced when wear becomes excessive.

Pistons

In simple terms, a *piston* is a long cylindrical forging or casting that is closed at the top and open at the bottom. It is

The Engine

connected at a central point to a wrist pin, which transmits the thrust power of the burning gases to the connecting rod and from there to the crankshaft.

The top surface of the piston is called the head. The *head* is that portion of the piston which receives the explosive force of the burning gases. A head may be flat, concave, convex, or have one of a great variety of other shapes. The design of the head promotes turbulence and helps control the combustion of the expanding gases.

The bottom of the piston is called the skirt. The *skirt* is actually the bearing surface of the piston. It supports the wrist pin bosses. The bosses are projections on the inside of the piston which hold and support the piston pin; they are cast or forged to be integral with the piston and help to strengthen it.

In many cases a piston is slightly tapered, with the top portion smaller than the skirt. This variation in size allows for differences of expansion in various parts of the piston. The top of the piston runs much hotter than the bottom and consequently expands more. This is particularly true of that portion above the top ring.

The piston appears to be a fairly simple part of the engine; it has, however, probably been subjected to more research than any other part. During normal operation a piston performs three functions:

   1. It converts heat energy into mechanical energy.
   2. It transmits the lateral thrust of the connecting rod to the cylinder walls.
   3. It carries off some of the excessive heat generated in the combustion chamber.

The piston undergoes tremendous mechanical and thermal stresses in performing these functions. A pressure of several tons is suddenly applied to the piston head during each power stroke. This occurs ninety or more times per second whenever the engine is running. Furthermore, the temperature within the combustion chamber climbs to approximately 5500° F. The effect of this rapid temperature increase is transmitted to the piston head.

Pistons are made of cast iron, semi-steel, or aluminum alloy. The particular material used and the shape of the piston depend upon the engine design. The piston must be capable of running with a minimum amount of clearance and must permit the least amount of oil to pass into the combustion chamber.

Piston clearance is the distance between the piston and the cylinder wall. Clearance provides for piston expansion and allows space for a thin film of oil. Too small a clearance results in excessive friction; too much clearance results in piston slap or knocking. To reduce friction the clearance fills with oil during operation, allowing the piston and the rings to move on a film of oil.

Piston slap is caused by tilting of the piston as it moves from one side of the cylinder to the other while it travels downward on the power stroke. This side-to-side movement occurs within all cylinders; however, knocking only occurs when the clearance is excessive or, sometimes, when there is inadequate lubrication between the piston and the cylinder wall. The movement of the piston from one side of the cylinder to the other comes from forces acting on the piston because of the angularity of the connecting rod.

Pistons must be carefully made and installed with the utmost care. All pistons within an engine must be of the same weight; otherwise, the unbalanced condition will cause noticeable vibration. They must be strong enough to withstand the tremendous pressures created during the explosive stage but light enough to reduce to a minimum the inertia caused by the constant reversal in the piston's direction. Inertia is the tendency of the piston to keep traveling in the original direction in which it was moving when the direction is reversed.

The material with which pistons are made has a direct influence on the amount of clearance required, the inertia created, and the ability of the piston to carry off the heat. Cast iron pistons have a rate of expansion the same as that of the cylinder and, therefore, can be fitted with a minimum of clearance. Although their expansion rate is the same as that of the cylinder walls, they do expand more than the walls due to their higher operating temperatures. Because of their weight, however, they will create more inertia than a similar size piston made of aluminum alloy.

Aluminum alloy pistons have less weight and, therefore, have the advantage of creating less inertia. They are also better heat conductors than iron. The disadvantage of these pistons is that they usually require greater clearances because of their higher expansion rates.

Piston Rings

Two important things must be done as the pistons move up and down in the cylinders. The tremendous pressures that are formed during the power stroke must be prevented from passing between the cylinder walls and the pistons and then on into the crankcase. Also, an adequate film of oil must be maintained between the pistons and the cylinder walls but at the same time prevented from entering the combustion chamber. Any passage of pressure of either the unburned air-fuel mixture or the burned gases from the combustion chamber to the crankcase is referred to as *blow-by*. Any passage of oil from the crankcase into the combustion chamber is referred to as *oil-pumping*. Both blow-by and oil-pumping are controlled by *piston rings*, which also help keep the piston cool by transmitting heat from the pistons to the cylinder walls.

Piston rings are fitted into the piston in grooves between the piston pin and the piston head or between the piston pin and the bottom of the piston (see Figure 1-9). There is some side clearance between the ring and groove to allow for expansion. The rings are classified in two general types: compression rings and oil-control rings. Most engines for fire apparatus have at least three compression rings and two oil-control rings (see Figure 1-9). For maximum effectiveness against blow-by, the ends of the rings are staggered around the piston during installation. The compression rings are fitted into grooves near the head of the piston. In some engines the oil-control rings are placed in grooves below the compression rings but above the piston pin; in others they are placed between the piston pin and the bottom of the piston.

Compression rings actually perform a dual function. Not only do they prevent blow-by; they also assist in oil control by scraping excess oil from the cylinder walls. Several compression rings are used, because a single ring cannot do the job effectively.

Oil on the pistons and cylinder walls is essential; however, much more oil is generally provided than is needed for lubrica-

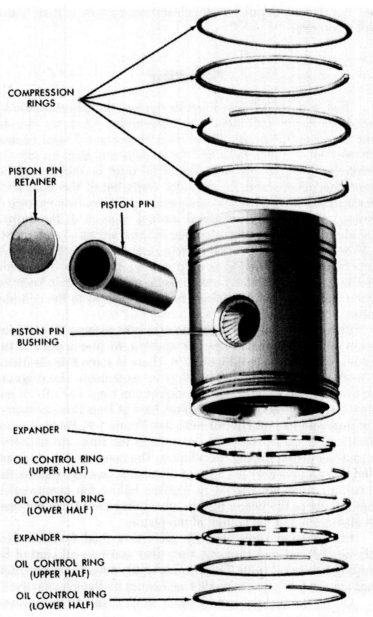

Figure 1-9 Typical Piston, Piston Rings, Pin, and Relative Location of Parts. *Courtesy of Detroit Diesel Allison, Division of General Motors Corporation*

The Engine

tion. This oil is supplied by several methods: by the fog that arises from the crankcase when the oil gets hot; by the connecting rods splashing oil from the oil troughs onto the cylinder walls; and, in full-force feed systems, directly from the oil pump through the piston pins.

The oil itself serves several purposes:

1. It creates a seal between the rings and the cylinder walls which helps prevent blow-by.
2. It lubricates the rings, pistons, and cylinder walls.
3. It carries away dirt that has come into the engine with the air-fuel mixture, since the rings scrape off the walls and drop the dirt particles into the oil pan, where they are eventually removed from the engine by the oil filter.
4. It helps cool the engine. It is estimated that as much as twenty percent of the engine cooling is accomplished by the oil, with the other eighty percent done by the cooling system.

The primary function of the oil ring is to control or meter oil for adequate lubrication of the rings and cylinder walls without causing excessive oil consumption. Modern oil control rings rely on increased unit pressure as one of the key ways to improve oil control. Such increased unit pressure can be achieved in two ways: increased ring tension and reduced cylinder wall contact area. Equally important in oil control is the conformability of the oil ring to the cylinder wall.

The number of oil control rings installed on a piston varies from engine to engine, depending upon engine design and operating characteristics. It is much more difficult to control the oil on a high-speed engine than on a low-speed engine. The engine turning at the higher speed not only splashes more oil on the cylinder walls, but it also causes the oil to get hotter, lowering the viscosity and permitting the oil to pass into the combustion chamber more easily.

All engines burn some oil, in spite of the design or operating speed. It may be possible to operate a new low-speed engine without adding any oil between oil changes, while it may be necessary to add a quart every few hundred miles with a new high-speed engine. Under normal circumstances, the actual oil consumption of an engine depends upon the engine wear and operating speed; however, oil consumption may be high in some engines even when the compression is good and the engine operates at a low speed. This effect is due to a condition in which the alternating vacuum and pressure in the combustion chamber may cause a ring to act as a pump while the oil helps seal excessive side clearance in the ring grooves, thereby preventing compression leaks.

Combustion Chamber

The *combustion chamber* is the air space above the top of the piston when the piston is at TDC in the cylinder. It is the space in which combustion of the air-fuel mixture takes place. The shape of the combustion chamber is extremely important to engine performance, having much to do with the proper mixing of the air and fuel to obtain maximum benefit from the combustion. The two general shapes of combustion chambers are the *wedge* and the *hemispheric*.

In the wedge-shaped chamber the flame has a relatively long distance to travel from its ignition point to the farthest point in the chamber. In a gasoline engine the spark plug is centrally located in a hemispheric chamber and the flame has a shorter distance to travel.

The *compression ratio* of an engine is a measurement of the compression of the air-fuel mixture in an engine cylinder. It is calculated by dividing the volume in a cylinder with the piston at BDC by the volume in the same cylinder with the piston at TDC. The air volume with the piston at TDC is called the *clearance volume*, since it is the clearance that remains above the piston when it is at TDC. The clearance volume is the same as the combustion chamber space.

An example of the calculation of the compression ratio of an engine is as follows. If the volume in a cylinder when the piston is at BDC is 80 cubic inches, and the volume in the cylinder when the piston is at TDC is 10 cubic inches, then the compression ratio would be 8 to 1 (80 divided by 10). This means that the air-fuel mixture when the piston was at BDC has been compressed to one-eighth its original size.

The compression ratio of an engine has a definite effect on its operating efficiency. In general, the higher the compression ratio,

The Engine

the greater the power and economy of the engine. Unfortunately, increasing the compression ratio also increases the problems. An example of one of the built-in problems is knocking or detonation in a gasoline engine. The higher the compression ratio, the greater the tendency of a gasoline engine to knock; this is due to the higher initial temperature of the fuel-air mixture caused by the heat of compression. With the higher initial pressure and temperature, the temperature at which detonation occurs is reached sooner. In designing an engine, a compromise must be made between the advantages of a higher compression ratio and the disadvantages of the additional problems encountered. Some engines are designed with lower compression ratios for particular reasons. The gasoline engines used in fire apparatus have a compression ratio of about 8 to 1, as compared with those used in some automobiles, which run as high as 11 to 1.

Valves

A *valve* is a device for opening and closing a passageway. The two types of valves used in internal combustion engines are intake and exhaust. An intake valve opens to allow the passage of the air-fuel mixture into the combustion chamber in a gasoline engine, or simply air in four-stroke-cycle-diesel engines. The exhaust valve opens to allow the discharge of the burning mixture from the combustion chamber. Only one intake valve and one exhaust valve are used for each cylinder in some engines; two of each are used in others. Some diesel engines employ four exhaust valves per cylinder (see Figure 1-10).

Valves used in internal combustion engines are referred to as *poppet valves*, receiving this name because they "pop" open and shut during operation.

In a gasoline engine, the air-fuel mixture enters the combustion chamber and the exhaust gases leave the chamber through ports. The ports have seats against which the heads of the valves rest. Passage of gas or vapor cannot take place when the valves are tight against the seats.

There are several types of valve controls. In some engines the valves are held against the seats by compression springs. The

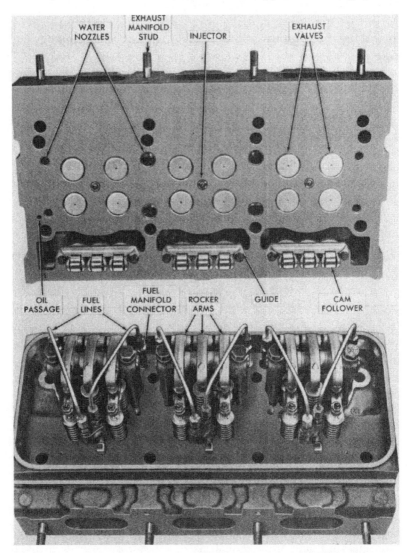

Figure 1-10 A Cylinder Head Assembly Using Four Exhaust Valves. *Courtesy of Detroit Diesel Allison, Division of General Motors Corporation*

The Engine

springs are constantly trying to expand, keeping the valves closed unless they are forced open. The lower end of the spring rests against a flat area of the cylinder head. The upper end rests against a washer, or spring retainer, which is held in place by a retainer lock. The retainer lock is attached to the valve stem.

Valves can be in the block, in the head, or in both. When both

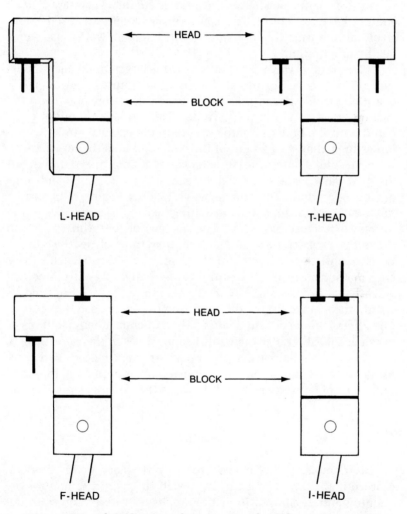

Figure 1-11   Identification of Engine Type by Valve Arrangement

the intake and exhaust valves are in the block on the same side of the cylinder, it is referred to as an L-head engine (see Figure 1-11). When both valves are in the block with the intake valve on one side of the cylinder and the exhaust valve on the other, it is referred to as a T-head engine. An F-head engine is one in which one valve is in the head and the other is in the block. When both valves are in the head, it is an I-head engine. Very few F-head engines have been built. The T-head arrangement was used extensively at one time. Most engines built today, however, use the I-head configuration.

The intake valve is usually a little larger than the exhaust valve. The passage of the air-fuel mixture into the cylinder is a result of a difference of pressure between the atmosphere and the vacuum created within the cylinder. This pressure is minor compared with the high pressure created on the exhaust stroke which forces the exhaust gases out of the chamber, thus the larger valve.

An intake valve and an exhaust valve look very similar, but the two valves are subjected to extremely different conditions during operation. The intake valve is only required to pass a relatively cool air-fuel mixture, while the exhaust valve is required to pass extremely hot gases. The passage of these hot gases can cause the exhaust valve to reach temperatures of nearly 1200° F. Consequently, special attention is given to cooling exhaust valves. One method is to use a sodium-cooled valve. A sodium-cooled valve has a hollow stem filled with sodium. The sodium moves within the stem, carrying the heat from the valve head to the stem. The effect is that sodium-cooled valves operate at temperatures as much as 200° F. cooler than solid-stem valves of the same design.

As mentioned earlier, the valves are kept closed by a compression spring and open only when a force is applied to the stem. The piece of equipment which induces this force is called a cam.

Cams

A *cam* is a device used to convert rotary action into reciprocating action. Cams are employed in engines to control the opening and closing of intake and exhaust valves. Cams are designed to lift valves at the correct moment and hold them open a

The Engine

sufficient length of time to fill and empty the cylinder most efficiently.

The cam is a circle with a triangular projection at one point which is referred to as a *ramp* (see Figure 1-3). The ramp is designed to open and close the valves smoothly and gradually, resulting in less shock to the valves, valve springs, and so forth. The exact shape of the ramp depends upon when the valve should open and close, and how long it should remain open. The design of the intake cam is slightly different from that of the exhaust cam. The nose on the ramp of the exhaust cam is flatter, causing the valve to remain open longer, thereby providing a better escape of the exhaust gases.

In most engines there are two cams for each cylinder, one for the intake valve and one for the exhaust valve. This is not true in two-stroke-cycle engines. Two-stroke-cycle gasoline engines use ports rather than valves. Two-stroke-cycle diesel engines use exhaust valves but are not equipped with an intake valve. Air is forced into the cylinder by a pump through a side port. This forced air is also used to scavenge exhaust gases. With two cams per cylinder, there are sixteen valve cams on an eight-cylinder engine and twenty-four on a twelve-cylinder engine, providing there is only one intake valve and one exhaust valve per cylinder. Additional cams are required when each cylinder uses more than one intake and exhaust valve.

The cams do not force open the valves directly but rather exert pressure on the valve lifters, which in turn exert pressure on a push rod or, depending on engine design, directly on the valve stem. There are two general types of valve lifters in use: the mechanical type and the hydraulic type. The mechanical type is designed with a valve clearance to allow for metal expansion; the hydraulic valve lifters provide for no valve clearance. With no need for valve clearance in engines using hydraulic lifters, the design of the cam is different from that in engines using mechanical lifters.

A hydraulic valve lifter operates quietly, with no tappet noise during normal operation. It is designed so that no adjustment is required during normal service, as any wear is compensated for hydraulically. The integral parts of the hydraulic valve lifter are operated by oil, which is forced in by the oil pump.

Camshafts

The role of the *camshaft* in proper engine operation is an important and difficult one. It must open and close the engine valves many times a minute and do this at the right instant against combustion pressures, valve spring loads, and high inertia forces.

The camshaft is a straight shaft, with cams an integral part of it. There may be one or two camshafts to a bank of cylinders, depending upon engine design. The cams for both the intake and exhaust valves are located on the same shaft in L-head and I-head engines. There is one camshaft for the intake valves and one for the exhaust valves on T-head and F-head engines.

There are usually as many cams on the camshaft in multiple-cylinder engines as there are valves to be operated. However, this is not the case in all V-type engines. Some of these engines use a single cam on the camshaft to operate a valve in each bank.

The camshaft is generally located above and to one side of the crankshaft, except in V-type engines, where it is usually located above the crankshaft.

A gear which is attached to one end of the camshaft is used to rotate the shaft. Power to turn the camshaft comes from the crankshaft. The two shafts are interconnected by one of three methods: (1) by direct meshing of gears at the end of the two shafts or through a gear train, (2) by a chain, or (3) by a toothed belt. When gears are used, they are referred to as *timing gears*. The chain is called a *timing chain,* and the belt is called a *timing belt.*

The gear on the end of the camshaft on four-stroke-cycle engines is twice as large as the gear on the end of the crankshaft. The result is that the camshaft will turn at half the speed of the crankshaft. Thus, each intake and exhaust valve opens and closes one time for every two revolutions of the crankshaft.

The gear on the end of the camshaft on two-stroke-cycle engines is the same size as the gear on the end of the crankshaft. The result is that the camshaft and crankshaft turn at the same speed. Thus, the exhaust valve opens once for every revolution of the crankshaft.

The camshaft generally does other duties in addition to operating the cams. In gasoline engines it may be used to drive the oil pump, the distributor, and the mechanical fuel pump, if one is used. The oil pump and the distributor are driven by a gear on the

The Engine

rear end of the shaft. The fuel pump is driven by an eccentric, which is an integral part of the shaft. In an eccentric, one circle is off-center from another, creating an action similar to the valve lob on the camshaft. On some diesel engines a separate gear in the gear train is used to drive the blower.

Crankshafts

The *crankshaft* is used to change the reciprocating motion of the pistons into rotary motion. It is often called the "backbone" of the engine because it handles the entire power output. The change from reciprocating power to rotary power is accomplished by *cranks* or *journals,* which are those portions of the crankshaft to which the connecting rods are attached (see Figure 1-12). These cranks or journals are commonly referred to as *throws*. The throw is actually the distance the journal is offset from the center line of the crankshaft.

The crankshaft is a one-piece casting or forging of heat-treated alloy steel which is hung on the bottom of the cylinder block. It is supported by large bearings which are called *main bearings*. The upper portion of the main bearing is counterbored into the cylinder block, the lower part being held in place with a bearing cap. These bearings are of the split or half type and are lubricated by positive forced feed from the oil pump.

The number of main bearings in an engine depends upon the

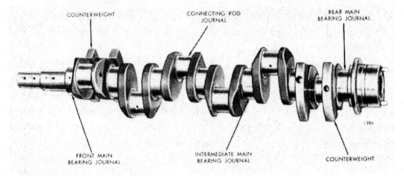

Figure 1-12 A Typical Six-Cylinder Crankshaft. *Courtesy of Detroit Diesel Allison, Division of General Motors Corporation*

engine design. Some apparatus engines are equipped with a main bearing between every throw. Other engines only use main bearings between every other throw. Supporting the shaft with bearings between every throw permits the use of a smaller, lighter crankshaft and provides better overall support. One of the main bearings is called a *thrust bearing*. It is designed to prevent excessive end play in the crankshaft.

The crankshaft is subjected to extreme forces from the downward push of the piston during the power stroke and from the fact that the throws are offset from the center line of the crankshaft and so receive the force from the connecting rods in an out-of-balance condition. A crankshaft commonly encounters problems of static and dynamic balance and torsional vibration. Several features are built into the crankshaft configuration to cope with these forces.

Crankshaft counterweights are usually forged or cast integrally with the shaft. These counterweights help balance the inertia and centrifugal loads imposed by the crankpin-piston-rod combination and help to dynamically balance the entire rotating assembly. The rotating assembly includes the crankshaft, the fan pulley, the timing gears, the vibration damper, the flywheel, and the clutch or converter parts attached to it.

Flywheels

The *flywheel* is a comparatively heavy wheel that is bolted to a flange at the rear of the crankshaft. The rear main bearing of the crankshaft is located near the flywheel. This bearing is longer and heavier than the others since it must carry the weight of the flywheel.

The primary purpose of the flywheel is to assist in providing smooth engine operation. Power to the crankshaft varies from second to second according to the timing of the power strokes of the various cylinders. Although there is a power overlap to some extent between the different cylinders, there are still variations in the amount of power being transmitted to the crankshaft. The crankshaft tries to speed up when the power is the greatest and slow down when it is less. The flywheel tends to resist any change in speed by storing power when the power is greatest and releasing it through its inertia during periods of less power.

The Engine

Engines with heavier flywheels generally idle more smoothly; however, due to the inertia in the flywheel, an engine with a heavier flywheel will accelerate and decelerate more slowly. Diesel engines generally use heavier flywheels than gasoline engines because of the more severe operating cycle and slower operating engine speeds.

A second function of the flywheel is to serve as a base for the clutch. One surface of the clutch is attached to the rear surface of the flywheel. The fluid flywheel or torque converter on those apparatus equipped with automatic transmissions is also attached to the flywheel.

The third function of the flywheel is to turn the crankshaft when the engine is started. The flywheel has gear teeth around its outer rim, which mesh with the teeth on the starting motor drive pinion when the starter is engaged (see Figure 1-13). The starting motor turns the flywheel, causing the pistons to move up and down.

Connecting Rods

The *connecting rod* serves as the link of power transmission between the piston and the crankshaft. It converts the reciprocating motion of the pistons into the rotary motion of the crankshaft. During the explosive cycle it is subjected to tremendous pressures. For example, with a Chevrolet Turbo-Thrust V-8 engine there is an explosion-pressure load of five tons transferred to the connecting rod.

The connecting rod not only transmits the power generated by the explosion of the cylinder fuel mixture from the pistons to the crankshaft, it also transmits stored-up energy in the crankshaft and flywheel back to the pistons to expel exhaust gases in the cylinder and to suck in and compress the cylinder fuel mixture.

The connecting rod is connected to the crankpin of the crankshaft by a bearing cap, rod bolts, and nuts. The bearing used on the crankpin is of the split type. See Figure 1-14 for a diagram of a connecting rod assembly. The orifice shown in the diagram is used for metering oil to the crankshaft bearing.

Connecting rods are forged from high-strength alloy steel. It is important that they be carefully balanced and that all con-

Figure 1-13 A Typical Flywheel Assembly. *Courtesy of Detroit Diesel Allison, Division of General Motors Corporation*

necting rods in the engine be of the same weight. Slight differences in weight would result in an off-balance engine and, consequently, engine roughness.

Wrist Pins

Connecting rods are connected to the pistons by means of *piston pins* (also referred to as *wrist pins*). The upper end of the connecting rod slips into the boss on the piston and the holes are lined up. The piston pin is inserted into one boss, through the connecting rod, and then into the other boss.

As previously mentioned, loads of five tons or more are transferred to the crankshaft from the combustion chamber during the compression stroke. These loads are carried from the pistons through the wrist pin bearings into the connecting rods to drive the crankshaft.

Cylinder Head

The *cylinder head* is a detachable portion of the engine which is bolted to the engine block. A gasket is used between the

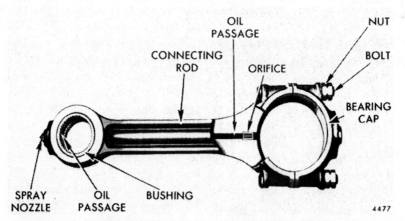

Figure 1-14 A Connecting Rod Assembly. *Courtesy of Detroit Diesel Allison, Division of General Motors Corporation*

head and the block. The cylinder head contains all or part of the combustion chamber. Most cylinder heads are made of cast iron. The head contains water jackets for cooling and passages that lead from the valve ports in the combustion chamber to the intake and exhaust manifolds.

There are two general types of cylinder heads in use: the I-head and the overhead camshaft head. The I-head has the valves in the head, while the overhead camshaft head houses the camshaft.

Manifolds

The *intake* and *exhaust manifolds* of gasoline engines are attached to the cylinder block or to the head, depending upon the valve arrangement. When a valve is located in the block, the manifold for that valve is normally attached to the block. When the valve is in the head, the manifold is normally attached to the head. The intake manifold conducts the air-fuel mixture from the carburetor to the engine intake valve ports. The exhaust manifold conducts the hot burned gases from the exhaust valve ports to the exhaust pipe or pipes.

Intake manifolds are designed to ensure as equal a distribution of air-fuel mixture to each cylinder as possible. Some cylinders usually receive less air-fuel mixture than others in spite of these design characteristics and consequently develop less power.

Another method used to increase the vaporizaton, and therefore to improve the distribution, is to heat the air-fuel mixture within the intake manifold. Various types of systems are used for the heating process, some using hot exhaust gases and some using hot water. The heating process is monitored by a thermostat, which provides more heat to the intake manifold when the engine is cold and reduces the heat as the engine warms up.

In spite of the attempts to provide better vaporization and distribution of the air-fuel mixture, some cylinders receive leaner mixtures than others, since so many factors affect the fuel distribution. These include: the amount of heat supplied to the intake manifold; the composition of the fuel; the characteristics and shape of the intake manifold; and the speed of the engine.

The location and attachment of the manifolds varies with engine types. The intake manifold is attached to the side of the engine block on T-head and L-head engines. It is situated between the two banks of cylinders and attached to the insides of the two blocks on I-head V-8 engines.

The exhaust manifold is also attached to the side of the engine block on T-head and L-head engines. Two exhaust manifolds are used on V-8 engines, one for each bank of cylinders.

Oil Pans

The *oil pan* is attached to the bottom part of the cylinder block. The pan and the inside of the lower portion of the cylinder block form the crankcase. A gasket is used between the cylinder block and the oil pan to form a seal and prevent oil from leaking from the pan.

ENGINE PERFORMANCE STANDARDS

It is important that an apparatus operator understand the meaning of engine specifications in order to comprehend fully the characteristics and capabilities of the engine. The objective of this section is to provide information about commonly used engine performance standards and general performance terminology.

Engine Displacement

Engine displacement is a measurement of the size of an engine. It is the sum total of the piston displacement of all cylinders within the engine. Engine displacement can be found by the formula:

$$\text{Engine displacement} = ASN$$

where A = area of a piston (sq. inches)
       S = stroke (inches)
       N = number of cylinders

The area of a piston can be determined by:

Introduction to Fire Apparatus and Equipment

$$\text{area} = .7854D^2$$

where D = piston diameter or bore

The piston stroke is the distance the piston moves from TDC to BDC.

QUESTION: What is the engine displacement of an eight-cylinder engine which has a 4½-inch bore and a 5-inch stroke?

ANSWER: Since

$$\text{piston area} = .7854D^2$$

where D = 4.5 inches

$$\text{piston area} = (.7854)(4.5)(4.5)$$
$$= 15.90 \text{ sq. inches}$$

And since engine displacement = ASN

where A = 15.90 sq. inches
        S = 5 inches
        N = 8

engine displacement
$$= (15.90)(5)(8)$$
$$= 636 \text{ cubic inches}$$

Horsepower

Power is the rate at which work is done. The unit for measuring power in apparatus engines is horsepower. One horsepower is defined as 33,000 foot-pounds per minute. Expressed in another way, it is the force required to raise 33,000 pounds at the rate of one foot in one minute. Horsepower can be found by the formula:

$$\text{hp} = \frac{\text{ft-lb/per min}}{33,000} = \frac{DW}{33,000T}$$

where D = the distance through which W must be moved
        W = the weight that is to be moved through D
        T = the time required to move W through D (in minutes)

The Engine

QUESTION: How much horsepower would be required to lift a tank of water weighing 2000 pounds a distance of 120 feet in two minutes?

ANSWER: Since

$$hp = \frac{DW}{33,000T}$$

where D = 120 feet
W = 2000 pounds
T = 2 minutes

$$hp = \frac{(120)(2000)}{(33,000)(2)}$$

$$= \frac{240,000}{66,000} = 3.636$$

Rated or SAE Horsepower

Rated or *SAE horsepower* is based upon a formula that was developed in the early days of the automotive industry. Today the formula no longer gives an indication of the power output of an engine; however, it is still used in some localities as a basis for licensing vehicles. The formula is

$$\text{rated hp} = \frac{D^2N}{2.5}$$

where D = piston diameter or bore (inches)
N = number of cylinders

QUESTION: What is the rated horsepower of an eight-cylinder engine with a 4-inch bore and a 4½-inch stroke?

ANSWER: Since

$$\text{rated hp} = \frac{D^2N}{2.5}$$

where D = 4 inches
N = 8

$$\text{rated hp} = \frac{(4)(4)(8)}{2.5}$$

$$= \frac{128}{2.5} = 51.2$$

Brake Horsepower

Brake horsepower is the actual power available at the crankshaft of an engine. On a fire apparatus it is the power available to propel the apparatus, drive the pump, raise the aerial, or perform any of the other numerous functions that are carried on beyond the flywheel. All of the power created within the cylinders which is used to drive the fan, water pump, power steering, and other equipment and is taken off ahead of the flywheel is not included when determining the brake horsepower.

The name brake horsepower was derived from a Prony brake, one of the first devices used to measure engine horsepower output. A Prony brake consists of a large drum and a brake. The drum is connected directly to the engine crankshaft when the power output test is taken. The brake is tightened, with the engine running at a predetermined rpm, until the speed drops off, at which time a reading is taken on a scale. The procedure is repeated with a number of readings taken at various rpm's. The brake horsepower at the various rpm's tested can be determined by the use of a formula.

It is also possible to determine the brake horsepower through the use of a dynamometer. Many diagnostic test centers use the dynamometer to evaluate improvements in engine performance.

Indicated Horsepower

Indicated horsepower is the power that is actually developed within the engine by the combustion process. It is determined by a formula after the mean effective pressure is found. The *mean effective pressure* is the average pressure during the power stroke minus the average pressures during the other three strokes. The mean effective pressure is determined by a special indicating device which makes a graph of the pressures developed during the

The Engine

four strokes. The formula used to determine the indicated horsepower is

$$\text{ihp} = \frac{PLANK}{33{,}000}$$

where P = mean effective pressure (psi)
L = length of the stroke (inches)
A = area of piston (sq. inches)
N = number of power strokes per minute (rpm/2)
K = number of cylinders

Friction Horsepower

Friction horsepower is the power required to overcome the friction within an engine. It is determined by bringing an engine up to operating temperature, shutting the engine off, then driving the engine through the use of an electric dynamometer. The throttle is wide open and there is no fuel in the carburetor during this process. The power required to drive the engine at various speeds is the friction horsepower at those speeds.

Friction within an engine develops when the numerous surfaces rub against one another. One major cause of horsepower loss in the engine occurs as a result of the rings moving along the cylinder walls. In some engines this can account for approximately 75 percent of the total loss of horsepower due to friction.

Horsepower loss through friction in an engine varies with speed. It is relatively low at low speeds, increasing rapidly as the engine speeds up.

Torque

Webster defines *torque* as "a force or combination of forces that produces or tends to produce a twisting or rotation motion." As a piston moves down on the power stroke it applies torque to the crankshaft through the connecting rod and crank on the crankshaft. The more power applied to the piston, the greater the amount of torque developed.

Engineers try to design as high a torque as possible throughout the normal operating range of an engine. They have in recent years begun to rely more on torque than on horsepower as a means of measuring engine performance.

Torque increases as engine speed increases until the point of maximum torque is reached; then it drops with any further increase in rpm's. This happens because there is less time for the air-fuel mixture to enter the cylinders at the higher speeds; hence there is less air-fuel mixture to burn. The result is that combustion pressures will not be as high, and the torque will not be as great.

Fire apparatus require engines that will perform adequately when maneuvering in heavy city traffic, when driving fire pumps, and when providing the power for the operation of aerial ladders and aerial platforms. Many fire departments favor high-torque low-speed engines because of the varied demands placed upon the engine during these operations. Apparatus operators should be aware of the point of maximum torque and the range of high torque, since it is within this range that extra power is available when needed.

Volumetric Efficiency

Volumetric efficiency is the ratio between the amount of air-fuel mixture that enters an engine cylinder and the theoretical amount which could enter the cylinder under ideal conditions. It would be possible to approach a 100 percent volumetric efficiency if it were possible to draw the air-fuel mixture into the cylinder very slowly; however, this is not the way a gasoline engine works. Engines turn over at a rapid rate, and valves open and close quickly. Cylinders do not completely fill due to the restrictions placed on the mixture movement by the intake manifold, valve operations, atmospheric pressure, and other factors. In fact, a volumetric efficiency of 80 percent is fairly good.

There are certain factors other than engine design that affect the volumetric efficiency. For example, an apparatus that is taken from a flatland operation to a mountainous operation will experience a drop in volumetric efficiency because of the reduction in atmospheric pressure at the higher elevation. This reduction of atmospheric pressure causes a smaller difference between the pressure outside the cylinders and that inside, providing, of course,

The Engine

that the engine is not equipped with a supercharger. The result is a slowdown of the mixture movement. This lowering of the volumetric efficiency results in a loss of approximately 3½ percent of the engine power for every 1000 feet of rise in elevation above sea level.

Another factor affecting the volumetric efficiency is the engine speed. It is more difficult to move the desired air-fuel mixture into the cylinder as the rpm's increase, since the valves remain open for a shorter period of time, and the mixture undergoes increased friction loss as it moves through the intake manifold, among other reasons. If the speed continues to increase, the point will eventually be reached where the engine begins to starve for air, resulting in a rapid drop in volumetric efficiency.

There is a good correlation between the maximum volumetric efficiency of an engine and the maximum torque. The two generally peak at about the same rpm.

Engine Efficiency

Efficiency is the relationship of what was done to what could have been done with the effort applied under ideal conditions. Engine efficiency is the relationship of the power delivered compared to the power that could have been delivered if there had not been any losses. The efficiency of an engine is judged from both a mechanical and a thermal viewpoint.

Mechanical efficiency is the relationship between brake horsepower and indicated horsepower. It is expressed as a formula as follows:

$$\text{mechanical efficiency} = \frac{\text{bhp}}{\text{ihp}} \times 100$$

QUESTION: The bhp of a certain engine at an rpm of 1800 is 142, while the ihp is 180. What is the mechanical efficiency of the engine?

ANSWER: Since

$$\text{mechanical efficiency} = \frac{\text{bhp}}{\text{ihp}} \times 100$$

where   bhp = 142
        ihp = 180

$$\text{mechanical efficiency} = \frac{142}{180} \times 100$$
$$= .7889 \times 100$$
$$= 78.89\%$$

This means that the other 21.11 percent of the power developed in the cylinders is consumed as friction horsepower.

Thermal efficiency is the relationship between the power output of an engine and the power capability in the fuel used to produce that power output. The thermal loss in gasoline engines, through the cooling system, exhaust system, friction, and other sources, is extremely high. A thermal efficiency of 25 percent is considered good.

Overall Efficiency

Overall efficiency is the comparison between the amount of power remaining to propel the apparatus and the power available in the fuel burned. The typical overall efficiency of a gasoline engine is about 15 percent. The remaining 85 percent is energy loss from the cylinders to the wheels. The general distribution of loss is as follows:

    35 percent loss in cooling water, air, and oil
    35 percent loss in exhaust gas
    15 percent loss in friction

REVIEW QUESTIONS

1. What is an internal combustion engine?
2. What is a reciprocating type of engine?
3. What are the two general types of internal combustion engines installed on fire apparatus?

The Engine

4. What is the general principle of operation of an internal combustion engine?

5. What events must take place in a gasoline engine in order to complete a power cycle?

6. What is the definition of a stroke?

7. What is meant by TDC and BDC?

8. What is the difference between a two-stroke-cycle engine and a four-stroke-cycle engine?

9. What are the four strokes of a four-stroke-cycle engine?

10. What causes the air-fuel mixture to move into the cylinder during the intake stroke?

11. What causes the burned gases to move out of the cylinder on the exhaust stroke?

12. What is meant by normal combustion?

13. What are two examples of abnormal combustion?

14. What is detonation?

15. What is preignition?

16. What are some of the advantages claimed by proponents of diesel engines?

17. What is the general operating principle of a diesel engine?

18. How do the compression ratios of diesel engines compare with those of gasoline engines?

19. How can the combustion of the diesel oil be controlled within the cylinder?

20. What is the difference between the general construction of a gasoline engine and that of a diesel engine?

21. How is the displacement of a cylinder determined?

22. What is a cylinder?

23. What is a piston?

24. What is the skirt of a piston?

25. What are the three functions a piston performs during normal operation?

26. What is piston clearance?

27. What causes piston slap?
28. What is blow-by?
29. What is meant by oil-pumping?
30. How are blow-by and oil-pumping controlled?
31. What are some of the purposes served by the oil on cylinder walls?
32. What is the combustion chamber?
33. How is the compression ratio of an engine determined?
34. What compression ratios are used on apparatus?
35. On gasoline engines? On diesel engines?
36. What is a poppet valve?
37. How many exhaust valves per cylinder do some diesel engines use?
38. Which is larger, the intake valve or the exhaust valve? Why?
39. What is a sodium-cooled valve?
40. What are the two general types of valve lifters in use?
41. What is a camshaft?
42. What drives the camshaft?
43. What is the function of the crankshaft?
44. What is the function of the counterweights on the crankshaft?
45. What is the function of the flywheel?
46. What is attached to the rear of the crankshaft?
47. What is the function of a connecting rod?
48. How is the connecting rod connected to the piston?
49. Which use heavier flywheels, gasoline engines or diesel engines? Why?
50. What formula is used to find engine displacement?
51. What is power?
52. What is horsepower?
53. What formula is used to determine horsepower?
54. What formula is used to find rated or SAE horsepower?
55. What is brake horsepower?

The Engine

56. How is the brake horsepower of an engine determined?

57. What is indicated horsepower?

58. What is meant by mean effective pressure?

59. What formula is used to find indicated horsepower?

60. What is friction horsepower?

61. What method is used to determine the friction horsepower of an engine?

62. What is probably the major cause of friction loss in an engine?

63. What is torque?

64. What is the relationship of torque to engine speed?

65. What is volumetric efficiency?

66. What is considered a good volumetric efficiency?

67. What are some of the factors, other than design, that affect volumetric efficiency?

68. What is engine efficiency?

69. What formula is used to find mechanical efficiency?

70. What is thermal efficiency?

71. What thermal efficiency is considered good?

72. What is overall efficiency?

73. What is the general distribution of losses in a gasoline engine?

2

Engine Systems

The objective of this chapter is to introduce apparatus operators to the various systems used on their apparatus. Fuel, ignition, lubrication, cooling, and electrical systems will all be explored.

FUEL SYSTEMS

Fire apparatus engines are primarily of the internal combustion type, which means that they burn a petroleum product in the cylinders in order to obtain power. The two primary fuels used are gasoline and diesel fuel.

Gasoline is a highly volatile liquid that gives off vapors three to four times heavier than air. While the flash points, ignition temperatures, and flammable limits of gasoline vary with different octane ratings, it is commonly considered that the flash point is approximately −45° F., the ignition temperature near 535° F., with the flammable limit ranging from 1.4 to 7.6.

Diesel engines burn diesel fuel instead of gasoline. While a wide range of fuels can be burned in diesel engines, it is important that the proper grades be used in fire apparatus engines. Not only is the quality of fuel used a very important factor in obtaining satisfactory engine performance, long engine life, and acceptable exhaust, but it is also a factor in the starting of an engine. The fuel must have a sufficiently high viscosity to lubricate the moving parts adequately, a low enough viscosity to flow easily through the fuel-pumping system, and a cetane number adequate for operational conditions. The cetane number is the measurement of the ignition value of a diesel fuel. The cetane number influences both ease of starting and the combustion roughness of an engine. High-cetane fuel should be required for low-temperature starts; however, there is no advantage to requiring a higher number than necessary. The fuel must also be relatively free of impurities so that harmful engine deposits and corrosion are kept to a minimum. While all diesel fuels contain a certain amount of sulfur, too high a sulfur content results in excessive wear of piston rings, valves, and cylinder liners.

The engine manufacturers' recommendations for cetane rating and sulfur content should be followed.

Gasoline Fuel Systems

The fuel system of a gasoline engine consists of the fuel tank, fuel pump, fuel lines, fuel filter, fuel gage, and a carburetor or fuel injection system.

Fuel Tank

The *fuel tank* on a heavy-duty fire apparatus should hold at least 20 gallons and be made of a corrosion-resistant material. The actual capacity of the tank depends upon the functions that must be performed by the apparatus. Fuel tanks of apparatus equipped

Engine Systems

with pumping equipment should be of sufficient capacity to permit the apparatus to move from the engine house to the fire and pump at 100 percent rated capacity for at least two hours.

The tank is normally near the rear of the apparatus. The tank fill opening should be adequately labeled as to the type of fuel required and be so located on the apparatus that it is adequately protected against mechanical injury. The tank should be vented and should be provided with a suitable drain.

Some tanks have a filtering element installed near the bottom of the tank at the fuel-line connection. The fuel-line connection should be at least a half-inch above the bottom of the tank. The tank also contains the sending unit for the fuel gage.

Fuel Pumps

Both mechanical and electrical *fuel pumps* are used on fire apparatus. The mechanical fuel pump consists of two chambers. One chamber contains an inlet valve and an outlet valve, both one-way valves that are opened or closed by pressure.

The other chamber houses a spring that pushes against a diaphragm that is clamped between the two chambers. The diaphragm is forced against the spring during operation by the action of a rocker arm. The outer end of the rocker arm rests on the camshaft eccentric and moves up and down as the cam rotates.

When the rocker arm forces the diaphragm away from the first chamber and against the spring, it causes a reduction of pressure in that chamber. The pressure differences between the reduced pressure in the chamber and the atmospheric pressure in the fuel tank forces the fuel against the inlet valve, opening it and allowing the chamber to fill with fuel. When the rocker arm releases the pressure on the diaphragm, the spring forces the diaphragm back to its original position, tending to compress the fuel in the first chamber. This creates a pressure that closes the inlet valve and opens the outlet valve, pushing the fuel from the chamber toward the carburetor.

A needle valve in the carburetor will not permit any fuel to enter the carburetor in the event the float bowl is full. In that case, the spring remains in a collapsed position until such time as the needle valve opens. The rocker arm will continue to move, sliding up and down on the diaphragm pull rod, but no fuel will leave the pump.

Electrical fuel pumps are of two basic types, external and internal. The external type is mounted outside the fuel tank (usually under the hood in a cool spot) and operates in a manner similar to that of the mechanical pump, using suction (reduced pressure) to draw fuel from the tank. The primary difference between the external electrical pump and the mechanical pump is that the electrical pump uses the movement of a bellows instead of a diaphragm to create the reduced pressure. The bellows is operated by the action of an electromagnet. The bellows is pulled down when the current is supplied to the electromagnet. The downward movement of the bellows by the armature pulls fuel into the pump in a manner similar to that of the diaphragm in the mechanical pump. A set of contact points are opened when the armature reaches its lower travel limit, shutting off the current to the electromagnet. A return spring collapses the bellows, forcing the fuel out of the pump.

The internal electrical pump is of the pusher type. The pump is located in the bottom of the fuel tank. It pushes fuel from the tank to the carburetor.

Both types of electrical pumps have certain advantages over the mechanical pump. First, they are less likely to vapor lock. Since they are mounted away from the heat of the engine, they operate at a much lower temperature than the mechanical pump. An additional advantage is that fuel is provided to the carburetor as soon as the ignition is turned on. The pumps deliver more fuel than required by the engine under maximum operating conditions, thus preventing the engine from ever becoming starved for fuel.

The internal pusher type appears to have two additional advantages. There are no valves between the pump and the carburetor; the fuel drains back into the fuel tank when the engine is shut off. This eliminates a pressure build-up, making it easier to start a hot engine. In addition, the fuel is delivered to the carburetor in a steady, non-pulsating motion.

Many of the fuel systems on fire apparatus employ two electrical fuel pumps. This not only provides a more positive supply of fuel to the carburetor but also acts as a safety factor in the event of failure of one of the pumps. The pumps are installed in a parallel fashion, each having its own line from the tank to the carburetor and separate check valves and filtering devices.

Fuel Filters

Filters are used to prevent dirt and water from entering the fuel pump and carburetor. Filters used on fire apparatus should be a type that can be serviced without disconnecting the fuel line.

Filters may be installed between the tank and the fuel pump, between the fuel pump and the carburetor, incorporated in the design of the fuel pump, or built into the carburetor.

Fuel Gages

Thermostatic and balancing coil gages are the two types of *fuel gages* in general use. The thermostatic gage has a sending unit located in the tank and a gage located on the instrument panel in the cab. The sending unit has a rheostat which varies its resistance by the movement of a float that travels up and down, depending upon the amount of fuel in the tank. Provisions are made to compensate for the splashing of fuel within the tank.

The balanced coil type has a sending unit in the tank that is similar to the unit in the thermostatic type. The instrument panel unit consists of two balancing coils that control an armature and a pointer. An armature dampening device prevents vibration of the needle.

Carburetors

Carburetion, in its simplest sense, is the mixing of gasoline and air in the proper proportions to form a combustible mixture. In doing this, the *carburetor* must meter the fuel and air and convert the fuel from a liquid into a gaseous state under a wide variety of operating conditions. For example, the air-fuel ratio required for starting is different from the one required for normal cruising.

The air-fuel ratio is the ratio of the weight of air to the weight of gasoline. For example, a ratio of 14 to 1 means that 14 pounds of air are required for every pound of gasoline.

The air-fuel demands of an engine vary under different operating conditions and vary slightly from one engine to another under the same operating conditions. In general, a ratio of 9 to 1 is required for starting. The mixture will lean out to about 12 to 1 while the engine is idling. The ratio at normal operating speeds approaches 15 or 16 to 1 but will enrich quickly to 12 or 13 to 1 when the engine is accelerating or at high speeds when maximum power is required. The carburetor must be capable of meeting these changing requirements under a wide variety of operating conditions. A carburetor contains five separate systems (or circuits). These consist of:

1. Float
2. Main metering
3. Idle
4. Accelerating pump
5. Power enrichment

It is important to understand a few basic principles of pressure reduction and air movement before examining each of these systems.

Air moves from a high-pressure area to a low-pressure area. The greater the difference in pressure, the stronger the movement. The normal atmospheric pressure at sea level is 14.7 pounds per square inch. A reduction of pressure occurs as the pistons within the engine move downward on the intake stroke, setting up a lower pressure than the outside atmospheric pressure. The amount of pressure reduction within the cylinder, and, consequently, the degree of potential air movement, depends largely upon the tightness of the piston within the cylinder, the piston displacement, and the engine speed. The air travels through the carburetor, the intake manifold, the intake valve port, and into the cylinder.

To secure an air-fuel mixture, there must be a method of drawing the fuel out of the fuel storage bowl in the carburetor and mixing it with the air as the air passes through the carburetor air horn. This is done with a venturi.

A *venturi* is a constriction in a flow path (see Figure 2-1). A reduction of pressure occurs in the venturi throat as the air moves through, the amount of reduction in a venturi of a given size depending upon the speed of the movement. This reduction of pressure is used to cause fuel to flow from the bowl into the air stream.

Engine Systems

The Float System   The *float system* consists of a float bowl, a float, and an inlet needle arrangement (see Figure 2-2). The float bowl stores the fuel which is fed into the intake manifold.

The objective of the float system is to maintain a ready supply of fuel at a constant level. The level at which the fuel is maintained is extremely important. Too high a level will result in an excessive amount of fuel reaching the discharge nozzle, and will cause fuel to drip from the nozzle when the engine is not running. Too low a level will result in poor engine performance, due to a lean mixture.

Fuel enters the bowl through the inlet needle arrangement. As the bowl fills, the float moves upward. As it does so, the arm on the float pushes against the needle valve, forcing it against the seat when the bowl is full. If fuel is pulled from the bowl through the discharge nozzle, then the float will drop, allowing more fuel to enter the bowl. Under operating conditions the float tends to hold the needle partly closed, balancing the incoming fuel with that being withdrawn.

The Main Metering System   The *main metering system* consists primarily of the main nozzle which is centered in the venturi. Figure 2-3 shows the position of the fuel in the main well when the engine is not running. At this time, the air pressure in the bowl and the air pressure in the venturi are identical.

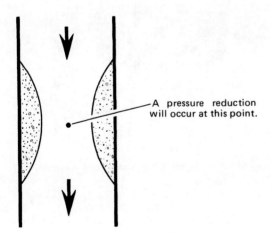

Figure 2-1   A Venturi

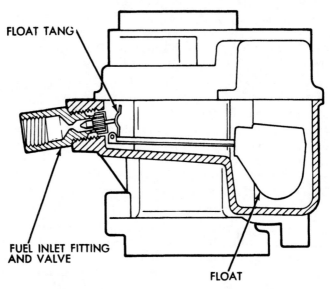

Figure 2-2   A Float System. *Courtesy of Colt Industries, Holley Carburetor Division*

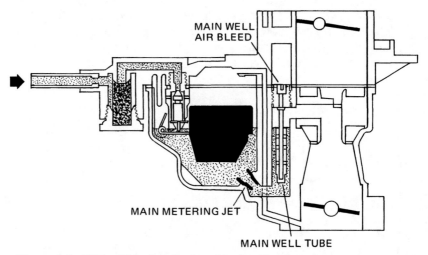

Figure 2-3   Main Metering System (Engine Not Running). *Courtesy of Colt Industries, Holley Carburetor Division*

Engine Systems

When the engine is first started with the throttle closed or mostly closed, the pressure within the venturi remains at or near atmospheric pressure, since the closed throttle restricts movement of air through the venturi. As the throttle is opened, the air movement through the venturi will cause a pressure reduction. Because of the difference in air pressure between the air in the bowl and that in the venturi, fuel will be forced out of the main discharge nozzle and into the air stream (see Figure 2-4). The fuel will be discharged at an increased rate as the pressure reduction in the venturi becomes greater. This discharge of fuel from the main nozzle maintains almost a constant air-fuel ratio, whether the throttle is partly open or wide open.

The Idle System   Little or no fuel flows out of the main nozzle when the throttle valve is closed and the engine is idling. At this time fuel is supplied to the engine through a separate unit called the *idle system* (see Figure 2-5).

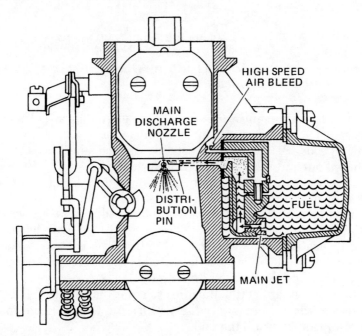

Figure 2-4   Main Metering System (Engine Running). *Courtesy of Colt Industries, Holley Carburetor Division*

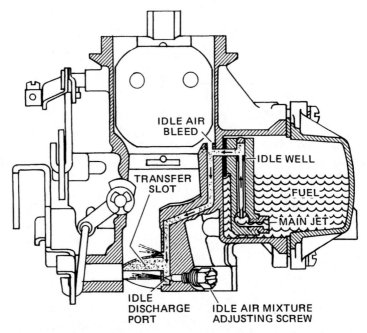

Figure 2-5   An Example of an Idle System. *Courtesy of Colt Industries, Holley Carburetor Division*

There are many varieties of idle systems in use. These are designed to provide an air-fuel mixture to the engine at idling and low speeds. As the throttle valve is opened, the idle system is phased out and the main metering system takes over.

The pressure in the venturi when the throttle value is closed is at or near atmospheric pressure; however, the pressure on the engine side of the throttle valve is reduced due to the vacuum action of the pistons. The idle system consists of a discharge hole on the engine side of the closed throttle valve. Fuel is discharged through the hole because of the difference in pressure between the float bowl and the point of discharge.

A secondary part of the idle system is the air bleed. A hole is inserted from a point above the venturi into the idle feed line, so that air can mix with the fuel.

The air-fuel mixture discharged from the idle system is very rich, since a rich mixture is needed for idling. An idle mixture

Engine Systems

needle allows for an adjustment of the amount of mixture discharged.

As the throttle is opened, the pressure at the idle discharge hole will increase and the pressure at the venturi decrease. This causes the idle system to start phasing out, and the main metering system to start taking over.

There is a point between the idle position and the time when the main metering system takes over where insufficient fuel is supplied from the idle jet to meet the demands of low-speed operation. An additional idle hole is provided slightly above the normally closed position of the throttle valve, to compensate for the loss of fuel from the idle discharge hole. As the throttle passes this point, additional fuel is supplied to meet the low-speed demands.

The Accelerating Pump System As with idle systems, a number of different types of *accelerating systems* are in general use. However, the various systems are all designed to meet the same objective—to compensate for the fuel lag when the throttle valve is opened quickly (see Figure 2-6).

The quick opening of the throttle causes air to rush in, creating a lean mixture and an instant demand for additional fuel. The accelerating pump system fills this demand.

Consider that the acccelerating pump system consists of a pump whose pump lever is linked to the throttle. The pump is built as an integral part of the carburetor. When the throttle is opened quickly, it forces the accelerator pump diaphragm to the right, which forces fuel out of the pump shooter jet and into the air passing through the carburetor horn. Measures are designed into the system to ensure that additional fuel is provided for a sufficient length of time to allow the full-power system to take over.

The Power Enrichment System An enriched mixture is required for a wide-open, full-power position. The main jet provides for economical operation through the normal range of operations but is not designed to provide for the extra fuel required at full power. This extra fuel is supplied by the *power enrichment system*.

There are two primary types of power enrichment systems, the mechanical and the vacuum (see Figure 2-7). Each system uses

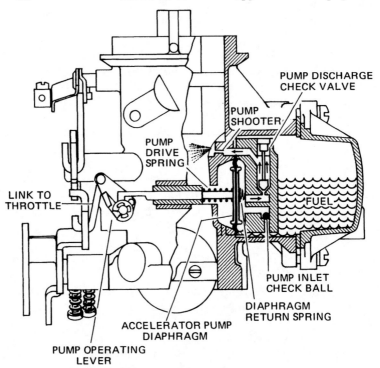

Figure 2-6 An Example of an Accelerator Pump System. *Courtesy of Colt Industries, Holley Carburetor Division*

a different method to open an extra passage to the main discharge jet that provides additional fuel and, hence, enriches the mixture.

Automatic Chokes   An extra rich mixture is required when starting an engine or when operating a cold engine. This mixture is required because of the lack of reduction of pressure in the venturi when the engine is turned over by the starter and poor vaporization of the fuel until the engine warms up. While a manual choke can be used to provide this rich mixture, most fire apparatus use an *automatic choke*.

Most automatic chokes operate on exhaust-manifold temperature and intake-manifold vacuum. The combination of these systems keeps the choke closed during starting operations, at

Engine Systems 55

intermediate positions as the engine warms up, and at the full open position under normal operating conditions.

Figure 2-8 is a schematic of an integral automatic choke system. A bimetal spring is installed inside the choke housing, which is a part of the cover. As the engine warms up, manifold heat transmitted by hot air to the choke housing relaxes the bimetal spring until it eventually permits the choke to open fully. The vacuum diaphragm housing is a part of the carburetor body. When the engine starts, manifold vacuum is applied to the choke diaphragm through a passage from the throttle body to the choke diaphragm assembly. The adjustment of the choke valve opening is called "vacuum kick" or "pull down." When the engine starts and vacuum is applied to the choke diaphragm, the choke valve is opened sufficiently to keep the engine running.

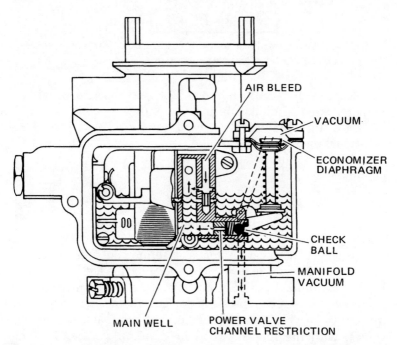

Figure 2-7 An Example of a Power Enrichment System. *Courtesy of Colt Industries, Holley Carburetor Division*

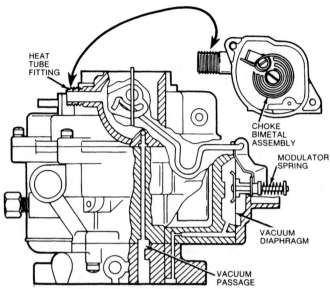

Figure 2-8 An Example of an Automatic Choke System. *Courtesy of Colt Industries, Holley Carburetor Division*

Manifold vacuum alone is not sufficient to provide the proper degree of choke during the entire choking period. The impact of inrushing air past the offset choke plate provides the additional opening force. A modulator spring permits correct initial choke opening after initial start, according to outside temperature.

Gasoline Fuel Injection Systems

In place of a carburetor, a *gasoline fuel injection system* injects fuel into the engine. The fuel nozzles are located in the intake manifold. Here only air enters the intake manifold, as a result of the reduced pressure formed in the cylinders on the intake stroke. The nozzles are located just back of the intake valves, with the fuel supplied by a pump system. The pump system includes a metering unit for supplying the correct amount of fuel at the proper time.

Engine Systems

Diesel Fuel Injection Systems

The fuel injection system for a diesel engine is quite different from a fuel injection system for a gasoline engine. The gasoline fuel injector nozzles inject fuel into the intake manifold where the pressure has been reduced; consequently, relatively low-pressure pumps can be employed in the system. The fuel in a diesel engine is injected directly into the cylinder at the end of the compression stroke, where ignition takes place as a result of the high temperature which has been created within the cylinder. The fuel is injected into the cylinder at the time of maximum compression. Consequently, a high-pressure pump system must be used in order to inject the fuel into the cylinders at pressures well in excess of those within the cylinder.

The fuel for a diesel engine must be supplied to the cylinder at the proper time and in the proper amount. The fuel injection system for a diesel engine must perform five separate functions:

1. Measure the amount of fuel injected
2. Provide the injection at the proper time
3. Control the rate of fuel injected.
4. Atomize the fuel.
5. Properly distribute the fuel within the combustion chamber

Three general types of systems are used to perform these functions. One is referred to as a *distribution system*. It operates much as does the ignition distributor in a gasoline engine. Filtered fuel is delivered from the fuel tank to the hydraulic fuel pump. The pump acts as a central distribution point, delivering fuel at high pressure to the various fuel nozzles in the proper firing order, proper amount, and at the proper time. Provisions are made for varying the amount of fuel according to the throttle positions, and for sending fuel to the cylinder earlier at high speeds so that it has sufficient time to ignite and burn.

A second system is called a *multiple pump system*. It uses individual plungers for metering and injecting the fuel to each cylinder. These pumps are mounted on the side of the engine and are connected to the injectors by long, high-pressure fuel lines.

The system was originally designed for slow-speed diesels but has been adapted to a number of high-speed models.

The third type of system is known as a *unit injector system*. It delivers fuel through low-pressure lines to individual fuel injectors at each cylinder. With this system no high pressure is developed in the pump or fuel lines outside of the engine. The unit fuel injectors develop the high pressure necessary to inject fuel into the cylinder. In the event of injector failure, only that cylinder is affected and the engine continues to operate.

IGNITION SYSTEMS

The objective of an ignition system is to provide a spark at the proper time to ignite the combustible mixture in the combustion chamber of a gasoline engine. A conventional ignition system consists of the ignition switch, battery, ignition coil, distributor, condenser, ignition cables, and the spark plugs.

Battery

Figure 2-9 shows a conventional ignition system. The battery provides current for the primary circuit. The current flows from the battery to the ignition switch, through the ignition resistor, and then through the primary windings in the coil, where it sets up a magnetic field in the core. It then continues through the distributor and on to ground.

Ignition Switch

The *ignition switch* activates the ignition system and energizes the starter motor. Note that there is a continuous draw of current from the battery if the ignition switch is left on and the points in the distributor are closed; however, there is no drain on the battery if the ignition switch is left on and the points are open.

Engine Systems 59

Ignition Coil

The purpose of the *ignition coil* is to step up the 12 volts from the battery to the approximately 20,000 volts needed to jump the gap in the spark plug. The coil consists of a primary winding, a secondary winding, and a core made of soft iron. The primary winding is made up of several hundred turns of a relatively heavy wire. The secondary winding consists of several thousand turns of fine wire. The primary windings are wound around the outside of the secondary windings, while the secondary windings are wound around the soft core (see Figure 2-10). The purpose of the iron core is to concentrate the magnetic field. As previously mentioned, the current from the battery flowing through the primary winding in the coil sets up a magnetic field. The current from the

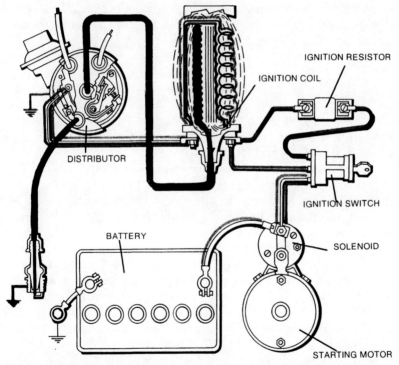

Figure 2-9 A Typical Conventional Ignition System. *Courtesy of Fram Division, Allied Automotive*

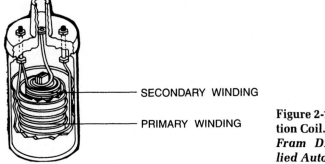

Figure 2-10 An Ignition Coil. *Courtesy of Fram Division, Allied Automotive*

battery continues through the coil to the breaker points in the distributor. When the distributor cam opens the points, the magnetic field collapses, developing high voltage in the coil's secondary windings. This voltage flows from the coil to the distributor cap's control tower (see Figure 2-9).

Distributor

Figure 2-11 shows a typical *distributor*. The distributor houses the *ignition points, condenser,* and *rotor*. The cap has a control tower in the center and terminals at the outer edge.

The distributor serves two basic functions. It controls the flow of primary current by opening and closing the ignition points, and it distributes the current in the high-voltage secondary circuit through the rotor, to the proper spark plug, in a carefully timed sequence.

The shaft of the distributor is turned at half crankshaft speed by the drive gear. The rotation of the shaft turns the cam and rotor. The lobes on the cam force open the points, causing the magnetic field to collapse and inducing the high voltage in the secondary circuit. The spring on the points closes the points when the flat portion of the cam is reached, again causing a magnetic field to develop.

It is extremely important that the points open at the proper time throughout the entire range of speeds and operating conditions. The requirement for the timing of the spark varies according to the load on the engine and the engine speed. Two methods are

Engine Systems

in general use for advancing or retarding the spark as conditions require. These are the *vacuum advance* and the *mechanical advance*. The vacuum advance diaphragm is shown is Figure 2-12. The centrifugal weights of the mechanical advance are not shown in the illustration, but would be located below the plate to which the points are attached. These two systems work at the same time under some conditions, and separately under others.

The vacuum advance shown in Figure 2-12 is controlled by the throttle position and the engine load. The intake manifold vacuum is high at low speeds. This compresses the diaphragm spring, which pulls the breaker plate against the cam rotation direction and advances the spark timing. The manifold vacuum drops when the engine load is high or during sudden accelerations, causing the spring to push the breaker plate in the direction of cam rotation and, therefore, reducing the spark advance.

The mechanical advance shown in Figure 2-13 is controlled by engine speed. The weights are thrown out by centrifugal force as the speed increases, causing the cam to move forward and the points to open sooner, advancing the spark. The weights are pulled back as the speed decreases, retarding the spark.

Condenser

The purpose of the *condenser* is to prevent arcing across the

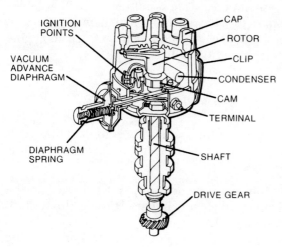

Figure 2-11 A Typical Distributor. *Courtesy of Fram Division, Allied Automotive*

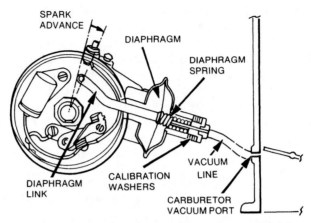

Figure 2-12 The Vacuum Advance. *Courtesy of Fram Division, Allied Automotive*

points when they are first opened. This is done by providing current storage space until the points are fully open.

Ignition Cables

The *ignition cables* serve a dual purpose They carry current from the coil to the distributor cap, and they carry current from the cap terminals to the spark plugs. Care should be taken that cables are not bent or that sharp tugs are not applied to them. Such action could result in internal damage, which might produce an engine miss similar to that resulting from a defective plug.

Spark Plugs

The primary function of the *spark plug* is to ignite the fuel-air mixture in a gasoline engine efficiently. In accomplishing this function, the spark plug:

    1. Conducts the high secondary circuit voltage into the combustion chamber of each cylinder. At times this voltage is in excess of 30,000 volts.

    2. Provides the gap for the high voltage spark to ignite the fuel-air mixture

Engine Systems 63

3. Helps dissipate combustion heat, which may exceed 3000° F

4. With its threads and seat in each cylinder head, it "seals" against the compression leakage in the combustion chamber.

Spark Plug Construction

Although it has no moving parts, the spark plug is exposed to more severe stresses than any other part of the engine. It must deliver a high-voltage spark with split-second timing, thousands of times a minute, and under a widely varying set of adverse conditions. Modern spark plugs combine ceramic engineering, sophisticated metallurgy, and dimensionally perfect assembly. Figure 2-14 shows a simple schematic of the basic spark plug components.

Engine Performance

Proper spark plug performance is important in helping to maintain gasoline mileage and engine efficiency. A spark plug that is functioning the way it should gives a hot, constant spark which provides power, quick starting, and responsive acceleration. A misfiring plug can result in poor mileage and inefficient operation.

A misfiring spark plug in a larger engine may not be detected as easily as in a smaller engine. An effective way to detect a spark plug misfire in larger engines is to use an electronic instrument designed for that purpose. Proper diagnosis and identification of spark plug problems can directly affect corrective action, from a simple engine tune-up to possible engine repair.

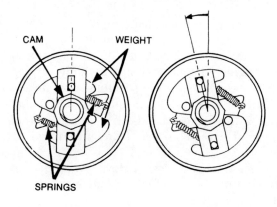

Figure 2-13 The Mechanical Advance. *Courtesy of Fram Division, Allied Automotive*

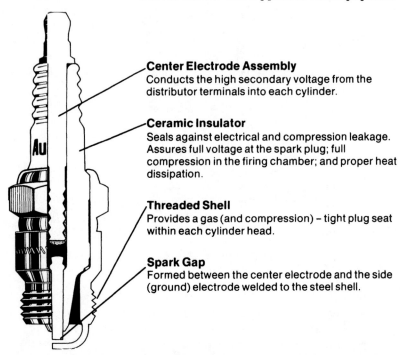

Figure 2-14 A Simplified Schema of Basic Spark Plug Components. *Courtesy of Fram Division, Allied Automotive*

Spark Plug Heat Range

Heat range is the measurement of a spark plug's capability to transfer heat received from the engine combustion chamber to the cylinder head. Spark plugs are designed in different ranges to accommodate different engine characteristics and widely varying driving and load conditions. The spark plug tip must operate at a high enough temperature to burn off combustion deposits that would otherwise collect at the firing tip and cause fouling, and at the same time remain cool enough to avoid preignition and/or electrode destruction.

The rate of heat transfer, whether fast or slow, is a product of spark plug design. It identifies the difference between a "hot" spark plug and a "cold" spark plug.

A "hot" plug transfers heat slowly and operates at a high

Engine Systems

How Spark Plugs Misfire

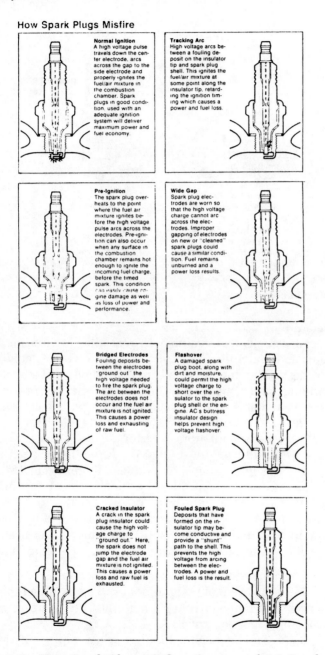

Figure 2-15 How Spark Plugs Misfire. *Courtesy of AC Spark Plug, Division of General Motors Corporation*

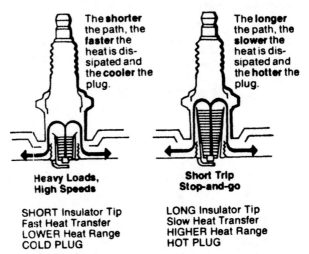

Figure 2-16 Heat Range of Spark Plugs. *Courtesy of Fram Division, Allied Automotive*

temperature. A "cold" plug has a faster rate of heat transfer and operates at a cooler temperature when installed in the same engine and operated under the same conditions. Therefore, a colder plug may be best suited for "full load" or continuous high-speed highway driving, while a hotter plug may be better for prolonged idling in typical stop-and-go city traffic. A plug with an "ideal heat range" operates between the preignition and fouling temperatures and delivers the best all-around performance.

Heat is conducted away from the firing tip, through the insulator and the spark plug shell, to the cylinder head. Thus, heat ranges are primarily controlled by the length of the insulator through which the heat must travel before escaping to the cylinder head (see Figure 2-16). The thermal characteristic of the insulator and the bond of the insulator to the shell also affect heat range. Good rules for selecting the correct spark plug are the following:

1. Follow the manufacturer's specifications.
2. In the event of preignition, use a spark plug in the next lower heat range (a colder plug) until the engine problems causing it can be corrected.
3. In case of engine fouling, select a plug in the next

Engine Systems

higher heat range (a hotter plug) until the engine problems causing it can be corrected.

How Spark Plugs Misfire

The proper servicing of spark plugs can help make the difference in good engine performance and fuel economy. An engine with adequate compression and the proper fuel-air mixture may fail to ignite the fuel charge properly in one or more of several ways, depending on the condition of the spark plugs. Misfire of the fuel charge could be caused by an improper fuel-air mixture in the cylinder. However, the problem is more commonly caused by faulty ignition. Figure 2-15 shows eight different spark plug conditions.

Spark Plug Conditions

The condition of a spark plug is a good indicator of the condition of a cylinder or perhaps the engine. Chart 2-1 shows a number of different spark plug conditions, together with possible causes and suggested remedies.

Diesel Engine Glow Plugs

The diesel engine glow plug aids cold-weather starting of the diesel engine. In a typical installation, the heated sheath of the glow plug extends into the pre-combustion chamber of the engine near the fuel injector. The glow plug is energized, prior to cranking the engine, for a period of time called "pre-glow." The pre-glow time period is dependent upon engine coolant temperature, which is measured by a thermistor. When engine coolant temperature is lowered, the electrical resistance of the thermistor increases.

Two types of glow plugs are commonly used. They are referred to as the fast-heat and the standard glow plug. Maximum pre-glow time for an A.C. fast-heat glow plug is seven seconds; 60 seconds, for the standard glow plug.

The maximum operating temperature of the A.C. glow plug is

Condition: Normal
Normal spark plug of the proper heat range as shown by the light coating of oil-ash deposits on the insulator.

Condition: Detonation
Possible Causes:
1. Overadvanced ignition timing
2. Improper fuel octane

Remedy:
1. Check ignition timing — set to recommended specifications
2. Use a higher octane fuel

Condition: Preignition or overheating
Possible Causes:
1. Spark plug too hot
2. Advanced engine timing
3. Glowing piece of carbon in combustion chamber
4. Plugged or partially plugged cooling system

Remedy:
1. Install colder spark plug
2. Check ignition timing — set to recommended specifications
3. Clean or clear cooling system

Condition: Carbon fouled
Possible Causes:
1. A sticky carburetor choke
2. Faulty ignition primary circuit
3. Defective spark plug wires
4. Rich fuel mixture
5. Spark plug too cold

Remedy:
1. Clean carburetor choke plate and shaft
2. Check for excessive resistance in the ignition primary circuit
3. Replace spark plug wires
4. Correct the fuel system problem
5. Use hotter spark plug

Condition: Oil fouled
Possible Causes:
Caused by defective piston rings, valve seals resulting in excessive oil consumption

Remedy:
Determine cause of oiling problem and repair

Condition: Dirt fouled
Possible Causes:
Excessive dirt and grit entering the combustion chamber caused by not using an air filter or operating the vehicle in a dusty atmosphere.

Remedy:
Install new Fram air filter

Condition: Glazed insulator
Possible Cause:
Normally occurs from a spark plug which is too hot for the application. Fuel deposits have melted from intense heat and center electrode is blistered.

Remedy:
Install colder spark plug

Condition: Burnt spark plug
Possible Cause:
1. Spark plug too hot for application
2. Over advanced ignition timing

Remedy:
1. Install colder spark plug
2. Check ignition timing — set to recommended specifications.

Condition: Lead Fouled
Possible Cause:
Due to the tetraethyllead used to boost octane in fuels

Remedy:
Install new Autolite spark plugs

Condition: MMT Fouled (Methycyclopentadienyl Manganese Tricarbonyl)
Possible Cause:
MMT is used to boost octane in some unleaded fuels

Chart 2-1 Spark Plug Condition Indicators of Engine Problems. *Courtesy of Fram Division, Allied Automotive*

Engine Systems

1800° F. (982° C.). Because the fast-heat glow plug is about half the electrical resistance of a standard glow plug, a device called a controller is used to prevent over-temperature destruction.

When operating, the standard glow plug remains on for up to a minute and a half (90 seconds) after the engine starts. This maintains ignition in all cylinders, improves throttle response, and reduces exhaust smoke. This glow plug operating period is called "after-glow."

Figure 2-17 is a typical A.C. diesel engine glow plug. The heater coil element is a resistance wire centrally positioned in the sheath. The heater coil element is welded to the center terminal conductor and to the sheath. The heater coil element and the center terminal conductor are supported and electrically insulated from the sheath by magnesium oxide. The entire assembly is joined to a threaded shell, which has a tapered seat and a hex section. Above the hex section are an insulator and electrical terminal blade.

Exhaust Oxygen Sensor

With the increased concern for energy conservation and cleaner air, better control of the engine combustion process has

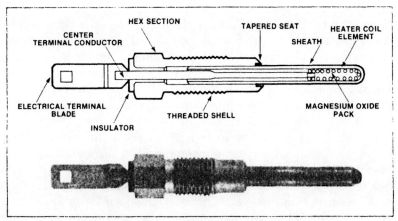

Figure 2-17 A Typical A.C. Diesel Engine Glow Plug. *Courtesy of AC Spark Plug Division of General Motors Corporation*

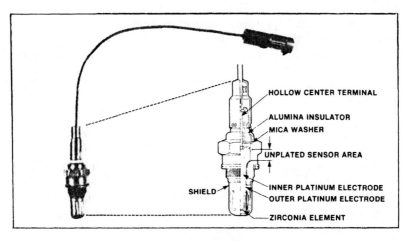

Figure 2-18 An Exhaust Oxygen Sensor. *Courtesy of AC Spark Plug Division of General Motors Corporation*

become necessary. AC Spark Plug Division produces an exhaust oxygen sensor for use in the internal combustion gasoline engine fuel control system.

The exhaust oxygen sensor adjusts and maintains desired air-fuel mixtures to better control exhaust emissions and fuel economy. In the general operation of the sensor (Figure 2-18), air enters the center of the zirconia element. When the zirconia element is activated by the heat of the exhaust gases passing around it, oxygen ions conduct through the element from the atmosphere to the outer electrode. The overall effect is to create a simple electrochemical cell which develops a voltage between the two electrodes. In use in an apparatus, as the air-fuel ratio becomes leaner, the oxygen concentration in the exhaust gas increases, and the output voltage drops to near zero. When the air-fuel ratio becomes richer, the oxygen concentration in the exhaust gas decreases, and the output voltage approaches one volt. The exhaust gas electrode is made of platinum to combine catalytically any unburned fuel and oxygen. This voltage signal is used, in combination with other components, to adjust and maintain desired engine air-fuel mixtures.

Engine Systems

Electronic (High-Energy) Ignition Systems

The primary current flow through the breaker points in a conventional ignition system is limited to about 2 amperes. This low amperage prevents burning of the points, but at times it may not be enough current for high-speed (or load) operations, particularly if the system is out of tune. This condition wastes fuel and greatly increases exhaust emissions.

The electronic ignition system (EIS) was developed to help reduce exhaust emissions. These systems offer the following advantages over a conventional system:

1. They use a higher secondary voltage through a vehicle's entire speed range.
2. They operate with a leaner fuel-air mixture.
3. They are not as subject to wear.
4. Engine dwell is pre-set and requires no adjustments.
5. Ignition maintenance is reduced to the inspection of wiring and connections, the distributor cap and rotor, and the lubrication of shaft and bushing areas.

Basic Electronic System Components

The principle of an electronic system is the same regardless of the manufacturer; however, the component parts vary from one manufacturer to another. Figure 2-19 shows a schematic based on the Chrysler system. This system is used simply for the purpose of illustration.

The starter, battery, and ignition coil in the electronic system are the same as in the conventional system except that the electronic ignition coil is designed to handle more primary current.

The EIS distributor contains a signal-generating device instead of points and a condenser. The signal-generating device induces a voltage signal in the pick-up coil and sends a timing pulse to the control unit.

Electronic Control Module

A switching transistor within the module receives the distributor timing pulse, which "tells" it to interrupt the coil's primary

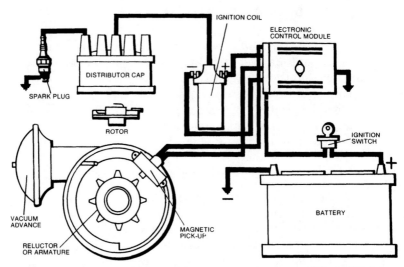

Figure 2-19 An Electronic Ignition System. *Courtesy of Fram Division, Allied Automotive*

circuit. The magnetic field in the coil's primary circuit collapses, inducing a high voltage in the coil's secondary windings.

Ballast Resistor

The ballast resistor regulates the primary current flow according to engine speed. It is bypassed during cranking to provide faster starting.

Ignition Cables

The ignition cables in an EIS are larger than those in a conventional system in order to prevent voltage leakage.

Spark Plugs

One feature of the EIS is its ability to ignite leaner fuel mixtures and thus further reduce hydrocarbon exhaust emissions. Some manufacturers specify spark plugs with wider electrode gaps, because of the inability of conventional plug settings to ignite the leaner fuel mixture continuously.

LUBRICATION SYSTEMS

The primary purpose of a lubrication system is to distribute an adequate amount of oil to the moving parts of an engine, in order to minimize wear and keep the loss of power due to friction to a minimum. However, the oil performs several other functions as it circulates throughout the engine.

1. It acts as a cooling agent, removing heat from the engine parts and carrying it back to the oil pan. The oil pan transfers the absorbed heat to the surrounding air.
2. It acts as a cleaning agent, washing dirt from the metal parts. The dirt is eventually filtered out by the oil filter.
3. It acts as a sealant, forming a good seal between the piston rings and the cylinder.
4. It acts as a shock absorber and noise reducer, absorbing and cushioning the hammerlike effects of the piston on the crankshaft, through the connecting rods.

Properties of Engine Oils

Perhaps the most important property of an engine lubricating oil is *viscosity*. Viscosity refers to an oil's internal friction or resistance to flow. The lower the viscosity, the more easily an oil will flow.

Oil viscosities have been assigned numbers by the Society of Automotive Engineers (SAE). The lower the number, the lower the viscosity. For example, SAE 10 oil flows much more easily than does one rated SAE 40. In general, low viscosity oil is recommended for cold-weather operation, while higher viscosity oil is recommended for warmer weather.

Special processing of oil permits use over a wide range of temperatures. Oils so processed are designated with a W. Thus, an SAE 10W oil can be used under more severe operating conditions and over a wider range of operating temperatures than can an oil with a simple SAE 10 rating.

Lubricating oils are also rated by a service designation. The service designation identifies the type of service for which an oil is best suited. There are six service ratings for gasoline engines and four for diesel engines. Ratings for gasoline engines start with an S; SA, SB, SC, SD, SE, and SF. Ratings for diesel engines start with a C; CA, CB, CC, and CD, with the later letters indicating oil approved for more severe operating conditions. For example, a CD oil is used for more severe operating conditions than is a CA oil.

Oil Changes

There is no hard-and-fast rule as to when oil should be changed that applies to all engines. Too much depends on the driving conditions, on the condition and type of engine, and on the quality and type of oil used.

Modern engines are sensitive to corrosion, rusting, damage from high engine compression, and other factors caused by increased speeds and bearing loads. Engine lubricating oil has been improved by additives. These include: corrosion and rust inhibitors; detergent-dispersant additives to prevent sludge and varnish deposits; pourpoint depressants to maintain pour characteristics in cold weather; foam inhibitors to prevent foaming; and oxidation inhibitors to reduce the possibility of oil oxidization at high temperatures.

Oil begins to lose its effectiveness from the first day it is put in the engine. The loss is primarily due to the wearing out of the additives. This breakdown of additives allows gum and varnish to form and corrosion and rust formations to begin. These contaminants start circulating with the oil. Some are removed by the oil filter, but some continue throughout the system. Over a period of time the buildup of contaminants will become so great that the contaminants will become a hazard to engine wear.

Different engine manufacturers recommend different periods of time between oil changes. In most cases, the oil in fire apparatus is changed even more frequently than is recommended by the manufacturer. Oil in fire apparatus should be changed in accordance with the operating procedures of individual departments. It is probably wise for apparatus operators to use the engine manufacturer's recommendations on oil changes in the absence of department policies.

Engine Systems 75

Oil Pumps

Oil pumps used in fire apparatus engines are of the *positive displacement* type. The theory of positive displacement pumps is discussed in Chapter 7. They are usually driven from the camshaft through the use of either gears or cams. The most common types in use are the gear and the rotor.

Oil Coolers

Some engines use an *oil cooler* to absorb some of the heat that is transferred from the hot engine parts to the oil. The oil cooler keeps the oil at a workable temperature.

Oil Filters

Every engine lubrication system has an oil filter. The filter removes dirt and abrasives from the oil, therefore keeping these contaminants from returning to the engine. The filters do not remove diluents such as gasoline and acids.

Oil filters are cartridges made of material which is capable of removing contaminants. The two types in general use are the bypass filter and the full-flow filter. Bypass filters filter part of the oil from the oil pump. Full-flow filters filter all the oil in circulation within the system.

COOLING SYSTEMS

The purpose of the *cooling system* is to keep the engine within its most efficient operating range at all engine speeds and during all driving conditions. Additional cooling methods may be employed during pumping operations.

While the individual construction of components of cooling systems varies from one engine to another, the principles involved are basically the same. The cooling system employed in Detroit V-6 and V-8 diesel engines will be used here for purposes of explanation.

A radiator and fan effectively dissipate the heat generated by the engine, in order to protect the engine and prevent severe damage.

The engine is equipped with a centrifugal water pump that circulates the coolant through the cylinder block water jackets, cylinder heads, and lubricating oil cooler. Thermostats maintain a normal engine operating temperature range of 160°F to 185°F.

Coolant is drawn from the lower portion of the radiator by the water pump and is forced through the oil cooler and into the cylinder block. From the cylinder block, the coolant passes up through the cylinder heads and, when the engine is at normal operating temperature, through the thermostat housing and into the upper portion of the radiator (see Figure 2-20). Then the coolant passes down a series of tubes, where its temperature is lowered by the air stream created by the revolving fan.

During the starting of a cold engine, or when the coolant is below operating temperature, the coolant is restricted at the thermostat housing, and a bypass provides water circulation within the engine during the warm-up period.

A properly maintained and cleaned cooling system will reduce engine wear and increase the satisfactory engine operating time between engine overhauls. This is accomplished by eliminating the hot spots within the engine. When they are operating within the proper engine temperature range and the recommended horsepower of the unit, all engine parts will be within their operating temperature ranges and at their proper operating clearances, as specified by the engine manufacturer.

Engine Coolant

The function of the *engine coolant* is to absorb the heat, developed as a result of the combustion process in the cylinders, along with the heat from the component parts such as exhaust valves, cylinder liners, and pistons, which are surrounded by water jackets. In addition, the engine coolant also removes heat absorbed by the oil in the oil-to-water oil cooler.

Engine Systems

Water Pump

The *centrifugal water pump* circulates the engine coolant through the cylinder block, cylinder heads, radiator, and oil cooler. The pump is mounted on the engine front cover and is driven by the front camshaft gear.

A bronze impeller is secured to one end of a stainless steel shaft by a lock nut. The water pump gear is pressed on the opposite end of the shaft. Two ball bearings carry the shaft. The larger bearing, at the drive gear end of the shaft, accommodates the thrust load. An oil seal is located in front of the smaller bearing

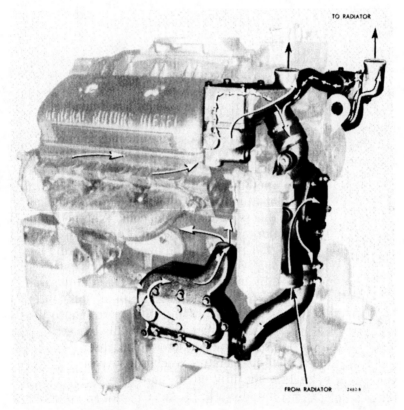

Figure 2-20 The Cooling System of a Detroit V-6 and V-8 Engine. *Courtesy of Detroit Diesel Allison, Division of General Motors Corporation*

and a spring-loaded face-type water seal is used behind the impeller. The pump ball bearings are lubricated with oil splashed by the camshaft gear and the water pump gear.

Water Manifold

The engines do not require external water manifolds. Coolant leaves the cylinder head through an opening directly over each exhaust port, and enters the water manifold, which is attached to the cylinder head with two nuts and lock washers at each of the water openings. A separate gasket is used at each attaching flange between the manifold and cylinder head.

Thermostat

The temperature of the engine coolant is automatically controlled by a *thermostat* located in a housing attached to the water outlet of each cylinder head. Blocking thermostats are used when a standard cooling system is employed; semi-blocking thermostats are used with the rapid-warm-up cooling system.

At coolant temperatures below approximately 170°F., the thermostat valves remain closed and block the flow of coolant to the radiator. During this period, all of the coolant in the standard system is circulated through the engine and is directed back to the suction side of the water pump via the bypass tube. In the rapid-warm-up system, enough coolant to vent the system is bypassed to the radiator top tank by means of a separate external deaeration line and then sent back to the water pump without going through the radiator cores. As the coolant temperature rises above 170°F., the thermostat valves start to open, restricting the bypass system, and permit a portion of the coolant to circulate through the radiator. When the coolant temperature reaches approximately 185°F., the thermostat valves are fully open, the bypass system is completely blocked off, and all of the coolant is directed through the radiator.

A defective thermostat that remains closed or only partially open will restrict the flow of coolant and cause the engine to

Engine Systems

overheat. A thermostat stuck in a full open position may not permit the engine to reach its normal operating temperature. The incomplete combustion of fuel due to cold engine operation will result in excessive carbon deposits on the pistons, rings, and valves.

Properly operating thermostats are essential for efficient operation of the engine. If the engine operating temperatures deviate from the normal range of 160°F. to 185°F., the thermostats should be removed and checked.

Radiator

The temperature of the coolant circulating through the engine is lowered by the action of the *radiator* and the *fan*. The radiator is mounted in front of the engine so that the fan will draw air through it, thereby lowering the coolant temperature to the degree necessary for efficient engine operation.

The life of the radiator will be considerably prolonged if the coolants used are limited to either clean, soft water with a rust inhibitor or a mixture of water and a high-boiling-point antifreeze. The use of any other type of antifreeze is not recommended.

A metal shroud is placed around (and slightly in back of) the fan to increase the cooling efficiency of the radiator. The fan shroud must be airtight against the radiator to prevent recirculation of the hot air drawn through the radiator. Hot air that is permitted to pass around the sides or bottom of the radiator will cause overheating of the engine.

Another cause of overheating is slippage of the fan drive belts. This is caused by incorrect belt tension, worn belts, worn fan belt pulley grooves, or the use of fan belts of unequal length when two or more belts are used. The belt tension and condition of the belts should be checked periodically.

A radiator that has a dirty, obstructed core, a leak in the cooling system, or an inoperative thermostat will also cause the engine to overheat. The radiator must be cleaned, the leaks eliminated, and defective thermostats replaced immediately to prevent serious damage from overheating. The external cleanliness of the radiator should be checked if the engine overheats and no other causes are apparent.

Engine Cooling Fan

The *engine cooling fan* is belt-driven from the crankshaft pulley. The three-groove pulley hub turns on a double-row ball bearing at the front of the hub and a single-row (shielded) ball bearing at the rear. On compact front-end engines the pulley hub turns on tapered roller bearings.

Thermo-Modulated Fan

A *thermo-modulated suction fan assembly* is provided on some engines. This fan assembly is designed to regulate the fan speed and maintain an efficient engine coolant temperature, regardless of the variations in the engine load or outside air temperature.

The entire fan drive assembly is a compact integral unit which requires no external piping or controls and operates on a simple principle. This system transmits torque from the input shaft to the fan by the shearing of a silicon fluid film between the input and output plates in a sealed, multiplate, fluid-filled clutch housing.

The thermostatic control element, which is an integral part of the fan drive, reacts to changes in engine temperature and varies the fluid film thickness between the plates, thereby changing the fan speed. The vehicle manufacturer selects the control element setting to maintain optimum cooling, and no further adjustment should be necessary.

The thermo-modulated fan is mounted and driven by the engine in the same manner as the conventional fan.

Coolant Filter and Conditioner

The engine cooling system *filter and conditioner* is a compact bypass unit with a replaceable element.

A correctly installed and properly maintained coolant filter and conditioner unit provides a cleaner engine cooling system, greater heat dissipation, and increased engine efficiency through improved heat conductivity and contributes to longer life of engine parts.

The filter provides mechanical filtration by means of a closely packed element through which the coolant passes. Any impurities suspended in the cooling system, such as sand and rust particles, will be removed by the straining action of the element. The removal of these impurities will contribute to longer water pump life and proper operation of the thermostat.

The filter also conditions the coolant by softening the water to minimize scale deposits, maintains an acid-free condition, and acts as a rust preventive.

Corrosion inhibitors placed in the element dissolve into the coolant, forming a protective rustproof film on all of the metal surfaces of the cooling system. The other components of the element clean and prepare the cooling passages while the corrosion inhibitors protect them.

ELECTRICAL SYSTEMS

The electrical system generally consists of a starting motor, an alternator (generator), a voltage regulator, a current regulator and cutout relay, a storage battery, and the necessary wiring. The radio, lights, indicating gages, and other electrically operated devices are usually considered to be accessory items, although they are, in reality, part of the overall electrical system.

Starting Motor

The *starting motor* is a direct-current electrical motor that is designed specifically for turning over the engine at a sufficient speed to permit starting. Power to turn the starting motor comes from the battery. Closing the ignition switch connects the motor to the battery.

Figure 2-21 is a cross-section drawing of a starting motor equipped with a Sprag heavy-duty clutch drive. A solenoid

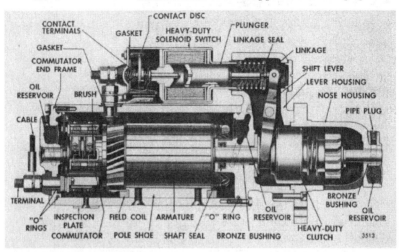

Figure 2-21 A Cross Section of a Starting Motor with a Spray Heavy-Duty Clutch Drive. *Courtesy of Detroit Diesel Allison, Division of General Motors Corporation*

switch, mounted on the starting motor housing, operates the current spray overrunning clutch drive by linkage and a shift lever. When the starting switch is engaged, the solenoid is energized. This shifts the starting motor pinion into mesh with the flywheel ring gear and closes the main contacts within the solenoid. Once engaged, the clutch will not disengage during intermittent engine firing. To protect the armature from excessive speed when the engine starts, the clutch "overruns," or turns faster than the armature, which permits the pinion to disengage itself from the flywheel ring gear.

The solenoid plunger and shift lever on this type of starting motor are totally enclosed to protect them from dirt, water, and other foreign material.

An oil seal, between the shaft and the lever housing, and a linkage seal prevent the entry of transmission oil into the main frame of the starting motor and solenoid case, allowing the motor to be used on wet clutch applications.

The starting motor drive pinion and the engine flywheel ring gear must be matched to provide positive engagement and to avoid clashing of the gear teeth. Flywheel ring gear teeth have either no

Engine Systems

chamfer, a Bendix chamfer, or a Dyer chamfer. *Chamfer* describes the type of bevel on the edge of the gear tooth.

Flywheel ring gears with no chamfer are used with starting motors equipped with an overrunning clutch drive. Ring gears with a Bendix chamfer are used with starting motors equipped with either a Bendix drive or an overrunning clutch drive. Ring gears with the Dyer chamfer, used with Dyer-type starting motors, may be reversed and used with the overrunning clutch drive starting motors.

Alternators (Generators)

The *alternator* (generator) provides electrical current to maintain the storage battery in a charged condition. It also supplies sufficient current to carry any other electrical load requirements up to the rated capacity of the alternator.

Direct-current alternators are manufactured in a wide range of sizes and types, but the basic design of all of them is the same. The size and type of alternator applied to a particular engine will depend on many factors, including maximum electrical load, type of service, ratio of engine idling time to running time, type of drive, drive ratio (engine speed to alternator speed), alternator mounting, and environmental conditions.

Most fire apparatus use an alternating current self-rectifying generator, generally referred to as an alternator. Figure 2-22 shows a cutaway of a typical alternator. These alternators are especially beneficial on apparatus with extra electrical accessories and those that must operate for extended periods at idle speeds. Diodes, built into the slip ring end frame, rectify the three-phrase A.C. voltage to provide D.C. voltage at the battery terminal of the alternator, thereby eliminating the need for an external rectifier. Alternators are available in a variety of sizes and types.

The proper selection of an alternator is extremely important. It must meet all the electrical demands of the apparatus and must be able to charge the batteries properly. Many apparatus are equipped with three-phase transformers and a rectifier unit which produces 110-volt D.C. current at special outlets.

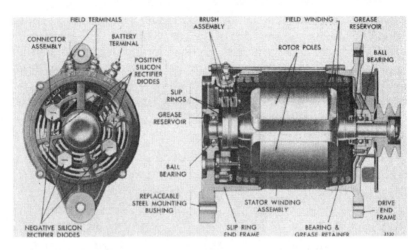

Figure 2-22 A Typical A.C. Self-Rectifying Alternator (Generator). *Courtesy of Detroit Diesel Allison, Division of General Motors Corporation*

Regulator (D.C. Circuit)

Several protective devices are employed to regulate the voltage and current output of the alternator (generator) and to maintain a fully charged storage battery, the type of which depends on the type of electrical system. While some alternators utilize an internal protection system, others use an external device. The most representative of these devices is the "three-unit" regulator.

The two types of three-unit regulators are identified as:

1. A "Circuit A" unit, in which the generator field circuit is connected to ground within the regulator. This regulator is used only with alternators having an externally grounded field circuit.

2. A "Circuit B" unit, in which the alternators field circuit passes through the regulator and returns to ground inside the alternator itself. This regulator must be used only with "Circuit B" alternators in which the field is internally grounded.

The regulators are dustproof and moisture-proof. On most applications, it is necessary to use shock mounts, which insulate the regulator against vibration but require the installation of a ground lead.

Engine Systems 85

The three-unit regulator consists of a *cutout relay*, a *voltage regulator*, and a *current regulator* mounted in a single assembly, as shown in Figure 2-23. These three units are basic and generally apply to most regulators in a D.C. alternator system.

Cutout Relay

The *cutout relay* (Figure 2-24) has two heavy windings assembled on one core: a series winding of a few turns of heavy wire and a shunt winding of many turns of fine wire. The relay core and windings are assembled into a frame. A flat steel armature is attached to the frame so it is centered just above the center of the core. The armature has two or more contact points, which are located just above a similar number of stationary contact points.

When the engine is not running, the armature contact points of the relay are held away from the stationary points by tension of a leaf spring. As the engine starts and the alternator (generator) speed increases, the current flowing through the shunt winding builds up until it reaches the value for which the relay has been set. At this point, sufficient magnetism overcomes the armature

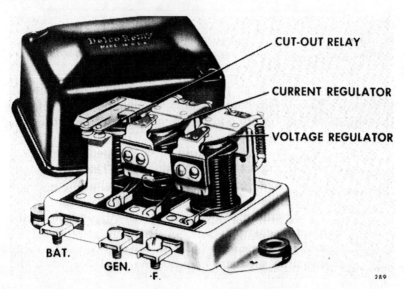

Figure 2-23 A Typical Regulator Assembly. *Courtesy of Detroit Diesel Allison, Division of General Motors Corporation*

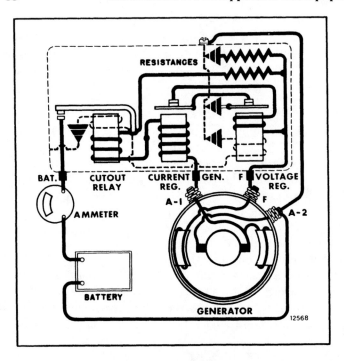

Figure 2-24 Wiring Circuit of a Typical Three-Unit Regulator. *Courtesy of Detroit Diesel Allison, Division of General Motors Corporation*

spring tension, the contact points close, and the current flows to the battery. The current that then flows through the series winding is in the right direction to add to the magnetic force holding the armature down and the points closed.

When the engine is slowed down or stopped, current will begin to flow from the battery to the alternator. This reverses the direction of the current flow through the series winding magnetic field. The magnetic field of the shunt winding does not reverse. Therefore, the two windings now oppose each other magnetically, and the resultant magnetic field is not strong enough to hold the armature down. The leaf spring pulls the armature away from the core and the points separate, opening the circuit between the alternator and the battery.

The regulator cutout relay contact points must never be closed by hand with the battery connected. This would cause a high current flow through the units and damage them.

Engine Systems

Voltage Regulator

The *voltage regulator* (Figure 2-24) has two windings on a single core. One is a shunt winding consisting of many turns of fine wire. In series with a resistor, it is shunted across the alternator at all times. The second winding is a field current winding connected between the alternator field circuit and ground whenever the regulator contact points are closed. In addition to the core frame, armature, and contact points, the unit has a spiral spring that holds the armature away from the core so the two contact points are touching when the voltage regulator is not operating.

When the alternator voltage reaches the value for which the regulator is adjusted, the combined magnetic field produced by the shunt winding and the field current winding overcomes the armature spring tension, pulls the armature down, and separates the voltage regulator contact points. This introduces resistance into the alternator field circuit so that the alternator field current and alternator voltage are reduced. The lowering of the output of the alternator causes the points to close again, thereby removing the resistance and increasing the alternator output. The complete cycle of opening and closing the points and alternately inserting and removing the resistance in the alternator field circuit is carried out rapidly, thus limiting the alternator voltage to a predetermined maximum value. With the alternator voltage limited, the alternator supplies varying amounts of current to meet the requirements of varying states of battery charge and electrical loads.

Current Regulator

Outwardly the current regulator looks like the voltage regulator. It contains two windings assembled on one core—a series winding and a field current winding. The series winding, consisting of a few turns of heavy wire, is connected into the charging circuit so that the full output of the alternator passes through it. The field current winding is connected in series with the alternator field circuit so that the field current flows through the field winding when the regulator contact points are closed.

When the current reaches the value for which the current regulator is adjusted, the magnetism produced by current flowing through the series winding overcomes the armature spring tension and the contact points open. This inserts a resistance into the alternator field circuit, resulting in a drop in alternator output. Immediately, the magnetic field of the series winding is weakened, the contact points close, the alternator output starts to increase, and the cycle is repeated. This action prevents the alternator from exceeding its rated output.

Therefore, when the load demand is heavy, alternator output will increase until it reaches the current value for which the current regulator is set; the current regulator will then begin to operate and regulate the current output from the alternator.

Storage Battery

The lead-acid storage battery is an electrochemical device for converting chemical energy into electrical energy. The battery has three major functions:

1. It provides a source of current for starting the engine.
2. It stabilizes the voltage in the electrical system.
3. It can, for a limited time, furnish current when the electrical demands exceed the output of the alternator.

There are two types of batteries in use today:

1. The dry charge battery contains fully charged positive plates and negative plates separated by separators. The battery contains no electrolyte unit; it is activated for service in the field and leaves the factory dry. Consequently, it is called a dry charge battery.

2. If the battery has been manufactured as a wet battery, it will contain fully charged positive and negative plates plus an electrolyte. This type of battery will not maintain its charged condition during storage and must be charged periodically to keep it ready for service.

Most heavy-duty fire apparatus are equipped with two banks of batteries, each having the capacity to serve the system individually. Most manufacturers supply two 6-volt batteries in each bank. A single 12-volt battery may be used in each bank, but this is not preferred. The batteries should be installed and wired so that both banks may be used simultaneously for starting the engine, and adequate switching arrangements must be provided so that either

Engine Systems

of the banks may be used for starting, ignition, or lighting; for ignition and lights alone; and for starting alone.

Servicing the Battery

A battery is a perishable item which requires periodic servicing. Only when the battery is properly cared for as described below can long and trouble-free service be expected.

    1. Check the level of the electrolyte regularly. When necessary, add water, distilled if local water is not suitable, but do not overfill. Overfilling can cause poor performance or early failure.

    2. Keep the top of the battery clean. When necessary, wash with a baking soda solution and rinse with fresh water. Do not allow the soda solution to enter the cells.

    3. Inspect the cables, clamps, and hold-down bracket regularly. The terminal posts and cable connections must be kept clean and free of corrosion. A thin layer of grease can be applied to retard corrosion buildup. Any corroded or damaged parts should be replaced.

    4. Use the standard battery test as the regular service test to check the condition of the battery.

    5. Check the electrical system if the battery becomes discharged repeatedly.

Many electrical troubles caused by battery failures can be prevented by systematic battery service. In general, the care and maintenance recommendations for storage batteries have traditionally remained constant.

Battery Safety Precautions

An explosive gas mixture forms beneath the cover of each cell when batteries are being charged. Part of this gas escapes through the holes in the vent plugs and may form an explosive atmosphere around the battery itself if ventilation is poor. This explosive gas may remain in or around the battery for several hours after it has been charged. Sparks or flames can ignite this gas, causing an internal explosion which could shatter the battery.

REVIEW QUESTIONS

1. What are some of the characteristics of gasoline?
2. What are some of the characteristics of the diesel fuel that is used in fire apparatus diesel engines?
3. What does the cetane number of a diesel fuel measure?
4. What are the units in a gasoline engine fuel system?
5. What is the minimum size of gasoline tank that should be used on fire apparatus?
6. How does a mechanical fuel pump for a gasoline engine work?
7. What are the two types of electrical fuel pumps?
8. What are some of the advantages of using an electrical fuel pump instead of a mechanical fuel pump?
9. What are the two types of fuel gages used on gasoline engines?
10. What is a simple definition of carburetion?
11. What is meant by an air-fuel ratio of 12 to 1?
12. What are the five separate systems (or circuits) in a carburetor?
13. What is a venturi?
14. What is the objective of a float system in a carburetor?
15. What happens if the fuel level in a carburetor bowl is too high? Too low?
16. At what throttle openings is fuel discharged from the main jet in a carburetor?
17. How is gasoline fed to an engine when the engine is idling or turning at a low speed?
18. What type of mixture is provided to an engine when it is idling?
19. What is the purpose of the accelerating pump system in a carburetor?
20. What is the purpose of the power enrichment system in a carburetor?
21. What are the two types of power enrichment systems used in carburetors?
22. Why is a choke required when starting a cold engine?

Engine Systems

23. On what principle do most automatic chokes work?
24. Where are the fuel nozzles in a gasoline fuel injection system located?
25. Where is the fuel nozzle in a diesel fuel injection system located?
26. What are the five separate functions which must be performed by a diesel fuel injection system?
27. What are the three general types of diesel fuel injection systems in use?
28. What is the objective of an ignition system?
29. What parts comprise a conventional ignition system?
30. What is the flow of current from battery to ground in an ignition system?
31. Will the battery run down if the ignition switch is left on?
32. What is the purpose of the ignition coil in an ignition system?
33. What is the composition of the primary coil windings? Secondary windings?
34. What causes the current in the secondary windings in an ignition system?
35. What two basic functions are served by the distributor in an ignition system?
36. What two methods are in general use for advancing or retarding the spark as conditions require?
37. What controls the vacuum advance? The mechanical advance?
38. What is the purpose of the condenser in an ignition system?
39. What dual purpose is served by the ignition cables in an ignition system?
40. What is the primary function of a spark plug?
41. What is meant by the heat range of a spark plug?
42. What is the difference between a "hot" spark plug and a "cold" spark plug?
43. For what type of operating conditions would a "hot" spark plug be best suited? A "cold" spark plug?
44. What is the purpose of a diesel engine glow plug?
45. What are the two types of glow plugs commonly used?

46. What is meant by "pre-glow" time?
47. What is the maximum pre-glow time for an A.C. fast heat glow plug?
48. What is the maximum pre-glow time for an A.C. standard glow plug?
49. What is meant by "after-glow" time?
50. How long does the standard A.C. glow plug remain on after the engine starts?
51. What is the purpose of an exhaust oxygen sensor?
52. How does an exhaust oxygen sensor achieve its objective?
53. Why was the electronic ignition system developed?
54. What are the advantages of an electronic ignition over a conventional ignition system?
55. What is the primary difference between an electronic ignition system and a conventional ignition system?
56. What is the primary purpose of an engine's lubrication system?
57. What are some of the functions performed by the oil in a lubrication system?
58. What is probably the most important property of an engine lubricating oil?
59. What are the meanings of the SAE numbers that are assigned to oils?
60. What is the special characteristic of an oil with a W rating?
61. What are the different service designations for oils used in gasoline engines? In diesel engines?
62. Why does oil have to be changed?
63. What usually drives the oil pump?
64. What is the objective of the oil cooler in the lubrication system?
65. What are the two general types of oil filters in use?
66. What is the purpose of an engine cooling system?
67. What is the normal operating temperature range of a Detroit V-8 engine?
68. What controls the temperature of the coolant within an engine?
69. What is the difference between a blocking-type thermostat and a semi-blocking-type thermostat?

Engine Systems

70. How does a thermostat valve control engine temperature?

71. What would be the result if a thermostat valve stuck closed? Stuck open?

72. How does a thermo-modulated cooling fan differ from a conventional cooling fan?

73. What parts normally make up the electrical system of an apparatus?

74. Is the starting motor a D.C. motor or an A.C. motor?

75. How is the armature of a starting motor protected from excessive speed when the engine starts?

76. What is the purpose of an alternator (generator)?

77. Why are alternators used on fire apparatus rather than generators?

78. What are the three parts of a three-unit regulator?

79. How does a cutout relay work?

80. What would happen if the cutout relay contact points were closed by hand with the battery connected?

81. How does a voltage regulator work?

82. How does a current regulator work?

83. What are the three major functions of a storage battery?

84. What precautions should be taken when charging a battery?

3

The Chassis and Component Parts

THE CHASSIS

Chassis is a French word meaning frame. Over a period of years the meaning has changed to include the springs, axles, wheels, and other parts of the apparatus. In the fire service the chassis is also thought to include the cab. To the chassis are added the pump, pump accessories, hose bed, compartments, water tank, and other component parts that make the apparatus functional.

Fire apparatus chassis are available in three different types. Some chassis are designed and custom-built from the ground up, in accordance with the purchase specifications of the buyer. Others are designed as a package, to which a body and other components are added to meet the purchaser's requirements (see Figure 3-1). Some fire apparatus are built on standard commercial

96   Introduction to Fire Apparatus and Equipment

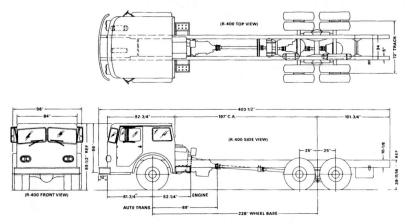

Figure 3-1 An Apparatus Chassis Available as a Package. *Courtesy of Nolan Company, Duplex Truck Division*

truck chassis. Regardless of the method of design and construction, the main member of the chassis is the frame.

The Frame

The *frame* is the backbone of the apparatus. It is to an apparatus what a foundation is to a house. The frame is the chassis member that must hold the whole set of components in such a way that they do not permanently sag or twist, while carrying the full load of the apparatus during operations.

A frame must be constructed properly in order to withstand the severe strains and shocks of acceleration, twisting, swaying, rocking, and braking to which the apparatus will be subjected. It consists of two main channels, running fore and aft, and transverse cross members, along with brackets, hangers, and other parts on which to mount the various chassis components.

Frames are generally made of either carbon steel or carbon manganese, an alloy steel. Carbon steel is more often used because of its adaptability. It can be straightened if bent, it can be welded, and it is comparatively easy to machine. Carbon manganese, on the other hand, has more strength, but it loses its physical properties when heated, it should not be welded, and it is more difficult to machine.

The Chassis and Component Parts

Gross Vehicle Weight

The frame is designed and engineered to support a maximum load. This maximum is designated as the rated gross vehicle weight (GVW). Included in this maximum is the weight of the chassis, body, all equipment carried on the apparatus, the water and water tank, the crew, and all other loads. During the design stage, a weight of 1200 pounds is usually allowed for the crew, along with a weight of 10 pounds for each gallon of water to be carried. The 10 pounds includes the weight of the tank.

To determine if the GVW is being exceeded, the apparatus should be weighed when it is fully loaded with all personnel aboard. Most apparatus with all equipment and personnel aboard will not weigh as much as the rated GVW. Apparatus operators, however, should remain alert to the total weight carried, as it is not unusual for equipment to be added from time to time with little thought given to the effects of the increased weight.

The GVW also affects the design of the suspension system, the rated axle capacity, the rated tire loading, and the distribution

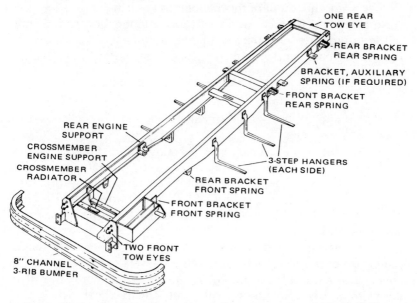

Figure 3-2 A Typical Frame Assembly. *Courtesy of Ward LaFrance Truck Corporation*

of weight between the front and rear wheels. An improper distribution of weight between the wheels will seriously affect the handling and riding characteristics of the apparatus.

Care should be taken by both the manufacturer and the user not to exceed the rated GVW of an apparatus. Exceeding the rated weight will make the apparatus more difficult to handle and will reduce the life of such component parts as the running gear, clutch, brakes, and transmisison. It will also increase the downtime and maintenance costs of an apparatus.

THE POWER TRAIN

The *power train* is a set of components and units that are used to transmit rotation force in the engine to the wheels. Some of the component parts or units of a power train are a clutch, a transmission, a driveline, and rear axles.

It is not the intent of this chapter to describe in detail all the various types and makes of power train components or units, but rather to provide the apparatus operator with a basic understanding of how each works, and how each contributes to the overall movement of the apparatus.

Clutches

The *clutch* is a device for connecting and disconnecting the engine from the transmission. In heavy-duty apparatus it usually consists of a pressure plate assembly, an intermediate pressure plate, and two discs (see Figure 3-3). For illustrating the principle of operation, it will be considered to consist of a pressure plate assembly and a single disc.

The pressure plate assembly consists of a *heavy pressure plate, coil* or *diaphragm springs, release levers* or *bearings,* and a *cover*. The assembly is bolted to the flywheel, with the disc mounted between the pressure plate and the flywheel.

The clutch is engaged when the clutch pedal is out. At this time, the heavy springs acting on the pressure plate compress the

The Chassis and Component Parts

clutch disc between the plate and the engine flywheel, thus making engine power available to the transmission.

The spring pressure must be released to disengage the clutch, through the use of clutch release bearings and levers, or fingers, depending upon the design of the clutch.

The engagement and disengagement of the clutch assembly is controlled by the clutch pedal, working through a set of linkages. Pushing down the clutch pedal activates the linkage, which moves the clutch release bearing or forks against the release levers on the pressure plate, thus relieving the spring pressure on the pressure plate, which in turn releases the clutch disc from the flywheel.

The heavy-duty springs return the pedal to the driving position when the driver takes his foot off the pedal. This, in turn, releases the clutch bearing or fork, allowing the springs to put pressure on the pressure plate, which, in turn, compresses the clutch disc between the pressure plate and the engine flywheel.

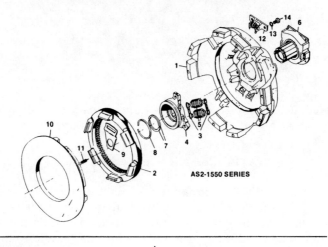

1 FLYWHEEL RING
2 ADJUSTING RING
3 SPRING PIVOT
4 RELEASE SLEEVE RETAINER
5 PRESSURE SPRINGS
6 RELEASE BEARING & SLEEVE ASSEMBLY
7 RELEASE SLEEVE RING
8 SNAP RING
9 LEVER
10 PRESSURE PLATE
11 RETURN SPRING
12 SELF ADJUSTER ASSEMBLY
13 LOCK WASHER
14 BOLT

Figure 3-3 An Exploded View of a Clutch. *Courtesy of Dana Corporation, Spicer Clutch Division*

Figure 3-4 A Knife-Edge-Design Clutch with a Self-Adjusting Mechanism. *Courtesy of Dana Corporation, Spicer Clutch Division*

Automatic Clutch Adjustment

Some clutches are equipped with an adjusting mechanism which automatically adjusts the clutch as wear takes place (see Figure 3-4). The adjusting mechanism, which is a replaceable section of the clutch cover, checks for facing wear each time the clutch is actuated. Once facing wear exceeds a predetermined amount, the adjusting ring is actuated by the adjusting mechanism and clutch free pedal dimensions are automatically returned to normal operating conditions.

The Coaxial Spring Damper

One of the improvements in clutch construction is incorporated in the *Spicer coaxial spring damper-driven disc assembly.*

Here is a brief explanation of the functions and design concept of this assembly.

Members in the truck driveline do not rotate at constant speeds. For example, the engine flywheel, while rotating at a specific rpm, will speed up and slow down several times in each revolution because of the periodic firing of each engine cylinder. Driveshafts (universal joints) also cause non-constant rotation in the drivetrain, because of geometry changes in a rotating joint set.

When the non-constant rotational response (NCR) of a member in the system occurs simultaneously with an NCR response of the remainder of the drivetrain system, resonance greatly increases the vibrational amplitude of the whole drivetrain. This type of vibration is known as a torsional vibration.

When transmissions of ten or more speeds were common, the resulting splits, occurring in the narrow range between 1800 and 2100 rpm, did not have critical torsional vibrations (resonance).

With the high-torque, fuel-efficient, lower-governed rpm engines, however, fewer speed transmissions are used, so that each gear change makes use of a broad range of engine speeds. It is impossible to operate over such a broad range of speeds without encountering a critical torsional area (resonance).

The resonance of a drivetrain and its resultant torsional vibrations can be moved out of the normal operating range of the drivetrain by changing the mass (weight), stiffness, or dampening of the drivetrain.

The coaxial damper (a spring within a spring) offers the dampening needed to tune the whole drivetrain system, so that these critical torsional vibrations are moved out of the operating speed range of the engine and the remainder of the drivetrain. By doubling the number of springs, the stress level in each spring is lowered in order to increase the life of the driven disc assembly. At the same time, torque capacity has been increased to accommodate engines with gross torque ratings up to 1400 foot-pounds.

Neglecting to use damper drive discs will cause a drivetrain to operate with torsional vibrations. These vibrations cause rapid wear of such components as transmission input shafts and subject the remainder of the transmission prop shaft and the axle to repeated high torque loadings, which, in turn, shorten their life as well.

Clutch Failure

The major cause of clutch failures can be summarized in two words: "excessive heat." Excessive heat is not the amount of heat a clutch can normally absorb and dissipate, but the amount of heat a clutch is forced to absorb and must attempt to dissipate.

Most clutches are designed to absorb and throw off more heat than is encountered in normal clutch operation without damage or breakdown of the friction surfaces. Clutch installations are engineered to last many thousands of miles under normal operating temperatures and, if properly used and maintained, they will give satisfactory service.

However, if a clutch is "slipped" excessively or asked to do the job of a fluid coupling, high heat quickly develops to destroy the clutch. Temperatures generated between the flywheel, driven discs, and pressure plates can be high enough to cause the metal to flow and the friction facing material to char and burn.

Heat or wear is practically nonexistent when a clutch is fully engaged, but considerable heat is generated during the moment of engagement, when the clutch is picking up the load. An improperly adjusted or slipping clutch will rapidly generate sufficient heat to destroy itself.

Clutch Brake

The rapidity with which an upshift can be made in any unsynchronized transmission is limited by the time it takes for the free-spinning clutch disc and countershaft assemblies to slow down to the slower rotating speed of the transmission mainshaft. Clashless sliding gear engagement is possible only at the moment these parts synchronize in speed. A *clutch brake* can be used to reduce this slow-down time and hold the apparatus momentum loss to a minimum.

Figure 3-5 shows a type of clutch brake that is available for use on Mack fire apparatus. The brake is assembled on the transmission lower countershaft. Since the countershaft is always connected to the clutch disc through the constant mesh driving pinion, action of the brake will overcome the tendency of the discs to continue to rotate at high speed when the clutch is disengaged.

When making an upshift, the driver depresses the clutch

The Chassis and Component Parts

pedal the normal amount, separating the clutch plates and allowing him to shift the transmission into neutral. He then presses down slightly farther on the clutch pedal and arm (1), which is located on the lower side of the throwout yoke that bears against the set screw (2), which pushes the brake pressure plate and lining assembly (3) into contact with the flange (4). This action applies a braking force to the freely spinning clutch disc and countershaft. Drivers quickly develop a sense of timing that enables them to apply just the right amount of braking effort, so that the gears can be meshed without the usual time delay. "Double-clutching" should be used on all downshifts.

Transmissions

The *transmission* is a device for controlling the speed and power, which go from the engine to the wheels. It is located directly behind the clutch in the power train. The power of the

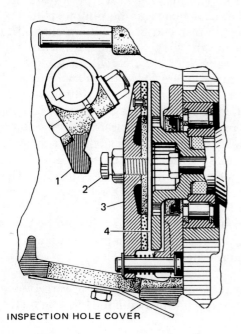

INSPECTION HOLE COVER

Figure 3-5 A Schema of a Clutch Brake. *Courtesy of Mack Trucks, Inc.*

engine goes to the transmission through the clutch. The manual transmission and the automatic transmission are the two basic types used on fire apparatus.

Manual Transmissions

A manual transmission enables the driver to select the amount of power transmitted to the rear wheels and the speed at which the rear wheels turn in relation to the engine speed. In the lower gears the engine turns faster than the output (drive) shaft. In high gear (direct drive or 1-to-1 ratio), the drive shaft turns at the same speed as the engine, and in overdrive, if the apparatus is so equipped, the drive shaft turns faster than the engine.

Most fire apparatus that have manual transmissions use transmissions that are equipped with at least four speeds forward, and one in reverse. However, a three-speed transmission and simplified drawings are used in this book to illustrate the principles involved.

Figure 3-6A illustrates the transmission with the gears in a neutral position. The clutch is attached to the pinion shaft. The pinion gear is attached to the rear end of the pinion shaft. The main shaft is attached to the drive shaft at the rear, while the forward end runs in the main shaft bushing within the pinion gear. Attached to the main shaft are gears that can be moved forward or to the rear, by the movement of the gearshift in the cab. These gears are referred to as the sliding gears. The rear gear on the main shaft is the low and reverse gear; the forward gear is second.

The shaft below is the countershaft. This shaft has a forward gear, which is in constant mesh with the pinion gear, causing the countershaft to rotate whenever the pinion shaft is turning. The gears on the countershaft, named from the rear forward, are reverse, low, and second.

In Figure 3-6A the countershaft is turning, but no power is being applied to the drive shaft, as no gear on the countershaft is in mesh with any on the main shaft.

Figure 3-6B illustrates the transmission in low gear. The low and reverse gear on the main shaft has been slid forward by the movement of the gearshift, and is meshed with the low gear on the

The Chassis and Component Parts 105

countershaft. Power is transmitted from the engine to the drive shaft as illustrated.

The transmission is in second gear in Figure 3-6C. The sliding gears on the main shaft have been moved to the rear so that the second gear on the main shaft and the corresponding gear on the countershaft are in mesh. Power is transmitted to the drive shaft as illustrated.

The reverse position is shown in Figure 3-6E. The low and reverse gear on the main shaft and the reverse gear on the countershaft are in mesh with the reverse idler gear. The reverse idler gear causes the drive shaft to turn in the opposite direction.

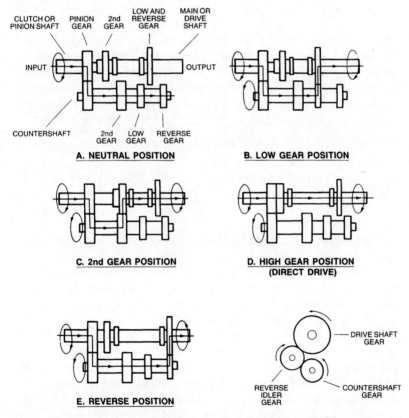

Figure 3-6   Typical Three-Speed Transmission Positions

Figure 3-6D shows the transmission in high gear. The sliding gears on the main shaft have been moved forward until the projections on the rear face of the pinion gear have engaged the matching indentations or notches on the forward face of the main shaft second gear. In this position there is a direct drive between the pinion shaft and the drive shaft. The countershaft is turning, but there is no connection between its gears and those on the main shaft.

Synchronizing Devices

Synchromesh transmissions are used to smooth out shifting operations. The synchronizing device controls the gears that are about to be meshed, so that the meshing teeth move at the same speed and therefore mesh without clashing. The device used is generally a clutch or synchronizing brake that slows down the faster-running of the two gears to be meshed, before the gears are actually engaged. This eliminates the clashing or scraping which might chip or break the edges of the gear teeth if an attempt is made to engage them before they are turning at the same speed.

Automatic Transmissions

An *automatic transmission* does automatically what the driver must do by hand with a manual transmission. It takes the guesswork out of gear selection, eliminating lugging and overspeeding. Proponents of automatic transmissions claim that such transmissions result in much better fuel economy and less apparatus downtime.

Several different types of automatic transmissions are used on fire appratus; however, the principles of operation of all of them are basically the same. The Allison model MT 644 is used in this manual for the purpose of explaining these principles.

Figure 3-7 shows a side view of an Allison model MT644 automatic transmission. This transmission can be used on apparatus having a gross vehicle weight of up to 73,280 pounds and having a pumping capacity of up to 1250 gallons per minute. It is designed for use with either gasoline or diesel engines. The transmission has four forward speeds and one in reverse. Shifting occurs automatically when the driver selects any range except low or reverse.

The Chassis and Component Parts

An automatic transmission is a complicated system of gears and hydraulic systems. The hydraulic system generates, directs, and controls the pressure and flow of hydraulic fluid within the transmission. The transmission consists of three assemblies—a torque converter, a constant mesh planetary gearing system, and an automatic hydraulic control.

The Torque Converter The torque converter multiplies engine torque during starts and acts as a hydraulic cushion between the engine and the gearing. The three primary components in the torque converter are the pump, a turbine, and a stator. The pump is the input element and is driven by the engine. The turbine is the output element and is driven by the pump. The stator is the torque-multiplying element.

The torque converter assembly is continually filled with oil, which flows through the converter to cool and lubricate it. When the converter is driven by the engine, the pump vanes throw oil against the turbine vanes. The impact of the oil against the turbine vanes rotates the turbine.

The turbine, splined to the turbine shaft, transmits torque to the transmission gearing. At engine idle speed, the impact of the

Figure 3-7 Model MT 644 Automatic Transmission (Side View). *Courtesy of Detroit Diesel Allison, Division of General Motors Corporation*

oil against the turbine vanes is not great. At high engine speed, the impact is much greater than at idle, and high torque is produced by the turbine.

Oil thrown into the turbine flows to the stator vanes. The stator vanes change the direction of oil flow (when the stator is locked against rotation), directing the oil to the pump, in a direction that assists the rotation of the pump. It is the redirection of the oil that enables the torque converter to multiply the input torque.

Greatest torque multiplication occurs when the turbine is stalled and the pump is rotating at its highest speed. Torque multiplication decreases as the engine rotates and gains speed.

When turbine speed approaches the speed of the pump, oil flowing to the stator begins striking the backs of the stator vanes. This rotates the stator in the same direction as the turbine and pump. At this point, torque multiplication stops and the converter becomes, in effect, a fluid coupling.

As indicated above, the torque converter accomplishes three functions. It acts as a disconnect clutch, since little torque is transmitted at engine idle speed. It multiples torque at low turbine and high pump speed to give greater starting or driving effort when needed. It acts as a fluid coupling to transmit engine torque efficiently to the transmission gearing during drive (other than starting or at idle).

The Constant-Mesh Planetary Gearing System   The constant-mesh planetary gearing system receives power from the torque converter. It is designed to provide a completely balanced torque loading and transmittal. Because the gears are in constant mesh, fast full-power shifts can occur, which result in quick acceleration.

The Automatic Hydraulic Control .   The *automatic hydraulic control* quickly senses load and speed conditions and selects the proper gear ratio. The system provides a far more accurate balance of engine power and load demand than most drivers can consistently do manually, thereby eliminating the danger of misshifts.

The Lockup Clutch   Another feature of the transmission is the *lockup clutch*. This unit consists of a piston, a clutch plate,

The Chassis and Component Parts 109

and a back plate. The piston and back plate are driven by the engine. The clutch plate, located between the piston and back plate, is splined to the converter turbine.

The lockup clutch automatically engages after the load is rolling and the torque demand is reduced. Engagement of the lockup clutch provides a direct drive from the engine to the transmission. The lockup clutch automatically releases at lower vehicle speeds. When the lockup clutch is not engaged, drive from the engine is transmitted hyrdaulically through the converter to the transmission gearing.

The Hydraulic System The hydraulic system generates, directs, and controls the pressure and flow of hydraulic fluid (transmission oil) within the transmission. Hydraulic fluid is the power-transmitting medium in the torque converter. Its velocity drives the converter turbine. Its flow cools and lubricates the transmission. Its pressure operates the various control valves and applies the clutches.

Torque Paths Through the Transmission Power is transmitted hydraulically through the torque converter. The engine drives the converter pump. The pump throws oil against the turbine vanes, imparting torque to the turbine shaft. From the turbine, oil flows between the vanes of the stator and reenters the pump, where the cycle begins again. When the engine is idling, impact of the oil upon the turbine blades is negligible. When the engine is accelerated, the impact is increased, and the torque directed through the turbine shaft can exceed the engine torque (by an amount equal to the torque ratio of the converter).

Automatic Upshifts When the transmission is operating in first gear, with the selector valve at drive (D), a combination of governor pressure and modulator pressure, or governor pressure alone, will upshift the transmission to second gear. At closed or part throttle, modulator pressure exists and will assist governor pressure. At full throttle, there is no modulator pressure. Thus, upshifts occur at a lower vehicle speed when the throttle is closed and are delayed by opening the throttle.

Governor pressure is dependent upon rotation speed of the transmission output. The greater the output speed (vehicle speed), the greater the governor pressure. When governor pressure is sufficient, the first upshift (first to second) will occur. With a further increase in governor pressure (and vehicle speed), the second upshift (second to third) will occur. A still further increase will cause a third upshift (third to fourth). Note that each of these shifts will be delayed or hastened by the decrease or increase, respectively, of modulator pressure.

In other drive selection positions, the same upshift sequence occurs until the highest gear attainable in that range selector is reached.

In any automatic upshift, the shift signal valve acts first. This directs a shift pressure to the relay valve. The relay valve shifts, exhausting the applied clutch and applying a clutch for a higher gear.

Automatic Downshifts Automatic downshifts, like upshifts, are controlled by rear governor and modulator pressures. Downshifts occur in sequence as rear governor pressure and/or modulator pressure decrease. Low modulator pressure (open throttle) will hasten the downshift; high modulator pressure (closed throttle) will delay downshift.

In any automatic downshift, the shift signal valve acts first. This exhausts the shift signal holding the relay valve downward. The relay valve then moves upward, exhausting the applied clutch and applying the clutch for the next lower gear.

Operating the Transmission The apparatus operator influences the automatic shifting of the transmission through the accelerator pedal. When the pedal is fully depressed, the transmission will automatically upshift near the governed speed of the engine. If the pedal is partially depressed, the upshifts will occur at a lower engine speed. The timing of the shifts is accomplished by using a cam and a cable from the throttle or by using the vacuum from the engine manifold. Either method provides the accurate shift spacing and control necessary for maximum performance.

The transmission can be downshifted or upshifted, even at full throttle, and although there is no speed limitation on upshifting, there are limits on downshifting and reverse. Downshifting should be avoided when the vehicle is above the maximum speed

The Chassis and Component Parts

attainable in the next lower gear. However, protection against improper downshifts and reverse shifts is inherent in the design of the hydraulic system. If a downshift or reverse shift is made at too high a speed, the hydraulic system automatically prevents the shift from taking effect until a safe, lower speed is reached.

To use the engine as a braking force, shift the range selector to the next lower range. If the vehicle is exceeding the maximum speed for a lower gear, use the service brakes to slow the vehicle to an acceptable speed, where the transmission may be downshifted safely.

Allison Transmission Electronic Control

Overall transmission performance, reliability, and durability can be improved by replacing hydraulic controls with electronic controls which incorporate microcomputer-based electronics for intelligence, but which continue to employ hydraulics for power to apply the clutches in the transmission.

Important benefits available with a microcomputer-based control system include: shift point optimization, improved shift point accuracy, transmission-application customization, mechanical linkage elimination, improved fuel economy, and overall powertrain optimization. As an example, by constantly measuring transmission speed and throttle position, the Allison automatic transmission electronic control (ATEC) automatically selects the most fuel-efficient gear ratio and torque converter mode. It also permits precision shift scheduling for fuel economy. ATEC's shifts are more accurate and positive than those obtained with hydraulic controls.

Figure 3-8 illustrates the major elements of the Allison automatic transmission electronic control. The electronic control unit (ECU) is the system's "brain." It receives input signals from switches and sensors, performs logic functions, and sends command signals to the appropriate solenoids on the valve body, providing positive, precise control of shift valves. One side of the ECU contains a PROM (Programmable Read Only Memory): a chip that tailors the transmission's operation to an individual department's needs, taking into account the apparatus, its engine, and the environment in which it operates.

The *throttle sensor* provides shift modulation by accurately

ATEC SYSTEM

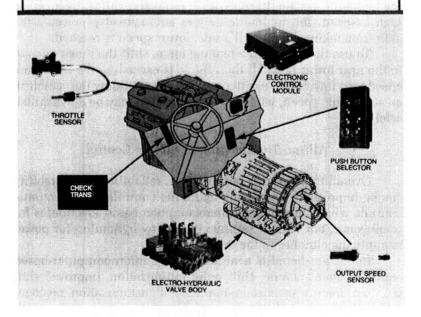

Figure 3-8 The Allison Transmission Electronic Control System. *Courtesy of Detroit Diesel Allison, Division of General Motors Corporation*

measuring throttle position. The sensor is self-calibrating and automatically adjusts for installation tolerances and wear.

The *output speed sensor* measures the transmission output speed which is used in calculating shift points and other control functions.

The *check transmission light* is part of the built-in diagnostic system. The light serves as a malfunction indicator, and when the system is put in the test mode will flash a coded signal indicating where a malfunction has occurred.

The *electro-hydraulic valve body* contains solenoids which provide positive, precise control of shift valves, thereby eliminating the need for complex hydraulic governors and mechanical control linkages.

The *pushbutton selector* is used to select the desired transmission gear. The ATEC system is available with either a pushbutton shift selector, as shown, or the conventional lever type.

The Chassis and Component Parts 113

Two other common elements of the ATEC system are a sump temperature sensor, which provides instant warning of elevated oil temperature due to oil overfill or other problem, and a lube pressure switch or an oil level sensor which monitors lube pressure or oil level in the transmission. If the lube pressure switch or oil level sensor comes on, the CHECK TRANSMISSION light is activated and the transmission inhibits high gear.

The built-in diagnostic features of the ATEC make it possible for the system constantly to monitor transmission components and check itself for proper operation. Upon detection of a problem in the control system or key areas such as oil pressure or sump temperature, the CHECK TRANSMISSION light will immediately indicate to the operator that a problem exists. Problems are detected early, before needless damage is done to any component. Maintenance efficiency is improved, since trouble areas are isolated and corrective action can be taken in exactly the right direction, thereby eliminating guesswork. The system can be interrogated to determine the source of the trouble in two ways: the first method involves manually activating a diagnostic mode switch located in the apparatus. With this switch on, the CHECK TRANSMISSION light will identify a two-digit trouble code by means of a flashing signal. The second method involves the use of a hand-held digital reader which has been developed for use as an optional maintenance tool to read out diagnostic information stored in the computer memory. Built-in diagnostics, digital maintenance tools, and the use of easily replaceable components help minimize maintenance expense.

In addition to the basic features of the system, there are several optional features which can be incorporated into fire apparatus to aid the apparatus operator in improving overall effectiveness. One is the reverse warning. With this feature, an output signal from the ECU identifies reverse range. The signal can be used to power backup lights and/or a warning horn. Another feature is the automatic retarder. The ECU can provide an output signal to apply the retarder automatically. This is beneficial where retarder usage is very high or for saving vehicle brakes.

A third feature is the engine overspeed protection. This is for transmissions equipped with a retarder. This feature will minimize inadvertent overspeed of the engine during downhill operations.

The most important feature of the system for fire apparatus is that the high-range lockup can be accomplished for split shaft operation without the necessity of making an expensive modification to the transmission.

The Driveline

The driveline connects the transmission to the driving axles. It consists of drive or propeller shafts, slip joints, universal joints, and the differential.

If there were no up-and-down motion of the apparatus, the driveline could consist of a simple driveshaft, which would extend from the transmission to the differential; however, such motion does occur. The rear springs compress and expand as the apparatus travels down the street. This compression and expansion causes the differential to move up and down, resulting in a change of the drive angle. Since the driving angle and the distance between the transmission and the differential are constantly changing, some means must be provided in the driveline to compensate for the changes. Compensation for the difference in the length is accomplished through the use of a slip joint. Compensation for the change in the driving angle is taken care of by the use of universal joints.

Slip Joints

Various types of *slip joints* are employed on apparatus. Their operation allows for the lengthening and shortening of the driveline. One type of slip joint uses an externally splined slip yoke and an internally splined shaft. The splines on the yoke slip into the splined shaft. The two must turn together because of the splines; however, the yoke can slip in and out of the shaft to compensate for distance changes.

Universal Joints

Universal joints are designed to provide for the change in driving angle as the differential moves up and down. Several different types are used on fire apparatus. Figure 3-9A illustrates a typical "U" bolt universal joint, Figure 3-9B a typical circular

The Chassis and Component Parts

flange universal joint, and Figure 3-9C a typical rectangular flange universal joint.

The Propeller Shaft

Propeller shafts and associated component parts provide a driveline, which transmits propelling forces between the transmission and rear axles. The propeller shaft arrangement may con-

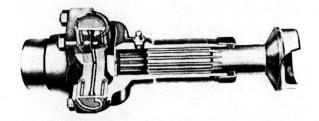

A   A Typical "U" Bolt Universal Joint

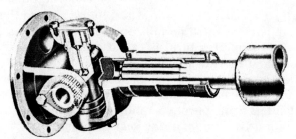

B   A Typical Circular Flange Universal Joint

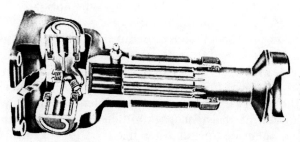

C   A Typical Rectangular Flange Universal Joint

Figure 3-9 Typical Types of Universal Joints. *Courtesy of Mack Trucks, Inc.*

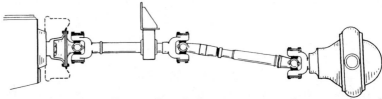

Figure 3-10 A Typical Propeller Shaft Arrangement. *Courtesy of Mack Trucks, Inc.*

sist of one or more propeller shafts and two or more universal joints. A typical arrangement on a fire apparatus uses two or more shafts and three or more universal joints. Figure 3-10 shows a typical propeller shaft arrangement that is used on a Mack truck.

The Differential

As an apparatus rounds a corner, the outside rear wheel has to travel a greater distance than the inside wheel. The outside wheel must turn faster than the inside wheel in order to do this. The unit on an apparatus which permits one wheel to turn faster than another, while making a turn, is called a differential. The term comes from the difference in wheel speeds.

The differential is found in the drive axle, which is the terminal point of the driveline, commonly referred to as the rear end. To understand its principle of operation, it is best to consider that it consists of a *drive pinion,* a *ring gear,* a *spider cross, four side pinions,* and *two side gears* (see Figure 3-11). The side gears are splined to the rear axle shafts. The ring gear is turned by the pinion gear attached to the end of the drive shaft, normally by means of a universal joint. The differential case, with the pinion gears and side gears enclosed, rotates as a unit when the apparatus is traveling in a straight-ahead direction. Resistance presses on the inside wheel's axle when the apparatus starts a turn. This resistance starts the pinion gears turning on their own axis (the spider), rolling around the side gear subjected to the resistance. This slows down the axle under the resistance, permitting the opposite axle to speed up to compensate for the difference in radius of turn of the two wheels.

The Chassis and Component Parts

Types of Rear Axles

There are two general classifications of axles used on fire apparatus. These are referred to as dead axles and live axles. A *dead axle* is basically a shaft upon which a wheel rotates. It supports load and may support braking.

On the other hand, a *live axle* is a driving axle. In addition to supporting load, it transmits power from the engine and transmission to the ground, provides differentiation, and is used to accommodate reaction to braking.

Gear reduction occurs as the drive pinion rotates the ring gear. Figure 3-12A shows a single reduction using a spiral bevel drive pinion and ring gear. Figure 3-12B shows a planetary double reduction gearing. In both cases the ring gear rotates the axle differential, the axle shafts, and then the wheels.

The Two-Speed Axle

Two-speed axles are designed to provide the performance of two axles combined into one unit: a high-speed *single reduction axle* and a low-speed *double reduction axle*. These axles provide a wide range of gear reductions within a single axle drivetrain, allowing for greater torque for climbing and moving heavy loads, and less torque for driving at higher speeds with lighter loads. Ratio changes are accomplished through a hand control button, mounted on the gearshift lever. Two basic types of axle shift systems are normally provided: electric, for a hydraulic brake chassis, and air, for an air brake chassis.

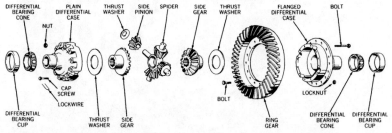

Figure 3-11 An Exploded View of a Differential. *Courtesy of Eaton Corporation, Axle Division*

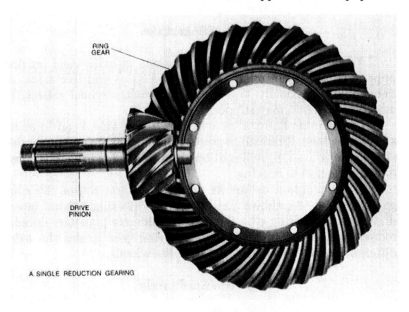

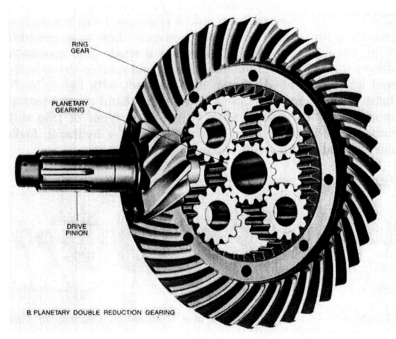

Figure 3-12 Two Types of Gear Reduction Components. *Courtesy of Eaton Corporation, Axle Division*

The Chassis and Component Parts

Figure 3-13 shows the power flow through an Eaton two-speed axle (in low ratio). The system includes a spiral bevel ring gear and drive pinion for the first reduction, and a planetary gearing unit for the second reduction. The planetary unit is confined within the internal teeth of the ring gear, permanently locked in alignment. A shifting unit and sliding clutch gear provide the means for selecting high or low axle ratios.

The driver has the option of many driving patterns. He can use low range through all transmission gears with only final axle

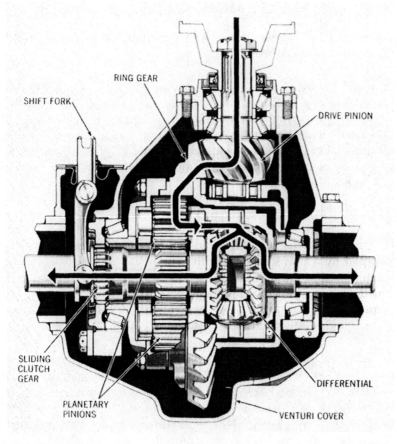

Figure 3-13 Power Flow Through a Two-Speed Axle (in Low Ratio).
Courtesy of Eaton Corporation, Axle Division

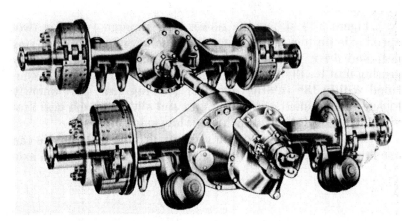

Figure 3-14 A Tandem Drive Axle Assembly. *Courtesy of Eaton Corporation, Axle Division*

shift to high range for cruising. He can split-shift, using the full range of gear selections or only those necessary to stay with the traffic. By observing the few simple rules for coordinating engine rpm with road speed, the driver with a two-speed axle can select a wide range of performance capabilities. Working with a combination of a five-speed transmission and a two-speed axle, he has ten gear selections.

Tandem Drive Axles

Tandem drive axles (see Figure 3-14) consist of two axle units coupled by a power divider unit. This combination drive offers inter-axle differentiation for easier steering and better roadability. The design prevents axle fight, and permits freer rolling for greater economy. Tandem drive axles are available in three different designs: single reduction, two-speed, and planetary double reduction.

Figure 3-15 shows the power flow through an Eaton single reduction "D" series tandem drive axle. The power divider unit equally distributes power from the vehicle transmission to both the forward and the rear axles. The inter-axle differential unit (mounted on the input shaft) operates the same way as a conventional axle differential, providing differential action between the forward and rear axles as follows:

The Chassis and Component Parts 121

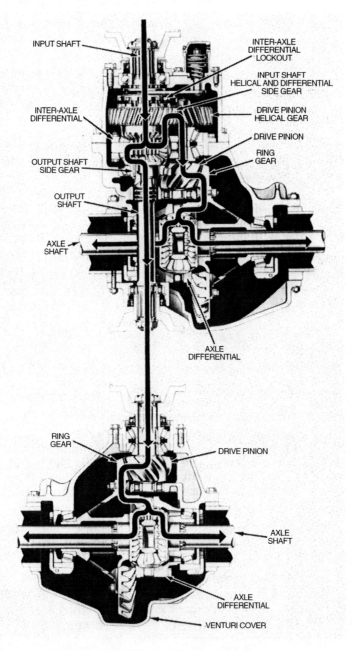

Figure 3-15 The Power Flow Through a Single Reduction Tandem Drive Axle. *Courtesy of Eaton Corporation, Axle Division*

1. Forward axle—power is transferred from the spider through a "floating" helical gear to the drive pinion and ring gears of the forward axle.

2. Rear axle—power is transferred from the spider through the output shaft side gear, the output shaft, and the inter-axle propeller shaft to the drive pinion and ring gears of the rear axle.

Thus, through use of the inter-axle differential as a power distribution point for the two axles, differential action is provided between the forward and rear axles.

When the inter-axle differential is not in operation (lockout mechanism engaged), the input shaft, helical side gear, and inter-axle differential rotate as one assembly. Power is then transferred to both axles without differential action. The forward axle drive pinion is driven from the input shaft, helical side gear to the drive pinion helical gear. The rear axle drive pinion is driven from the output shaft side gear, through the output shaft and the inter-axle propeller shaft.

Lockout Mechanism Tandem axles will rotate at different speeds when cornering, driving over uneven road surfaces, or equipped with tires of different sizes. The inter-axle differential is simply a mechanism that lets one axle rotate faster or slower than the other. With the *lockout mechanism*, a driver can lock out the inter-axle differential whenever extra traction is needed under adverse road conditions. When the lockout is engaged (LOCK position), the inter-axle differential acts as a solid shaft and does not compensate for differences in axle speed. When the lockout is disengaged (UNLOCK position), the inter-axle differential operates normally. A lockout selector valve, located in the apparatus cab, controls the inter-axle differential lockout.

The lockout mechanism consists of a *sliding clutch* (mounted on the power divider input shaft), *shift fork* and *push rod*, and control system. The shift fork may be controlled by an air-operated shift cylinder or a vacuum-operated control.

When the lockout mechanism is engaged by the driver, the sliding clutch engages the helical and differential side gear and locks out the inter-axle differential action. This provides a positive drive to both axles.

When the lockout control is released, a spring on the shift

The Chassis and Component Parts

fork push rod operates, disengaging the sliding clutch from the helical and differential side gear, and inter-axle differential action is restored.

The lockout should be used only on, or when approaching, poor traction surfaces, such as ice, snow, wet surfaces, and mud, or in loose terrain. Continuous unnecessary use of lockout must be avoided, as such use causes abnormal tire wear.

To operate the lockout, the driver temporarily releases the accelerator pedal. This interrupts the torque applied to the inter-axle differential, allowing easy engagement or disengagement at any speed. The lockout control can then be flipped to lock (engage) or unlock (disengage) the lockout.

BRAKING SYSTEMS

A braking system is designed to slow or stop a moving vehicle or to keep a stopped vehicle from moving. The terminal end of most braking systems is the same, regardless of the type of system (see Figure 3-16). The difference in braking systems is in the method of forcing the brake liners against the drums.

When the brakes are applied, the brake linings are forced into contact with the brake drums. The brake drum is an integral part of the wheel. The pressure of the brake linings against the drums slows the rotation of the wheels. This causes friction between the tires and the road surface, which slows the speed of the vehicle.

Brake Linings

Brake liners vary considerably in both size and content, and the type used on a fire apparatus varies according to the service to which the apparatus will be subjected.

Most brake linings are made of heat-resistant fiber. Materials such as zinc, brass, copper, ceramics, and so on are added to provide the ability to withstand higher temperatures and to give the desired coefficient of friction.

124 Introduction to Fire Apparatus and Equipment

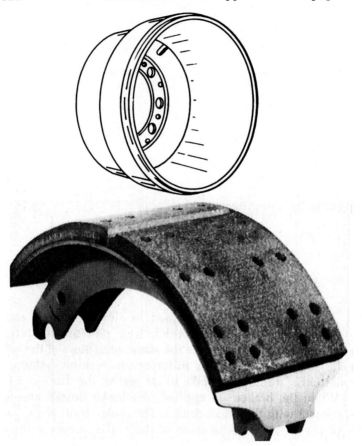

Figure 3-16 The Terminal End of a Braking System. *Courtesy of Rockwell International*

Two brake shoes with attached liners are installed in each wheel. The shoe that faces the front of the vehicle is called the primary shoe. The shoe that faces the rear is called the secondary shoe.

Hydraulic Braking Systems

A *hydraulic braking system* consists of a master cylinder containing a piston and brake fluid. The master cylinder is con-

The Chassis and Component Parts

nected with tubing to smaller cylinder assemblies in each wheel. Each of these wheel cylinder assemblies contains two pistons that face one another with a space between them. The outsides of the pistons are attached to push rods, which rest against the brake lever assembly (see Figure 3-18).

When the foot pedal is applied, the piston in the master cylinder forces brake fluid through the tubing to each wheel cylinder assembly (see Figure 3-17). The fluid forces the two wheel cylinder pistons to opposite ends of the cylinder, causing the push rods to move the brake liners against the wheel drums. When the driver's foot is released from the pedal, springs return the linings to their original positions and force the extra brake fluid out of the wheel assemblies back to the master cylinder.

As the brake liners wear, the liner assembly must be pushed a greater distance before contact is made with the drum. Some brake assemblies have brake adjusting bolts as shown in Figure 3-18 for adjusting the distance between the liners and the drums. Other brake assemblies have self-adjusting mechanisms.

Apparatus using hydraulic brake systems employ a dual master cylinder. The dual cylinder is referred to as a split system and is designed to help prevent total brake failure. In operation, the front and rear brakes, in a diagonal arrangement, perform as separate hydraulic systems. Should a leak occur in one system, the other will continue to operate. Total braking effort will be re-

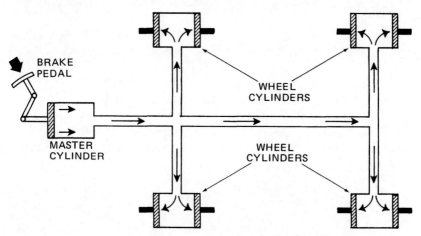

Figure 3-17   A Schema of a Hydraulic Braking System

Figure 3-18 An Exploded View of a Hydraulic Brake Layout. *Courtesy of Rockwell International*

duced, but it will still be possible to stop the vehicle by a heavy application of the brake.

Air Brake Systems

An *air brake system* uses compressed air to actuate the brakes. The three primary elements in an air brake system are the compressor, with its governor and reservoir, the brake application valve, and the brake chamber; however, other important components are also needed to increase utility and performance.

The Air Compressor

The *air compressor* is the source of energy for the brake system. It is driven by the vehicle engine, either by belt or drive gear, and on most vehicles utilizes the vehicle's lubrication and cooling systems; however, some compressors are self-lubricated and air-cooled. Compressors feature automatic inlet and discharge valves and unloading mechanisms.

The Governor

The *governor* operates in conjunction with the compressor unloading mechanism and maintains reservoir air pressure between a predetermined maximum and minimum pressure. Some

The Chassis and Component Parts

governors are designed to permit adjustment of minimum and maximum pressures, while others have built-in settings. The governor can be mounted either on the compressor or in a remote location.

The Reservoir

The *reservoir* serves the air brake system as a storage tank for a volume of compressed air. The size of the reservoir is determined by the vehicle manufacturer to provide an adequate volume of air for the braking system and auxiliary control devices. Generally more than one reservoir is used in the air brake systems. A secondary function of reservoirs is to provide a location where the air, heated by compression, may be cooled and the water vapor condensed. The condensed water vapor must be periodically drained from the reservoir.

Reservoir Draining Devices

Draining devices are installed on the reservoir. They allow the contaminants collected in the reservoir to be drained off to the outside. Both manual and automatic devices are available. Manual draining devices consist of drain cocks which require direct manual operation.

The Bendix DV-2 automatic reservoir drain valve is an example of a completely automatic draining device. It is installed directly into the end or bottom drain port of the reservoir and does not require any additional control lines except wires for the heating units. It is controlled automatically by ascending and descending reservoir pressures.

The Air Dryer

As the complexity of the air brake system has increased, so has the need for clean air. Many of the valves of later design contain small orifices and passages and thus are more susceptible to contaminants. In addition, the prevention of freeze-ups in the system has become important. The invention of the air dryer has been a valuable development in providing clean, dry air, and its use is highly recommended.

Air dryers are in-line filtration systems that remove both oil and water vapor from air coming from the compressor, before that air reaches the reservoir. This causes only clean, dry air to be supplied to the system, aiding in the prevention of air system freeze-ups. Alcohol evaporators also serve the system by helping to prevent freeze-ups.

Air dryers are equipped both with automatic drain valves controlled by the system governor and with integral heating elements.

Brake Application Valves

The *foot-operated brake valve* is the control point of the vehicle's air brake system. It provides the driver with an easily operated and graduated means of applying and releasing the brakes on a single vehicle or on vehicles in combination (tractor and trailer).

Brake valves are available in various mounting configurations and, generally, are mounted either on the floor or on the fire wall. Actuation of the valve can be by treadle, pedal, or, in some cases, by a lever/linkage arrangement.

The "feel" or sensitivity of the valve will vary, depending upon the method of actuation and the design of the valve. All brake valves are designed to provide a gradual means of applying air in the 5 psi to 80 psi range, with the capability of delivering full reservoir pressure.

Brake Chambers

Brake chambers and *slack adjusters* convert the energy of compressed air into mechanical force and motion. They move the brake camshaft, which in turn operates the foundation brake mechanism, forcing the brake shoes against the brake drum.

Brake Lining Actuation

There are two general methods of actuating the brake linings of air brakes used on heavy-duty fire apparatus. One method is referred to as a wedge system while the other is referred to as an S-cam system. The two systems derive their names from the devices, used to spread the brake shoes, which force the brake linings against the drum.

The Chassis and Component Parts

With the wedge system, a wedge-shaped push rod is forced between the brake shoe actuators by the action of the brake chamber push rod. (See Figure 3-19.) As the driver increases the pressure on the foot pedal, the wedge-shaped push rod is forced farther between the actuaters, resulting in increased brake lining pressure against the drum. Some systems use a single-chamber wedge while others use a double-chamber wedge.

In the S-cam system, a brake cam made in an "S" shape is rotated by a shaft, by the action of the brake chamber. The rotation of the S-cam forces the brake shoes apart, resulting in the brake lining moving against the drum. (See Figure 3-20). The shape of the S-cam results in increased brake lining pressure with an increased rotation of the S-cam shaft.

Slack Adjusters

The *slack adjuster* is the link between the actuator and the foundation brake camshaft. It transforms and multiplies the force developed by the chamber into a torque which applies the brakes via the brake camshaft. Slack adjusters are equipped with an adjusting mechanism, providing a means of adjusting for brake lining wear. Slack adjuster models are designated by a number

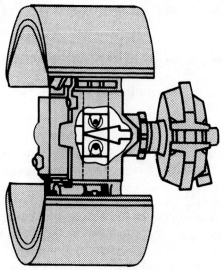

Figure 3-19 A Wedge Brake System

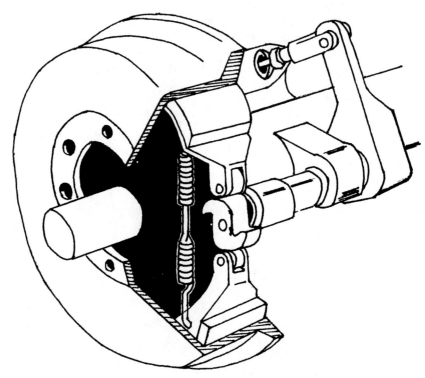

Figure 3-20   An "S" Cam Brake System

representing the maximum torque rating of each (a Bendix type 20 unit is rated for a maximum of 20,000 inch-pounds of torque).

Air Brake Maintenance

An ideal braking system can be defined as one in which the braking pressures reach each actuator at the same time and at the same pressure level. One factor that affects this pressure buildup is application and release time. Vehicle manufacturers carefully select tubing and hose sizes in order to establish this performance. Air application and release performance are partially dependent upon the size and volume of the chambers, vehicle weights, and locations of the valves and chambers or the distance that the air must travel. The vehicle is designed by the manufacturer to meet certain performance standards. The role of the apparatus operator and the mechanic is to preserve that performance.

The Importance of Reservoir Draining

The contaminants that can collect in air brake reservoirs consist of water condensed from the air and a small amount of oil from the compressor. The water and oil normally pass into the reservoir in the form of vapor, because of the heat generated during compression.

There is probably no more simple yet more important maintenance than reservoir draining. All reservoirs not equipped with automatic draining devices or moisture removal devices should be drained daily unless otherwise specified by the vehicle manufacturer or department policy. This is especially true on those days when apparatus are in use, since there is an increased amount of water present. All automatic drain valves and moisture-removing devices should be checked periodically for proper operation.

Maintenance Intervals

Maintenance intervals vary, because no two vehicles are operated under identical conditions. Experience is a valuable guide in determining the best maintenance interval for any one particular operation. Apparatus operators should follow the maintenance schedules as established by their departments; however, they should remember that scheduled intervals are the maximum allowable, not the minimum. Apparatus operated under severe or adverse conditions may require more frequent service.

Disc Brakes

Disc brakes depart from the tradition of forcing brake liners against the drum to stop the vehicle. Disc brakes are designed with a rotor which turns with the wheel. The rotor is straddled by a housing called a *caliper assembly.* With hydraulic disc brakes, the caliper assembly contains pistons on each side of the rotor that are similar to those used in a conventional hydraulic braking system. When the driver steps on the brake pedal, hydraulic fluid forces the piston and the *friction pads* (liners) against both sides of the machined surfaces of the rotor. The pinching action of the pads against the rotor causes friction, which slows or stops the apparatus.

Heavy-duty fire apparatus also use air-actuated disc brakes. While the basic principle of operation is the same as in a hydraulic system, air instead of hydraulic fluid is used to force the friction pads against the rotor.

The primary advantage of disc brakes is the reduced brake fading and, consequently, a shorter stopping distance; however, disc brakes also have the advantage of running cooler than drum brakes and are not affected by water as are drum brakes, since the water is thrown off the braking surface of the rotor. Disc brakes are used only on the front wheels of some vehicles, while others have them on all wheels.

Parking Brakes

The *parking brake* provides a secondary means of stopping an apparatus. The parking brake system is also commonly referred to as an *emergency brake* or *hand brake system*. The parking brake can be used to bring the apparatus to a safe stop, should the normal foot brake fail; however, the driver should understand the disadvantages of using the system indiscriminately.

The parking brake on most heavy-duty fire apparatus is located on the driveshaft. It is separate and distinct from the foot brake system. This separation provides two systems for stopping the vehicle. Some apparatus have a double-face brake on the driveshaft, while others have a single-face brake. With both systems, direct pressure is applied to the facing by cable, rod, or a combination of the two. A ratchet allows the brake handle to remain in the "on" position.

Normally, the parking brake should only be applied after the apparatus has been brought to a full stop. Engaging the parking brake while the apparatus is underway will cause:

    1. Glazed surfaces to build up on the lining surface, resulting in brake failure
    2. Burning of the brake surface due to friction
    3. Terrific stresses to be placed on the drive shaft, as well as on the rear-end assembly
    4. Stretching of cables, rods, and linkages

The foot brake should be used to stop the apparatus, and then the parking brake should be applied. Should the foot brake fail when no immediate danger exists, the driver should attempt to shift down, using the gear box to slow the apparatus and eventually bring it to a complete stop, if possible. However, the parking brake should be applied without hesitation whenever a serious emergency exists during a brake failure.

Automatic Transmission Retarders

Some fire apparatus automatic transmissions have built-in retarders. While a transmission retarder is not in itself a braking device, its assistance in braking situations is sufficient for it to be considered a secondary braking system. It is used to best advantage when traveling through cities where stop-and-go traffic is encountered. Engagement of the retarder through the service brake pedal slows down the apparatus, absorbing much of the braking load normally handled by the service brakes, thus preventing resultant heat buildup with possible brake fade and premature brake wear.

A different approach to transmission retardation is the output retarder developed by Detroit Diesel Allison. It is adaptable to the Allison automatic MT and HT series transmissions. The retarder is a two-stage unit which incorporates a hydraulic fluid coupling brake for high-speed retardation and a hydraulic, multiple-disc brake for low-speed retardation. This combination provides effective, sustained braking at high speeds as well as at very low speeds. Under most operating conditions, the output retarder covers a majority of the apparatus braking demand without the necessity of using the service brakes. In cases of emergency the service brakes therefore are not overheated and full braking is available. The output retarder brakes only the drive wheels, which leaves the front wheels free for steering when slippery road conditions are encountered. For downhill runs, the apparatus can be controlled even at low speeds by the sole use of the retarder.

The unit is mounted to the rear of the transmission, hence the name output retarder. The rear-end mounting provides braking effort directly to the drive shaft. This configuration eliminates associated transmission component wear and reduces the need to make gear changes in order to maintain retardation.

The output retarder can be engaged automatically through a vehicle's service brake system or with a separate hand lever in the case of downhill speed control situations. If the retarder is used for most of the vehicle braking, the service brakes can be kept cooler and can perform more efficiently in emergency stops. It is estimated that use of the retarder can extend the life of the service brakes approximately four to seven times.

The retarder consists of a double torus fluid coupling with a vaned rotor. The rotor revolves between cast stator vanes in the retarder housing and is splined to the transmission output shaft. Retarder braking is generated by filling the retarder housing with transmission oil. The oil is circulated through a transmission oil cooler which dissipates the heat. The apparatus cooling system must be capable of cooling the transmission oil to recommended temperature limits in all expected duty cycles.

The integral output retarder absorbs horsepower as a function of rotor speed (vehicle speed) and oil fill pressure. Increased oil fill pressure and/or high speed results in increasing retarder horsepower capacity. Since the retardation force acts directly on the transmission output shaft, there is less need to make gear changes to increase braking force, as compared to non-retarder transmissions.

The clutch portion of the retarder is the most useful to an apparatus operator. It provides the fast initial response and maintains the required braking capacity when the hydraulic retardation diminishes as transmission output speed is slowed. This multi-plate clutch pack consists of six non-metallic friction plates, which are splined to the rotating rotor hub, and six steel reaction plates, which are externally splined to the retarder housing and front stator. The dual torus hydraulic rotor is positioned just outside the clutch pack with four friction plates on the transmission side of the rotor and two friction plates to the rear of the rotor hub. The friction plates were designed with a "three-pass, multi-parallel" grooving pattern to allow maximum oil flow through the plates for lubrication. In addition, minimum restriction results in the oil "auto-flowing" through the clutch pack and rotor as it absorbs the power.

The Allison retarder provides the driver with up to 600 braking horsepower at 2000 rpm steady rate. On adverse driving surfaces such as ice, snow, or even wet pavement, the added braking capacity may be difficult to "feather" or control. When encountering these conditions, apparatus operators should test the retarder to determine when to turn it off. Tractor-trailer ladder trucks are even more sensitive than pumpers when adverse driving conditions are encountered.

STEERING

At the speeds common today, any apparatus operated with defective steering or defective brakes is a potential instrument of destruction. In addition, steering, as well as improper wheel and frame alignment, contributes to tire wear. Proper steering alignment must, therefore, be considered an economic factor as well as a safety factor. The trend toward smaller-diameter tires, lower tire pressures, and higher speed limits has caused the abrasive action between the tire tread and road surface to increase. Accurate wheel alignment has become imperative to extended tire life.

Stable steering and normal tire life depend on maintaining balanced forces in the front end and the steering system. Deflection; the weight of the vehicle on the tire; the vertical, lateral, and torsional vibrations of the chassis and steering parts caused by irregularities in the road surface; centrifugal force; and gyroscopic action of the wheels will tend to produce forces of variable frequencies and magnitude. To prevent the undesirable effects of these forces, it is essential that the various parts of the steering system be properly aligned or balanced. Additionally, all units encompassing the front end and wheel suspension system must be maintained in good mechanical condition and proper alignment.

Although this section deals with front end geometry, it must be noted that the steering characteristics also depend on proper rear wheel geometry, frame alignment, and maintenance of wheels and tires.

An apparatus operator should possess knowledge of the basic factors that affect the steering system so that he can logically trace the cause-and-effect relationship in any trouble.

The following glossary covers terms relating to steering components, alignment factors, and troubleshooting. Major components of the front axle are shown in Figure 3-21.

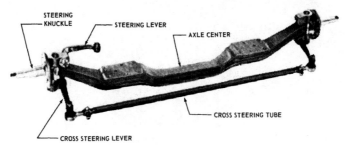

Figure 3-21 Front Axle Components. *Courtesy of Mack Trucks, Inc.*

Steering Components

1. *Cross Steering Levers*—The arms attached to the right- and left-hand steering knuckles
2. *Cross Steering Socket Assembly*—A ball socket assembly installed into each end of the cross steering tube and attached to the cross steering levers
3. *Cross Steering Tube Assembly*—An adjustable tube assembly connecting the cross steering levers
4. *Drag Links*—The mechanical connection between the steering lever and the steering gear lever
5. *Knuckle Pins*—The pins used to attach the steering knuckle to the axle center
6. *Pivot Center*—A theoretical point on the knuckle pin axis where it is intersected by the axis of the steering knuckle spindle
7. *Steering Gear Lever (Pitman Arm)*—The arm or lever on the end of the steering gear shaft and connected to the drag link. A timing mark is scribed on the arm or lever and must be aligned with a similar mark on the steering gear shaft.
8. *Steering Gear Shaft*—The output shaft of the steering gear with which the steering gear lever is aligned
9. *Steering Knuckle*—The moveable portion of the axle which pivots on the knuckle pins
10. *Steering Knuckle Spindle*—The part of the steering knuckle upon which the wheel revolves
11. *Steering Lever*—The lever attached to the steering knuckle and connected to the drag link, for turning the front wheels

Alignment Factors

1. *Camber*—The amount the front wheels incline outward at the top, expressed in degrees. If the wheels are inclined outward at the top, this is known as positive camber. If the wheels are inclined inward at the top, this would be described as negative or reverse camber.
2. *Caster*—The amount, in degrees, that the top of the knuckle pin is tilted toward the front or rear of the vehicle. Caster may be positive, negative, or zero.
3. *Caster, Positive*—The inclination of the top of the steering knuckle pin toward the rear
4. *Caster, Negative (or Reverse)*—The inclination of the steering knuckle pin toward the front
5. *Caster, Zero*—A perfectly vertical position of the steering knuckle pin
6. *Included Angle*—The total angle of camber and the knuckle pin inclination
7. *Knuckle Pin Inclination*—The amount, in degrees, that the knuckle pins are inclined inward at the top
8. *Steering Error*—The degree of error in the angular relation between the two front wheels when the vehicle is operating through right- or left-hand turns
9. *Steering Geometry*—The analysis of the steering mechanism that connects the front wheels and the steering gear
10. *Toe-in*—The distance by which the front wheels are closer at the front than they are at the rear of the front axle
11. *Turning (Steering) Angle*—The maximum number of degrees, both left and right, through which the front wheels can be turned from a straight-ahead position
12. *Turning Radius*—The radius of the arc described by the center of the track made by the outside front wheel of the vehicle when making its shortest turn
13. *Wheel Balance*—A condition in which the weight of the wheel and tire together is so evenly distributed that the tire will stand in any position in which it may stop, when free from friction. This is *static balance*. A wheel and tire that have been statically balanced should also be balanced centrifugally to attain a *dynamic balance*.

Troubleshooting

1. *Diving or Darting*—The tendency of the vehicle to dive or dart suddenly from the direction of travel, particularly when one wheel hits a hole or obstruction
2. *Road Shock*—A kick felt in the steering wheel as the front wheels travel on uneven roads. It is not noticeable on smooth roads.
3. *Leading*—The tendency of the vehicle to pull in the same direction at all times
4. *Shimmy*—A rhythmic oscillation of the front wheels, right and left, usually between speeds of 10 and 50 mph
5. *Tramp*—A rhythmic and alternate bouncing of the front wheels, usually noticeable at high speeds
6. *Wandering*—Straying indiscriminately from side to side, sometimes at both high and low speeds
7. *Wheel Fight*—A continuous jerking of the steering wheel in either direction, on rough roads, at all speeds; not violent, but extremely annoying and tiring to the driver.

Purpose

The purpose of the alignment factors included in the design of the front axle and steering component is to provide reliable steering, convenience in handling, maximum tire life, and minimum component wear. Steering geometry analysis depends entirely upon an understanding of these alignment factors, whether built-in or adjustable, along with a knowledge of possible corrective measures. A summary of the alignment factors is presented below to provide a basic knowledge of the front axle steering system.

Camber

Camber is built into the front axle to allow the weight of the vehicle to be carried directly below the spindle, easing the steering and reducing road shock. Too much or too little camber may cause tire wear. The tire wear caused by too much or too little camber will appear on the inner or outer half of the tire tread and the wear will be smooth, not feather-edged. Wear on the outer half of the tire tread indicates too much camber. Wear on the inner half of the tire tread indicates too little camber. Incorrect or unequal camber may also cause wandering. On flat road surfaces, the vehicle will usually lead to the side that has the highest positive camber.

All vehicles should have camber that is equal on both sides. Differences in camber between the left and right sides may be attributed to unequal tire size, tire wear, improper tire inflation, worn steering knuckle pin bushings, a deformed wheel, or a bent axle.

Caster

Caster is built into the front axle to assist in maintaining directional stability and to aid the front wheels in returning to a straight-ahead position after completing a turn. Caster is provided to compensate for the rearward lateral forces that arise when the vehicle is moving forward.

Insufficient caster will cause low-speed wandering or weaving and high-speed instability. When cornering, the wheels will be slow in returning to the straight-ahead position. Too much caster will be indicated by hard steering and impact shimmy. Excessive caster may also be indirectly related to road shock and high-speed wander and may cause power steering chatter. Unequal positive caster will cause leading of the vehicle to the side having the lesser caster. It is important to note that incorrect caster will not affect tire wear.

Knuckle Pin Inclination

Knuckle pin inclination influences the directional stability of the vehicle. Slanting the knuckle pin out at the bottom tends to minimize the' pulling that comes from unequal brake action and irregular road surfaces.

When the apparatus rounds a corner, the spindle remains in a level plane and therefore, because of the inclination of the knuckle pin, the front end of the vehicle is raised slightly. After the turn is completed and upon release of the steering wheel, the vehicle weight will tend to force the wheels to their normal forward position.

Knuckle pin inclination will not affect tire wear. Incorrect knuckle pin inclination may, however, cause a variance in camber, which will contribute to tire wear.

Toe-In

Toe-in is adjustable to compensate for caster, to reduce tire wear, and to prevent wander. Due to the outward tilt at the top of the wheels, the wheels tend to pull outward or away from the vehicle. Because both front wheels normally have positive camber, they will tend to pull in opposite directions. This pull in opposite directions causes the wheels to fight each other, thereby developing steering wander. The outward pull of the wheels also tends to compress the cross steering tube and cross steering tube ball studs, especially if the steering tube sockets are worn. Front wheels, therefore, are toed-in slightly to allow the wheels to run in a straight line when the vehicle is in motion.

Too much or too little toe-in will cause the wheels to fight each other and contribute to tire wear. Wear may appear on the inner or outer half of both front tires as a feathered edge. A feathered or sawtooth edge, pointing inward, will indicate excessive toe-in; too little toe-in, or toe-out, will be indicated by a feathered or sawtooth edge pointing outward.

Toe-Out

The purpose of *toe-out* is to compensate for the different radii that the front wheels are forced to travel when the vehicle is in a turn. This characteristic is built into the steering linkage and is not adjustable. Because of toe-out, it is mandatory that toe-in be measured while the front wheels are in a straight-ahead position.

Pivot Centers

The dimension between the right- and left-hand pivot centers depends on the design of the specific axle.

Turning (Steering) Angle and Turning Radius

Correct setting of the turning angle is important, as too large an angle will cause an interference between the tire and the chas-

The Chassis and Component Parts 141

sis or the steering linkage. If the turning angle is too small, the maneuverability of the vehicle will be reduced and the turning radius will increase.

REVIEW QUESTIONS

1. What three types of chassis packages are available to fire department purchasers?
2. What is the main member of the chassis?
3. Of what material are frames generally made?
4. What is the definition of gross vehicle weight?
5. When an apparatus is designed, how much weight is usually allowed for the crew? For the water to be carried?
6. What are some of the component parts of the power train?
7. What is the purpose of the clutch?
8. What component parts are usually found in a clutch assembly?
9. Explain how a clutch works.
10. What is the major cause of clutch failures?
11. What is the purpose of a clutch brake?
12. What is the purpose of the transmission?
13. What turns the countershaft in a transmission?
14. What is the purpose of synchronizing devices?
15. What are the three basic assemblies an Allison automatic transmission?
16. What is the purpose of the torque converter in an automatic transmission?
17. What is the purpose of the automatic hydraulic control in an automatic transmission?
18. What does the lockup clutch do in an automatic transmission?

19. What happens if a driver downshifts in an Allison automatic transmission when the vehicle is above the maximum speed attainable in the next lower gear?

20. What are some of the benefits available with a microcomputer-based control system for an automatic transmission?

21. What is considered the "brain" of the ATEC?

22. What are the major elements of the ATEC?

23. What do the built-in features of the ATEC make it possible for the system to monitor?

24. What does the driveline do?

25. What are the component parts of a driveline?

26. What is used to compensate for the differences in the length of the driveline as the differential moves up and down?

27. What is used to compensate for the change in the driving angle as this movement occurs?

28. What are some of the types of universal joints used on fire apparatus?

29. What is the purpose of the differential?

30. What are the two general classifications of rear axles used on fire apparatus?

31. Where does rear end gear reduction take place?

32. What is the purpose of a two-speed axle?

33. What are tandem drive axles?

34. What are the component parts of the terminal end of most braking systems?

35. Describe what happens when the brakes on an apparatus are applied.

36. What material is used in the construction of most brake linings?

37. What is the name given to the brake shoe that faces the front of the apparatus?

38. How does a hydraulic brake system work?

39. Explain the dual master cylinder concept as used in the hydraulic brake system.

The Chassis and Component Parts

40. What are the three primary elements in an air brake system?

41. What is the control point of a vehicle air brake system?

42. What is used in an air brake system to convert the energy of the compressed air into mechanical force and motion?

43. What are the two general methods of actuating the brake shoes in an air brake system?

44. What is the definition of an ideal braking system?

45. For an apparatus operator, what is probably the most important maintenance element in the air brake system?

46. What is the principle of operation of a disc brake?

47. What is the primary advantage of disc brakes?

48. Where is the parking brake on most heavy-duty fire apparatus located?

49. What might possibly be the result of engaging the parking brake while the apparatus is underway?

50. What is the purpose of an automatic transmission retarder?

51. How is a transmission retarder engaged?

52. Where is an output retarder located?

53. Where is the best use made of a transmission retarder when driving?

54. What precautions are necessary when driving an apparatus equipped with an automatic transmission retarder on adverse driving surfaces?

55. What is camber?

56. Define caster.

57. Define toe-in.

58. Define turning radius.

59. What is meant by diving or darting?

60. What is the difference between a shimmy and a tramp?

61. What is meant by wheel fight?

62. What is the purpose of camber?

63. Where will the tires wear as a result of too much or too little camber?

64. What is the purpose of caster?

65. What would be the result of insufficient caster?

66. What effect will improper knuckle pin inclination have on tire wear?

67. What would be the result of too little or too much toe-in?

68. What is the purpose of toe-out?

69. What is most likely to cause feathered-edge tire wear?

70. What would be the effect of too much caster?

4

Engine and Systems Troubleshooting

Apparatus operators are not expected to know as much about the apparatus and engine as a mechanic; however, they should be familiar with the component parts so that they have some idea of the possible causes when problems develop. The objective of this chapter is to introduce the operator to some of the common problems in an apparatus, and their probable causes. While it was not possible to list every problem that might develop, those an operator is most likely to encounter are included, in either a specific or a general way.

Diesel Engines

Problem	Possible Cause
1. Engine will not rotate during starting operations.	a. Low battery voltage, loose starter connections, or faulty starter. b. Defective starting motor switch. c. Internal seizure.

Diesel Engines
(Continued)

Problem	Possible Cause
2. Low cranking speed.	a. Improper lubricating oil viscosity.
	b. Low battery output, loose starter connections, or faulty starter.
3. Low engine compression.	a. Exhaust valves sticking or burned.
	b. Compression rings worn or broken.
	c. Cylinder head gasket leaking.
	d. Improper valve clearance adjustment.
	e. Blower not functioning.
4. No fuel to engine.	a. Out of fuel.
	b. Air leaks.
	c. Flow obstruction.
	d. Faulty fuel pump or faulty installation.
	e. Injector racks not in full-fuel position.
5. Uneven running or frequent stalling of engine.	a. Low coolant temperature.
	b. Insufficient fuel.
	c. Faulty injectors.
	d. Low compression pressures.
	e. Governor instability.
6. Lack of engine power.	a. Improper engine adjustments (tune-up) and gear train timing.
	b. Insufficient fuel.
	c. Insufficient air.
	d. Engine horsepower too little for vehicle.
	e. High return fuel temperature.
	f. High ambient air temperature.
	g. High altitude operation.
7. Engine detonating.	a. Oil picked up by the air stream.
	b. Low coolant temperature.
	c. Faulty injectors.
8. No fuel or insufficient fuel caused by air leaks.	a. Low fuel supply.
	b. Loose connections or cracked lines between fuel pump and tank or suction line in tank.
	c. Damaged fuel oil strainer gasket.
	d. Faulty injector tip assembly.
9. No fuel or insufficient fuel caused by fuel obstruction.	a. Fuel strainer or lines restricted.
	b. Temperature less than 10° F. above pour point of fuel.

Engine and Systems Troubleshooting

Diesel Engines
(Continued)

Problem	Possible Cause
10. No fuel or insufficient fuel caused by faulty fuel pump.	a. Relief valve not seating. b. Worn gears or pump body. c. Fuel pump not rotating.
11. No fuel or insufficient fuel caused by faulty installation.	a. Diameter of fuel suction lines too small. b. Restricted fitting missing from return line. c. Inoperative fuel intake line check valve. d. High fuel return temperature.
12. High lubricating oil consumption.	a. External leaks. b. Internal leaks. c. Oil control at cylinder.
13. External oil leaks.	a. Oil lines or connections leaking. b. Gasket or oil seal leaks. c. High crankcase pressure. d. Excessive oil in air box.
14. Internal oil leaks.	a. Blower oil seal leaking. b. Oil cooler core leaking.
15. Excessive oil consumption caused by oil control at cylinder.	a. Oil control rings worn, broken, or improperly installed. b. Piston pin retainer loose. c. Scored liners, pistons, or oil rings. d. Piston and rod misalignment. e. Excessive installation angle. f. Excessive oil in crankcase.
16. Excessive crankcase pressure.	a. Cylinder blow-by. b. Breather restriction. c. Air from blower or air box. d. Excessive exhaust back pressure.
17. Cylinder blow-by.	a. Cylinder head gasket leaking. b. Piston or liner damaged. c. Piston rings worn or broken.
18. Excessive exhaust back pressure.	a. Excessive muffler resistance. b. Faulty exhaust piping.
19. Excessive crankcase pressure caused by air from blower or air box.	a. Damaged blower-to-block gasket. b. Cyinder block end plate gasket leaking.
20. Low oil pressure.	a. Lubricating oil insufficient. b. Pressure gauge faulty. c. Poor circulation. d. Oil pump faulty.

Diesel Engines
(Continued)

Problem	Possible Cause
21. Lubricating oil causing low oil pressure.	a. Suction loss. b. Lubricating oil viscosity.
22. Pressure gage causing low oil pressure.	a. Faulty gage. b. Gage line obstructed. c. Gage orifice plugged. d. Electrical instrument panel sending units faulty.
23. Oil pump causing low oil pressure.	a. Intake screen partially clogged. b. Relief valve faulty. c. Air leak in pump suction. d. Pump worn or damaged. e. Flange leak (pressure side).
24. Poor circulation causing low oil pressure.	a. Cooler clogged. b. Cooler bypass valve not functioning properly. c. Pressure regulator valve not functioning properly. d. Excessive wear on crankcase bearings. e. Gallery, crankshaft, or camshaft plugs missing.
25. Engine coolant operating temperature above normal.	a. Insufficient heat transfer. b. Poor circulation.
26. Engine coolant operating temperature below normal.	a. Improper circulation. b. Excessive leakage at thermostat seal.
27. Black or gray exhaust smoke.	a. Incompletely burned fuel. b. Excessive fuel or irregular fuel distribution. c. Improper grade of fuel.
28. Blue exhaust smoke.	a. Lubricating oil not burned in cylinder (blown through cylinder during scavenging period).
29. White exhaust smoke.	a. Misfiring cylinders.

Source: Courtesy of Detroit Diesel Allison, Division of General Motors Corporation.

Carburetion Maintenance

Proper operation of the carburetor requires the following:
1. Fuel supply.
2. Linkage and emission control systems.
3. Engine compression.
4. Ignition system firing voltage.
5. Ignition spark timing.
6. Secure intake manifold.

Engine and Systems Troubleshooting

Carburetion Maintenance
(Continued)

7. Engine temperature.
8. Carburetor adjustments.

Any problems in the above areas can cause the following:

1. No start or hard starting (hot or cold).
2. Rough engine idle and stalling.
3. Hesitation on acceleration.
4. Loss of power on acceleration and top speed.
5. Uneven running or surging of engine.
6. Poor fuel economy.
7. Excessive emissions.

Note: The above items should be checked before proceeding with the carburetor. Also make sure all emission control units are installed and operating properly. This includes all emission system solenoids and connecting hoses. Carburetor problems cannot be isolated effectively unless all other engine systems are operating correctly and the engine is properly tuned.

Source: Courtesy of Holley Replacement Parts Division, Colt Industries.

Carburetor Problems

Problem	Possible Cause
1. Engine cranks (turns over) but will not start or is hard to start when cold.	a. Improper starting procedure used. b. No fuel in gas tank. c. Choke valve not closing sufficiently when cold. d. Choke valve or linkage binding or sticking. e. No fuel in carburetor. f. Engine flooded. g. Carburetor flooding.
2. Engine hard to start when hot.	a. Choke valve not opening completely. b. Engine flooded. c. Carburetor flooding. d. No fuel in carburetor. e. Leaking float bowl. f. Fuel percolation.
3. Engine starts and stalls.	a. Engine does not have fast enough idle speed when cold. b. Choke vacuum diaphragm unit not adjusted to specification or defective. c. Choke coil rod out of adjustment. d. Choke valve and/or linkage sticking or binding.

Carburetion Problems
(Continued)

Problem	Possible Cause
	e. Idle speed setting in need of adjustment.
	f. Not enough fuel in carburetor.
	g. Carburetor flooding.
4. Engine idles roughly and stalls.	a. Idle mixture in need of adjustment.
	b. Idle speed setting in need of adjustment.
	c. Manifold vacuum hoses disconnected or improperly installed.
	d. Carburetor loose on intake manifold.
	e. Intake manifold loose or gaskets defective.
	f. Hot idle compensator not operating (where used).
	g. Carburetor flooding.
5. Engine runs unevenly or surges.	a. Fuel restriction.
	b. Dirt or water in fuel system.
	c. Fuel level too low.
	d. Main metering jet defective, loose, or carrying incorrect part.
	e. Power system in carburetor not functioning properly.
	f. Vacuum leaks.
6. Engine hesitates on acceleration.	a. Defective accelerator pump system.
	b. Dirt in pump passages or pump jet.
	c. Fuel level too low.
	d. Leaking air horn to float bowl gasket.
	e. Carburetor loose on manifold.
	Note: A quick check of the pump system can be made as follows: With the engine off, remove the air cleaner, look into the carburetor bores, and observe the pump stream, while briskly opening the throttle valve. A full stream of fuel should be emitted from the pump jet and strike near the center of the venturi area.
7. No power on heavy acceleration or at high speed.	a. Carburetor throttle valves not opening wide. (Check by pushing accelerator pedal to the floor.)
	b. Dirty or plugged fuel filters.
	c. Power system not operating.

Engine and Systems Troubleshooting

Carburetion Problems
(Continued)

Problem	Possible Cause
	d. Fuel level too low.
	e. Float not dropping far enough into float bowl.
	f. Main metering jets dirty, plugged, or carrying incorrect part.
8. Poor fuel economy.	a. Engine in need of complete tune-up.
	b. Choke valve not opening fully.
	c. Fuel leaks.
	d. Main metering jet defective, loose, or carrying incorrect part.
	e. Power system in carburetor not functioning properly; power valve or piston sticking in up position.
	f. Fuel level in carburetor too high or carburetor flooding.
	g. Fuel being pulled from accelerator system into venturi through pump jet.
	h. Air bleeding or fuel passages in carburetor dirty or plugged.
9. Engine backfires.	a. Binding or sticking choke valve.
	b. Accelerator pump not operating properly.
	c. Old or dirty (fouled) spark plugs.
	d. Old or cracked spark plug wires.
	e. Partially clogged fuel filter.
	f. Defective air pump diverter valve (backfire on deceleration).

Source: Courtesy of Holley Replacement Parts Division, Colt Industries.

Clutches

Problem	Possible Cause
1. Poor release.	a. Intermediate plate sticking on drive lugs.
	b. Pressure plate not retracting.
	c. Driven disc distorted or warped.
	d. Spines worn on main drive gear of transmission.
	e. Internal clutch adjustment not correct.
	f. Flywheel pilot bearing fitting too

Clutches
(Continued)

Problem	Possible Cause
	tightly in flywheel or on end of drive gear.
	g. Facings gummed with oil or grease.
	h. Damaged clutch release bearing.
	i. Clutch release shaft projecting through release yoke.
	j. Release yoke contacting cover assembly at full release position.
	k. Release yoke not properly aligned with bearing.
	l. Broken intermediate plate.
2. Clutch slippage.	a. Weak pressure springs.
	b. No free pedal.
	c. Worn clutch facings.
	d. Release mechanism binding.
	e. Grease or oil on facings.
3. Noisy clutch.	a. Clutch release bearings dry or damaged.
	b. Flywheel pilot bearing dry or damaged.
	c. Clutch release bearing housing striking flywheel ring.
	d. Excessive clearance between drive slots and drive lugs on intermediate and pressure plates.

Source: Courtesy of Dana Corporation, Spicer Clutch Division.

Drive Train

Problem	Possible Cause
1. Whining noise at all speeds; pitch varies directly with road speed.	a. Rear axle trouble.
2. Grinding or grating noise at all speeds.	a. Chipped or scored center bearing.
3. Vibration at any road speed.	a. Universal flange or yoke nut loose.
	b. Universal joint journal cross worn.

Source: Courtesy of Mack Trucks, Inc.

Engine and Systems Troubleshooting

Manual Steering

Problem	Possible Cause
1. Hard steering.	a. Lack of lubrication or wrong lubricant. b. Front tires underinflated. c. Excessive caster. d. Front axle overloaded. e. Binding in knuckles, linkage, or steering gear. f. Axle thrust bearing worn or seized. g. Steering column misaligned (full-length column).
2. Excessive free play.	a. Steering gear worn or in need of adjustment. b. Steering linkage ball socket ends worn or out of adjustment. c. Steering gear mounting bolts loose. d. Steering shaft universal joints or slip joints worn. e. Steering knuckle bushings worn. f. Wheel bearings worn or out of adjustment.
3. Shimmy.	a. Incorrect or uneven tire pressure. b. Wheel bearings worn or out of adjustment. c. Caster or toe-in incorrect. d. Tire size not uniform. e. Wheels out of balance or not true. f. Steering linkage loose. g. Steering gear mounting bolts loose. h. Steering gear adjustment loose. *Note:* Shimmy may also be caused by factors not directly related to the front axle or steering system, such as an unbalanced propeller shaft or loose engine mounts.
4. Wandering.	a. Tire pressure low or unequal. b. Caster or toe-in incorrect. c. Steering linkage or knuckle pins loose. d. Wheel bearings worn or incorrectly adjusted. e. Steering gear mounting bolts loose. f. Steering gear adjustment incorrect. g. Steering knuckle spindle bent. h. Sagging or broken spring.

Manual Steering
(Continued)

Problem	Possible Cause
	i. Incorrect load distribution.
	j. Rear axle shifted.
5. Pulling to one side.	a. Caster unequal or incorrect.
	b. Toe-in incorrect.
	c. Tire pressure low or unequal.
	d. Tires not of uniform size.
	e. Brake dragging on one side.
	f. Wheel bearings tight or not lubricated.
	g. Spring sagging or broken.
	h. Steering knuckle spindle bent.
	i. Rear axle shifted.
	j. Frame out of alignment.
	k. Steering gear not centered.
6. Uneven or excessive tire wear.	a. Tire inflation pressure incorrect.
	b. Wheel alignment incorrect.
	c. Tire and wheel assembly out of balance.
	d. Steering knuckle bushings worn.
	e. Steering linkage loose.
	f. Harsh or unequal brake action.
	g. Overloading.
	h. Poor driving practices.
7. No recovery after turn.	a. Caster insufficient.
	b. Binding in knuckles or linkage.

Source: Courtesy of Mack Trucks, Inc.

Hydraulic Power Steering

Problem	Possible Cause
1. Abnormal noise.	a. If the power steering pump is belt-driven, a "squealing" noise during steering may indicate that the belt(s) should be tightened or replaced.
	b. A "clicking" noise, heard when initiating a steering maneuver or when changing directions or turning, may indicate that some component is loose and is shifting under load.
	c. A change in the normal noise from

Engine and Systems Troubleshooting

Hydraulic Power Steering
(Continued)

Problem	Possible Cause
	the power steering pump may indicate that an excessive amount of air has been trapped in the fluid, or that the fluid level is low.
	d. Excess noise from the power steering pump may indicate that the hydraulic system has dirt, sludge, or other impurities trapped or that the power steering pump is worn or defective.
2. Wandering.	a. Components in the steering linkage, such as ball sockets on drag link or axle arm, loose or worn.
	b. Tire pressure incorrect or unequal left to right.
	c. Wheel bearings improperly adjusted or worn.
	d. Steering gear mounting bolts loose on frame.
	e. Front end alignment out of specification.
	f. Steering gear center adjustment improperly adjusted.
	g. Dry fifth wheel or poor finish on fifth wheel or trailer plate.
	h. Steering geometry incorrect (pitman arm and/or axle arm ball position).
3. No recovery after turn.	a. Front end alignment incorrect.
	b. Front end components binding.
	c. Pump flow insufficient.
	d. Tire pressure low.
	e. Fifth wheel dry.
	f. Control valve spool or sleeve sticking.
	g. Tight front axle spindles.
	h. Steering column binding.
4. Shimmy.	a. Front end alignment incorrect.
	b. Air in hydraulic system.
	c. Wheels out of balance.
	d. Components in steering linkage, such as ball sockets on drag link or axle arm, loose or worn.

Hydraulic Power Steering
(Continued)

Problem	Possible Cause
	e. Unevenly worn tires.
	f. Wheel bearings improperly adjusted or worn.
5. External oil leakage.	a. No external leakage is acceptable at the steering gear. However, finding the location of a leak may be difficult, since oil may run away from the leak and drip from a low point on the gear or chassis.
6. Oversteer or darting.	a. Front end components binding.
	b. Sector shaft adjustment too tight.
	c. Control valve spool or sleeve sticking.
	d. Steering geometry incorrect (pitman arm and/or axle arm ball position).
	e. Steering column binding.
7. High steering effort in one direction.	a. Vehicle overloaded.
	b. Valve sleeve non-stick seal worn and allowing internal leakage.
	c. Auxiliary cylinder lines crossed.
	d. Inadequate flow.
	e. Inadequate pressure.
	f. Leaking piston relief valve.
8. High steering effort in both directions.	a. Vehicle overloaded.
	b. Low flow or pressure from the pump, which may be due to loose drive belts, sticky flow control or relief valve, or high internal leakage in pump.
	c. Components of steering system binding. High back pressure caused by clogged return line filter.
	d. Restriction in return line, or return line too small.
	e. Excessive internal leakage in steering gear.
	f. Low hydraulic fluid level.
	g. Control valve spool or sleeve sticking.
	h. Low tire pressure.
	i. Restriction in supply line, or supply line too small.

Engine and Systems Troubleshooting

Hydraulic Power Steering
(Continued)

Problem	Possible Cause
	j. Leaking relief valve.
9. Lost motion at steering wheel.	a. Loose ball socket connections or linkage connections.
	b. Loose steering gear adjustments, side cover adjustment, and/or worm adjustment.
	c. Pitman arm loose on output shaft.
	d. Steering gear loose on frame.
	e. Loose connections between steering column and steering gear.
10. Excessive heat.	a. Undersize replacement hose or line.
	b. Hose or line restricted by a kink or severe bend.
	c. Poppets not adjusted properly to relieve pressure at ends of steering gear travel.
	d. Hydraulic system fluid overheated, causing seals in the steering gear to dry out, harden, and lose their sealing ability.
	Note: If the hydraulic system fluid has been overheated, it will have a rancid odor. If it has been highly overheated, the internal parts of the steering gear may show some discoloration.

Source: Courtesy of Mack Trucks, Inc.

Air Brakes

Problem	Possible Cause
1. Brakes insufficient.	a. Mechanical parts damaged.
	b. Brakes in need of adjusting, lubricating, or relining.
	c. Low air pressure in the brake system.
	d. Brake valve delivery pressure below normal.
2. Brakes apply too slowly.	a. Brakes in need of adjusting or lubricating.
	b. Low air pressure in the brake system.

Air Brakes
(Continued)

Problem	Possible Cause
	c. Brake valve delivery pressure below normal.
	d. Excessive leakage when brakes are applied.
	e. Restricted tubing, hose line, or valves.
3. Brakes release too slowly.	a. Brakes in need of adjustment or lubrication.
	b. Brake valve not returning to fully released position.
	c. Restricted tubing or hose line.
	d. Exhaust port of brake valve, quick release valve, or relay valve restricted or plugged.
	e. Defective brake valve.
4. Brakes do not apply.	a. No air pressure in system.
	b. Restricted or broken tubing or hose line.
	c. Defective brake valve.
	d. Defective brake chamber diaphragm.
5. Brakes do not release.	a. Brake linkage binding.
	b. Brake valve not in fully released position.
	c. Defective brake valve.
	d. Restriction in tubing or hose line.
6. Brakes grab.	a. Grease on brake lining; reline brakes.
	b. Brake drum out of round.
	c. Defective brake valve.
	d. Brake linkage binding.
7. Compressor knocks continuously or intermittently.	a. Excessive backlash in drive gears or drive coupling.
	b. Worn or burned-out bearings.
	c. Excessive carbon deposits in compressor cylinder head.
8. Safety valve "blows off."	a. Safety valve out of adjustment.
	b. Air pressure in the system above normal.
	c. Governor out of adjustment.
	d. Air line blocked.
9. Excessive oil or water in the system.	a. Reservoirs not being drained daily.
	b. Compressor passing excessive amount of oil.

Engine and Systems Troubleshooting

Air Brakes
(Continued)

Problem	Possible Cause
	c. Compressor air strainer dirty.
	d. Automatic drain valve not functioning.
10. Brakes uneven.	a. Brakes in need of adjustment, lubrication, or relining.
	b. Grease on brake lining; reline brakes.
	c. Brake shoe release spring or brake chamber release spring broken.
	d. Brake drum out of round.
	e. Brake chamber diaphragm leaking.
11. Air pressure will not rise to normal.	a. Defective air gage.
	b. Excessive leakage.
	c. Reservoir drain cock open.
	d. Governor out of adjustment.
	e. No clearance at compressor unloader plungers.
	f. Defective or worn compressor.
12. Air pressure rises to normal too slowly.	a. Defective air gage.
	b. Excessive leakage.
	c. Clogged compressor air strainer.
	d. Engine speed too low.
	e. Compressor discharge valves leaking.
	f. Worn compressor.
	g. Excessive carbon in the compressor cylinder head or discharge line.
	h. Compressor belt loose.
13. Air pressure rises above normal.	a. Defective air gage.
	b. Governor out of adjustment.
	c. Defective governor.
	d. Restriction in line between governor and compressor unloading mechanism.
14. Air pressure drops quickly with engine stopped and brakes released.	a. Leaking brake valve.
	b. Leaking tubing or hose line.
	c. Compressor discharge valves leaking.
	d. Governor leaking.
	e. Excessive leakage elsewhere in the system.

Source: Courtesy of Mack Trucks, Inc.

Automatic Transmission (Allison)

Problem	Possible Cause
1. Automatic shifts occur at too high speed.	a. Governor valve stuck. b. Shift signal valve spring adjustment too tight. c. Valves sticking. d. Modulator valve spring adjustment too loose.
2. Automatic shifts occur at too low speed.	a. Governor valve stuck. b. Governor valve weak. c. Shift signal valve spring adjustment too loose. d. Modulator valve stuck. e. Modulator valve spring adjustment too tight.
3. Main pressure low in all ranges.	a. Low oil level. b. Oil filter element clogged. c. Seal ring at oil intake pipe (filter output) leaking or missing. d. Main pressure regulator valve spring weak. e. Control valve body leakage. f. Valves stuck (trimmers, relays, and main pressure regulator). g. Oil pump worn or damaged.
4. Main pressure low in one operating range, normal in other ranges.	a. Leakage in clutch apply circuits for specific range. b. Excessive leakage in clutch piston seals for specific range.
5. Excessive creep in first and reverse gears.	a. Idle throttle setting too high.
6. Low lubrication pressure.	a. Oil level low. b. Excessive internal oil leakage. c. Cooler lines restricted or leaking. d. Lubrication valve spring weak.
7. Oil leaking into converter housing.	a. Converter pump hub seal worn. b. Converter pump hub worn in seal area. c. Engine rear seal worn.
8. Transmission overheating in all ranges.	a. Oil level low. b. Oil level high. c. Cooler restricted (oil or coolant side).
9. No response to shift lever movement.	a. Range selector linkage unhooked. b. Range selector linkage defective or broken.

Engine and Systems Troubleshooting

Automatic Transmission (Allison)
(Continued)

Problem	Possible Cause
	c. Main pressure low.
	d. Range selector not engaged at control valve.
10. Rough shifting.	a. Manual selector linkage out of adjustment.
	b. Control valves sticking.
	c. Modulator valve sticking; spring adjustment too tight.
	d. Modulator actuator cable kinked or out of adjustment.
11. Dirty oil.	a. Failure to change oil at proper interval.
	b. Heat excessive.
	c. Clutch failure.
	d. Damaged oil filter.
12. Oil leaking at output shaft.	a. Oil seal at output flange worn or damaged.
	b. Flange worn at seal surface.
13. High stall speed.	a. Oil level low.
	b. Clutch pressure low.
	c. Forward clutch slipping (forward).
	d. First clutch slipping.
	e. Fourth clutch slipping (reverse).
	f. Low clutch slipping (five-speed).
14. Low stall speed.	a. Engine not performing efficiently (may be due to high altitude).
	b. Broken converter parts.
15. Clutch slippage in all forward gears.	a. Oil level low.
	b. Clutch (main) pressure low.
	c. Forward clutch slipping.
	d. Seal rings on front support hub worn or broken.
16. Vehicle moved in neutral.	a. Range selector or linkage out of adjustment.
	b. Forward clutch not releasing.
	c. Fourth clutch not releasing.
17. Oil thrown from filler tube.	a. Dipstick loose.
	b. Oil level too high.
	c. Breather clogged.
	d. Dipstick gasket worn.
	e. Improper dipstick marking.

Source: Courtesy of Detroit Diesel Allison, Division of General Motors Corporation.

Introduction to Fire Apparatus and Equipment

REVIEW QUESTIONS

Following is a list of problems that could occur on a fire apparatus. Without referring to the chapter, give some of the possible causes.

Diesel Engines

1. Engine will not rotate during starting operations.
2. Low cranking speed.
3. Low engine compression.
4. No fuel to engine.
5. Uneven running or frequent stalling of engine.
6. Lack of engine power.
7. Engine detonating.
8. No fuel or insufficient fuel caused by air leaks.
9. No fuel or insufficient fuel caused by fuel obstruction.
10. No fuel or insufficient fuel caused by faulty fuel pump.
11. No fuel or insufficient fuel caused by faulty installation.
12. High lubricating oil consumption.
13. External oil leaks.
14. Internal oil leaks.
15. Excessive oil consumption caused by oil control at cylinder.
16. Excessive crankcase pressure.
17. Cylinder blow-by.
18. Excessive exhaust back pressure.
19. Excessive crankcase pressure caused by air from blower or air box.
20. Low oil pressure.
21. Lubricating oil causing low oil pressure.
22. Pressure gage causing low oil pressure.
23. Oil pump causing low oil pressure.
24. Poor circulation causing low oil pressure.
25. Engine coolant operating temperature above normal.
26. Engine coolant operating temperature below normal.

Engine and Systems Troubleshooting

27. Black or gray exhaust smoke.
28. Blue exhaust smoke.
29. White exhaust smoke.

Carburetor

1. Engine cranks (turns over) but will not start or is hard to start when cold.
2. Engine hard to start when hot.
3. Engine starts and stalls.
4. Engine idles roughly and stalls.
5. Engine runs unevenly or surges.
6. Engine hesitates on acceleration.
7. No power on heavy acceleration or at high speed.
8. Poor fuel economy.
9. Engine backfires.

Clutches

1. Poor release.
2. Clutch slippage.
3. Noisy clutch.

Drive Train

1. Whining noise at all speeds; pitch varies directly with road speed.
2. Grinding or grating noise at all speeds.
3. Vibration at any road speed.

Manual Steering

1. Hard steering.
2. Excessive free play.
3. Shimmy.
4. Wandering.

5. Pulling to one side.
6. Uneven or excessive tire wear.
7. No recovery after turn.

Hydraulic Power Steering

1. Abnormal noise.
2. Wandering.
3. No recovery after turn.
4. Shimmy.
5. External oil leakage.
6. Oversteer or darting.
7. High steering effort in one direction.
8. High steering effort in both directions.
9. Lost motion at steering wheel.
10. Excessive heat.

Air Brakes

1. Brakes insufficient.
2. Brakes apply too slowly.
3. Brakes release too slowly.
4. Brakes do not apply.
5. Brakes do not release.
6. Brakes grab.
7. Compressor knocks continuously or intermittently.
8. Safety valve "blows off."
9. Excessive oil or water in the system.
10. Brakes uneven.
11. Air pressure will not rise to normal.
12. Air pressure rises to normal too slowly.
13. Air pressure rises above normal.
14. Air pressure drops quickly with engine stopped and brakes released.

Automatic Transmission (Allison)

1. Automatic shifts occur at too high speed.
2. Automatic shifts occur at too low speed.

Engine and Systems Troubleshooting

3. Main pressure low in all ranges.
4. Main pressure low in one operating range, normal in other ranges.
5. Excessive creep in first and reverse gears.
6. Low lubrication pressure.
7. Oil leaking into converter housing.
8. Transmission overheating in all ranges.
9. No response to shift lever movement.
10. Rough shifting.
11. Dirty oil.
12. Oil leaking at output shaft.
13. High stall speed.
14. Low stall speed.
15. Clutch slippage in all forward gears.
16. Vehicle moves in neutral.
17. Oil thrown from filler tube.

5

Apparatus Testing

Apparatus should be tested to ensure that they meet minimum standards of performance and safety. Tests should be conducted prior to delivery, upon delivery, and at least annually once apparatus have been placed in service. Apparatus should also be tested after any extensive repairs that might affect pumping or operating performance or contribute to lower safety standards. Standards have been established for certain types of apparatus testing; however, standardized testing procedures are still lacking in some important aspects of apparatus testing, aspects that affect safety. This chapter contains information on pumper testing, aerial ladder testing, elevating platform testing, and some testing safety procedures.

PUMPER TESTING

Most pumpers built prior to 1939 conform to the standards established for Class B pumpers. These standards were developed by the National Board of Fire Underwriters, now the Insurance Services Office, and the International Association of Fire Engineers, now the International Association of Fire Chiefs. Class B standards required that a pump deliver rated capacity at 120 pounds per square inch net pump pressure, 50 percent of rated capacity at 200 psi net pump pressure, and 33⅓ percent of rated capacity at 250 psi net pump pressure. Testing consisted of running the pump at capacity for 6 hours, half capacity for 3 hours, and one-third capacity for 3 hours.

Between 1939 and 1947, national authorities on fire apparatus toyed with the concept of a Class A pumper that would deliver rated capacity at 150 psi net pump pressure, 70 percent of rated capacity at 200 psi net pump pressure, and 50 percent of rated capacity at 250 psi net pump pressure. Both the Class A standards and those for a Class B pumper were outlined in the 1947 edition of the *Suggested Specifications for Motor Fire Apparatus*, issued by the National Board of Fire Underwriters. By 1956 the specifications no longer included those for a Class B pumper.

As a result of this gradual change, most pumpers built prior to 1939 are Class B, those built from from 1939 through 1956 are Class A or B, and those built since 1956 are Class A.

Seven pump capacities are now considered standard for the fire service: 500, 750, 1000, 1250, 1500, 1750, and 2000 gallons per minute. Pumper requirements are outlined in the National Fire Protection Association Standard 1901, *Automotive Fire Apparatus*.

Fire department purchase contracts should specify the required capacity of the fire pump. Pumps manufactured today conform to NFPA Standard 1901 and are designed to meet the following discharge requirements:

> 100 percent of rated capacity at 150 psi net pump pressure
> 70 percent of rated capacity at 200 psi net pump pressure
> 50 percent of rated capacity at 250 psi net pump pressure.

Pumpers are subjected to three different types of tests: certification, delivery, and service. Certification tests are conducted

Apparatus Testing

prior to delivery, delivery tests are conducted upon delivery, and service tests are conducted after an apparatus has been placed in service.

Certification Tests

Certification tests are designed to ensure that a new pumper meets minimum standards of construction and performance prior to delivery to a purchaser. They are conducted at the manufacturer's factory and supervised by engineers from Underwriters Laboratories Inc. The apparatus manufacturer is required to furnish all equipment needed for the tests and a sufficient number of personnel to conduct the tests.

The certificate for apparatus tested at elevations between sea level and 2000 feet will indicate "Performance certified to 2000 feet elevation." For apparatus tested at elevations above 2000 feet, the certificate will indicate elevation of the test site.

Each pumper delivered to a purchaser is not subjected to the certification test, but the certification confirms that at least one identical pumper has successfully passed the test.

The certification program consists of two parts. The first part, the pumping test, ensures that the pump and accessories meet minimum standards of performance. The second part is an inspection to ensure that the pumper complies with certain nationally accepted standards.

Pump Test

The pump test consists of the following six parts:

1. Engine governor test
2. Pumping test
3. Overload test
4. Automatic pressure control test
5. Water tank flow test
6. Vacuum test

Some general rules apply to all of the testing procedures. These include:

This page includes copyrighted material of Insurance Services Office with its permission. Copyright, Insurance Services Office, 1975.

1. Tests are to be conducted while pumping from draft, with the water source at least one foot below the surface on which the pumper is located. The pumping water source is to be sufficiently large, clean, and free of debris to minimize clogging of the strainers.
2. Tests must be made with all accessories and power-consuming appliances attached.
3. The only stops permitted during the tests are those necessary for changing hose and nozzles. During and after the tests, the engine, pump, transmission, and all parts of the machine must show no undue heating or excessive strain or vibration, and the engine must show no loss of power, overspeed, or other defect.

Engine Governor Test

The engine governor test is conducted first and then repeated after the vacuum test. The tests are conducted on the pumper assembly with the road transmission in neutral.

Method   The pumper is warmed up, and the suction and discharge hoses are laid out. The engine speed is increased steadily—with the hand throttle at the operator's position in the first test, and with the accelerator pedal in the cab in a second test—until the throttle is at its maximum setting or the speed of the engine is in excess of the specified no-load speed.

The engine speed may be determined by reading the permanently installed engine tachometer at the pump operator's position, or a separate (calibrated) test tachometer may be used instead of the pump panel tachometer. If the pump panel tachometer is connected to the pump rather than to the engine, the engine speed should be determined by reading the cab panel tachometer or a separate test tachometer. In any case, the accuracy of the pump panel tachometer or the cab panel tachometer should be verified by a hand counter. Avoid prolonged engine operation at high no-load speed.

Basis for Acceptability   The maximum no-load speed is that speed specified by the engine manufacturer on the certified brake horsepower curve or stamped on the engine block.

The maximum speed of the engine should not exceed the maximum no-load speed by more than 2 percent.

Pumping Test

The pumping test consists of drafting water and pumping at rated capacity at a net pump pressure of 150 psi for a continuous period of 2 hours, followed by two half-hour periods of continous pumping.

Method   During the first period, at least 70 percent of the rated capacity shall be delivered at a net pump pressure of 200 psi; during the remaining half-hour, 50 percent of the rated capacity shall be delivered at a net pump pressure of 250 psi.

Basis for Acceptability   The pumping test demonstrates the ability of the equipment to deliver rated capacity at 150 psi net pressure, and the required minimum capacity at higher pressures for the prescribed periods, while maintaining the engine temperature within the operating range, with no drop in engine oil pressure, no overheating in the engine or transmission, and no other apparent malfunction of the engine or pump.

Overload Test

The overload test and the automatic pressure control test may be conducted immediately following the 2-hour pumping test.

Method   The apparatus should be given an overload test of 10 minutes' duration. The test consists of discharging rated capacity at 165 psi net pump pressure. Test data shall be recorded as prescribed under the pumping test.

Basis for Acceptability   The overload test demonstrates the ability of the engine to develop 10 percent excess power.

Automatic Pressure Control Test

The pressure control test determines the efficiency of the automatic means of controlling the pump pressure.

Method The automatic pressure control test should be conducted while pumping from draft, as previously described, with the operating pressure set at various levels between 90 psi and 250 psi, and while pumping at specified capacity, with the automatic pressure control set at not more than 10 psi higher than the operating pressure. In each case, the hose outlet valve is slowly closed.

Basis for Acceptability Under the operating conditions outlined above, the pump pressure should not increase to more than 30 psi above the operating pressure.

Water Tank Flow Test

The weighted capacity of the water tank is to be that given by the manufacturer on the *Manufacturer's Record of Pumper Construction Details.*

The test is to be conducted with the truck on level grade.

Method Connect the discharge hose layout and select the proper nozzle. Use a nozzle diameter of 1 inch for 250 gpm, and 1½ inches for 500 gpm. At this point the discharge valves should be closed.

Fill the water tank to overflowing.

Close all inlets. The draft inlets are closed by either capping or by closing the gate valves in the hose line or on the apparatus. Close the tank fill and/or the bypass line.

Fully open the tank-to-pump supply line.

Fully open the discharge valves to the hose lines.

Adjust the engine throttle for the required flow within the range of 0 to +5 percent.

After the flow has been adjusted and determined to be within the acceptable range, note the discharge pressure on the test gage. Shut off the discharge valves (the tank-to-pump valve or the bypass line may be opened to cool the pump) without disturbing the throttle setting. The automatic pressure control device may be used, if set for a pressure greater than the noted discharge pressure.

Refill the water tank to overflowing. Open the tank-to-pump supply line fully.

Apparatus Testing

After the tank has been refilled, close the lines as before, and open the tank-to-pump supply line fully. The tank fill or bypass line should be closed. The test time period starts when the hose discharge valves are fully opened. The engine revolutions may be adjusted to maintain the discharge pressure at the previously noted pressure.

The test time period ends when the discharge pressure on the test gage drops and remains at least 5 psi below the previously noted pressure.

The amount of water discharged from the tank is determined by multiplying the actual flow rates in gallons per minute by the length of the test time period in minutes.

Basis for Acceptability The connections between the water tank and the pump should permit a flow of at least 250 gpm for tanks of 300- to 750-gallon capacity, and 500 gpm for tanks with a capacity over 750 gallons.

The specified rate of flow must be maintained until not less than 80 percent of the capacity of the tank has been discharged.

The purchaser may desire a different tank-to-pump flow rate. In this case, the Underwriters Laboratories inspector should be provided with a copy of the letter (or other document) in which the desired flow rate is specified.

Vacuum Test

The vacuum test should be the last test.

Method As a condition of the test, all water must be drained from the pump and all the internal suction and discharge lines. This is done by opening all the drain valves and removing all caps on the suction and discharge lines.

To set up the remaining conditions of the test, the pump and lines are then closed. Close all valves on the suction and discharge side. If more than one valve can control an inlet or outlet, the valve farthest from the pump is closed, and all the others in the line are opened. If no valves are in the line, such as in the case of drafting suction inlets, the protective caps are put tightly in place at the inlet.

Take care to close the tank-to-pump valve tightly.

After the pump and system are dry and sealed, the priming device is used to develop the required vacuum. During the vacuum test, no part of the apparatus is to be in operation.

The required vacuum is 22 inches of mercury (Hg). At test site elevations over 1000 feet, this is decreased by 1 inch for every 1000 feet. For example, at a 2000-foot test site, the required vacuum is 21 inches Hg; at 3000 feet, it is 20 inches Hg, and at an elevation of 1500 feet, it is 21.5 inches Hg.

When the last operative device is turned off, the vacuum inside the pump is the "maximum vacuum"; the test time period is to start at this instant. At the end of 10 minutes, the manometer is read; the difference between the reading at the start and that after the 10-minute test period is the "maximum drop."

Basis for Acceptability The priming device must develop a vacuum equal to a minimum of 22 inches of mercury (as corrected for altitude) and maintain a drop of not more than 10 inches in 10 minutes.

Test Inspection

The second part of the certification program is an inspection to determine whether the apparatus conforms to those portions of the standards listed in NFPA 1901, *Automotive Fire Apparatus*, that, in the judgment of Underwriters Laboratories Inc., apply to its Certification Program on Fire Department Pumpers. The following list is only a sample of the many factors that are considered:

1. An engine governor should be installed. This will limit the speed of the engine under all conditions of operation to that speed established by the engine manufacturer; this is the maximum no-load speed.
2. At least one gated suction inlet should be provided.
3. When a water tank is installed on the vehicle, the tank should be connected to the suction side of the pump, with the valve controlled from the pump operator's position.
4. All outlets should have national (American) standard fire hose coupling threads.
5. Adequate illumination should be provided for all gages and controls that are located at the pump operator's position.

6. Provisions should be made for quickly and easily placing the pump in operation. The lever or other device must be marked to indicate when it is in pumping position.

7. Any control device used in the power train between the engine and the pump must be so arranged as to prevent it from being accidentally knocked out of operating position.

8. With parallel-series centrifugal pumps, the positions for parallel operation (volume) and series operation (pressure) should be clearly indicated. The control for changing the pump from series to parallel, and vice versa, should be controlled from the pump operator's position.

9. A hand throttle controlling the fuel supply to the engine should be of a type that will hold its set position and should be so located that it can be manipulated from the operator's position with all gages in full view.

10. All gages and instruments should be so located as to be readily visible from the pump operator's position.

11. A pumper should be equipped with one suction gage, located on the left side of the gage panel. The gage should be not less than 4½ inches in diameter and should read from 30 inches of vacuum to at least 300 psi pressure but not more than 600 psi pressure. The pumper must also have one discharge gage located to the right of the suction gage, also not less than 4½ inches in diameter, and reading from 0 to at least 300 psi pressure but not more than 600 psi pressure, except in cases where pump maximum working pressure is above 600 psi.

12. An oil pressure gage and an engine-coolant temperature gage should be located at the pump operator's position, in addition to those on the vehicle's instrument panel.

13. A test panel plate at the pump operator's position should give the rated discharges and pressures, together with the speed of the engine, as determined by the manufacturer's test for each unit.

The Certificate of Inspection Form

Those pumpers that satisfactorily comply with both the pumping and inspection portions of the certification program are issued a Certificate of Inspection for Fire Department Pumper. A specimen of the form is shown in Figure 5-1.

176   Introduction to Fire Apparatus and Equipment

UNDERWRITERS LABORATORIES INC.
333 PFINGSTEN ROAD · NORTHBROOK, ILLINOIS 60062

an independent, not-for-profit organization testing for public safety

CERTIFICATE OF INSPECTION FOR FIRE DEPARTMENT PUMPER

The scope of the UL Certification Program is limited to witnessing the conduct of the tests reported below and the checking of only those controls, instruments, and equipment necessary to accomplish this testing against the requirements of the Edition of the NFPA Standard No. 1901 in effect as of the date of this certificate unless otherwise noted.

RATED CAPACITY .... gpm   Date ....

This certifies that the pumper described below has performed acceptably and is provided with items of equipment as shown.

Manufacturer: .... Model No. .... Serial No. ....
For: .... Location ....
Chassis: Mfr. .... Model No. .... Serial No. ....
Engine: Mfr. .... Model No. .... Serial No. ....
Pump: Mfr. .... Model No. .... Serial No. ....
Test Conditions: Barometric Pressure .... in. Hg. (corrected to Sea Level); Temp. .... F; Elevation .... ft.;
Suction Hose: Size .... in.; Length .... ft.; Pump elev. above water source .... ft.;
Performance Certified to .... ft. Elevation above Sea Level.

TEST CONDITIONS	PUMP CONTROL POSITION	FLOW GPM	DISCHARGE PRESSURE PSIG	SUCTION ALLOWANCE PSIG (NEG.)	NET PUMP PRESSURE PSI	PUMP SPEED RPM	ENGINE SPEED RPM	GEAR RATIO ENGINE TO PUMP
Capacity 150 psi—2 Hrs.								
Overload 165 psi—10 Min.				SPECIMEN				
70% Capacity 200 psi—30 Min.								
50% Capacity 250 psi—30 Min.								

Automatic Pressure Control Test: Max. Increase .... psi. Vacuum Test: .... in. Hg. drop in 10 Min.
No Load Governor Speed .... rpm; Specified .... rpm. Pump Location ....
Water Tank Capacity .... Gal;   Tank Flow Test: .... Gal. discharged at .... gpm
.... In. Suction gauge .... In. Hg. to .... psi   .... In. Discharge gauges .... to .... psi

REMARKS

Not Valid Unless Countersigned   Signed: *D. L. Breting*

UNDERWRITERS LABORATORIES INC.   D. L. BRETING
Vice President
Follow-Up Services

Figure 5-1   Certificate of Inspection for Fire Department Pumper. *Courtesy of Underwriters Laboratories, Inc.*

Delivery Tests

Every new pumper, on delivery and before acceptance, should be put through a test that will determine its ability to meet contract specifications. The test should be conducted at the delivery site by the apparatus manufacturer's representative, with the fire chief or a department representative present and responsible for supervision of the test. In addition to determining whether the pump satisfactorily fulfills contract requirements, delivery tests provide pump operators with the opportunity to operate their pumpers at capacity before the pumper is placed in service.

Delivery tests run 3 hours. They consist of drafting water and pumping rated capacity against a net pump pressure of 150 psi for a continuous 2-hour period, followed by two half-hour periods of continuous pumping. During one period, at least 70 percent of the rated capacity is delivered at a net pump pressure of 200 psi, and during the remaining half-hour, 50 percent of rated capacity is delivered at a net pump pressure of 250 psi.

The following three tests should also be included in the delivery test procedure:

1. The apparatus should be given a short overload test to demonstrate its ability to develop 10 percent excess power. This test should consist of discharging rated capacity at 165 psi net pump pressure.

2. A test should be conducted when operating at draft, to determine the efficiency of the means provided for automatically controlling the pump pressure. With the pressure control device set at 10 psi higher than any operating pressure over 75 psi, and the pump discharging through one or more hose lines, the pump pressure should not increase more than 30 psi when discharge valves are closed slowly.

3. A vacuum test, with a capped suction at least 20 feet long, should show the apparatus's ability to develop 22 inches of vacuum and hold the vacuum for 10 minutes with a drop not in excess of 10 inches.

The delivery test should be made with all accessories and

This page includes copyrighted material of Insurance Services Office with its permission. Copyright, Insurance Services Office, 1975.

power-consuming appliances attached, using a grade of fuel recommended by the apparatus manufacturer.

Service Tests

Pumpers that are in service should be tested after any extensive repairs that might affect their pumping capabilities, and at least annually to determine their condition. Investigation has shown that where regular and systematic tests of pumpers are not made, undetected defects may continue for considerable periods, under the light demands at ordinary fires, and only become apparent at a large fire where the pumper is called upon to perform at or near full capacity. Furthermore, regular tests are a valuable drill for pump operators, since only a few departments give sufficient training in operating pumpers at capacity and at draft. The breakdown of a pumper at a fire, or the inability of the crew to operate it properly, may cause needless loss of life and property.

If possible, service tests should be conducted by the crew normally assigned to the apparatus. Off-duty crews should be encouraged to attend the tests in order to obtain valuable experience in drafting operations and operating the pump at full capacity. The test gives both on- and off-duty personnel an opportunity to become more familiar with their pumper's performance capabilities and with test procedures.

The test should be supervised by the mechanic or other person responsible for the maintenance of the apparatus. This individual should be prepared to instruct personnel on proper pump operation procedures and to make any necessary adjustments required during the test.

Service tests are based on the capacity specified in the original purchase specifications. The usual service test consists of pumping rated capacity at 150 (or 120) psi for at least 20 minutes, a pressure test at 70 percent (or one-half) rated capacity at 200 psi for about 10 minutes, and a pressure test at one-half (or one-third) rated capcity at 250 psi for about 10 minutes.

Tests in which the pressures developed and/or the quantities

This page includes copyrighted material of Insurance Services Office with its permission. Copyright, Insurance Services Office, 1975.

Apparatus Testing

discharged are less than those specified by the original contract do not show the true condition of the apparatus and are of little value other than for operator training.

Data submitted at the time of the delivery test and all results of service tests should be maintained in a permanent file so that the condition of the pumper can be compared over years of operation.

Test Facilities

If possible, pumpers should be tested at draft, with a vertical lift of not more than 10 feet. The vertical lift is the vertical distance from the surface of the water to the center of the pump suction inlet. It is important that an adequate site be selected for testing. A pit constructed especially for testing purposes is ideal.

Pumper test pits greatly facilitate test operations. These pits are small reservoirs of clear water where pumpers can be tested at draft, using the water over and over. A pumper test pit is usually a simple concrete tank at least 10 feet deep and twice as long as it is wide. It generally contains a central partition to keep the turbulence of the discharge compartment from affecting the suction compartment.

A source that provides an adequate supply of clean static water may be used if a test pit is not available. Locations such as reservoirs, swimming pools, or docks generally meet this criterion.

The site chosen should, if possible, be located along an improved roadway or on solid ground where the water level is from 4 to 8 feet below the grade. It should be possible to reach the water from the pumper intake inlet with not more than 20 feet of hard suction hose, with the strainer submerged at least 2 feet, and with no humps in the hose. The water should be a least 4 feet deep where the strainer is located, to provide adequate clearance below the strainer and sufficient depth above it. A special basket or container should be used to prevent the suction in the hose from drawing in particles from the stream bed, if drafting from shallow water is necessary. Clean fresh water is desirable, but where salt

This page includes copyrighted material of Insurance Services Office with its permission. Copyright, Insurance Services Office, 1975.

water is drafted, the pump, pipe, fittings, and pressure-regulating governors should be thoroughly flushed out after testing.

Tests should be made at draft if at all possible; however, it may sometimes be necessary to run service tests by connecting to a hydrant. This procedure does not test the ability of the priming system or of the pump to maintain prime and creates an additional problem of getting rid of the large amount of water that is discharged. Water pumped during the test may vary from 11,000 gallons to over 50,000 gallons, depending on the pump capacity. The hydrant used should be one at which the initial pressure is not too high—below 40 pounds is preferable—yet should be connected to a large enough water main to assure a sufficient supply of water.

During pumping at draft, the net pump pressure is the sum of pump discharge gage pressure, corrected for any gage error, plus lift (the vertical distance from water level to the gage), plus suction losses. Allowances for lift and suction losses are usually given in feet and must be divided by 2.3 to convert the figures into pounds per square inch. Table 5-1 shows the allowance for friction loss in the suction hose. The net pump pressure is equal to the pumper discharge pressure, less the pressure on the suction line at the pump, when water is being drawn from a hydrant.

During the test the pumper should be parked as close as possible to the water's edge. It is usually more convenient to have the pump control panel side of the pumper away from the water. Front or rear intake inlets should be avoided, as the piping between the pump and inlet is usually restricted by twists and bends.

Required Equipment

Equipment items are the same for any type of test. On days when more than one pumper test is to be made, nozzles and hose layouts are usually left at the test site until completion of all tests. Essential equipment is listed below:

1. *Hard Suction Hose.* Twenty feet or more of good 4- to 6-inch hard suction hose should be available for drafting.

This page includes copyrighted material of Insurance Services Office with its permission. Copyright, Insurance Services Office, 1975.

Apparatus Testing

The size of the suction hose to be used will depend on the altitude and the lift, as well as the rated capacity of the pump to be tested. The suction hose should be of the following minimum sizes for the pumps indicated:

500-gpm pumpers, 4-inch
750-gpm pumpers, 4½-inch
1000-gpm pumpers, 6-inch or, at low altitudes, 5-inch

Table 5-1 Allowances for Friction Loss in Suction Hose

Rated Capacity of Pumper gpm	Diameter of Suction Hose in inches	Allowance (feet)	
		For 10 ft. of Suction Hose	For each additional 10 ft. of Suction Hose
500	4	6	plus 1
	4½	3½	plus ½
750	4½	7	plus 1½
	5	4½	plus 1
1000	4½	12	plus 2½
	5	8	plus 1½
	6	4	plus ½
1250	5	12½	plus 2
	6	6½	plus ½
1500	6	9	plus 1
1500	4½ (dual)	7	plus 1½
1500	5 (dual)	4½	plus 1
1500	6 (dual)	2	plus ½
1750	6	12½	plus 1½
1750	4½ (dual)	9½	plus 2
1750	5 (dual)	6½	plus 1
1750	6 (dual)	3	plus ½
2000	4½ (dual)	12	plus 2½
2000	5 (dual)	8	plus 1½
2000	6 (dual)	4	plus ½

Note: The allowance computed above for the capacity test should be reduced by 1 psi for the allowance on the 200-psi test and by 2 psi for the allowance on the 250-psi test.

Example: 1000-gpm pumper, 9-foot lift, 20 feet of 5-inch suction.

$$\text{Pressure Correction} = \frac{9 + 8 + 1\frac{1}{2}}{2.3} = 8 \text{ psi (capacity test)}$$

This page includes copyrighted material of Insurance Services Office with its permission. Copyright, Insurance Services Office, 1975.

1250-gpm pumpers, 6-inch

1500-gpm pumpers or larger, 6-inch or two suction lines (5-inch minimum size) in a siamese

Although a 4½-inch suction hose is large enough for even a 1000-gpm pumper when taking water from a hydrant, departments having 1000-gpm or larger pumpers should keep at least two sections of 5- or 6-inch suction hose for testing purposes.

Chafing pads should be provided for suction hose to prevent damage to the hose when it is in contact with sharp edges of docks, manholes, walls, and rocks.

2. *Strainer.* An adequate strainer is needed for attaching to the intake end of the hard suction hose.

3. *Nozzles.* A set of nozzles suitable for testing should be provided. Suitable nozzles are usually found in the regular equipment of every fire department. Mounted turrets (deck guns) offer more safety than deluge sets when testing pumpers. For safety reasons, all nozzles, deluge sets, and turrets must be securely tied down. Only smooth-bore nozzles should be used. Care should be taken that washers or gaskets do not protrude into the pipe, as a perfectly smooth waterway is essential. Particular care must be taken to see that shut-off nozzles do not have any projections or breaks in the waterway and that the shut-off handle does not become partially closed.

4. *Nozzle Tips.* For various capacity and pressure tests, nozzle tips with inside diameters from ¾ to 2½ inches are best. They should be free from nicks and scratches to ensure a smooth stream. Tips should be measured to ensure that there is no mistake about the size used; the measurements should be taken after the tips are attached and ready for the test.

5. *Hose.* From 50 to 350 feet of 2½-inch hose should be used, the length depending on the hose layout and the capacity of the pump to be tested. It should be double-jacket, rubber-lined hose. That commonly used in the department is

---

This page includes copyrighted material of Insurance Services Office with its permission. Copyright, Insurance Services Office, 1975.

Apparatus Testing

satisfactory. The hose should be in good condition and have been tested recently to at least 250 psi.

6. *Strong Rope or Light Chain.* At least one length of strong rope or light chain is required, in order to secure the nozzles in position.

7. *Pitot Tube with Air Chamber and Pressure Gage.* A pitot tube with an air chamber and a pressure gage is necessary for checking the velocity pressure of the water at the nozzle. The pitot tube may be one of several types. The type shown in Figure 5-2 is relatively easy to construct. It should

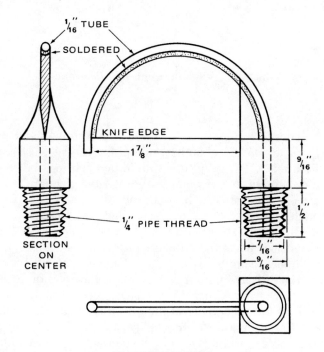

SCALE FULL SIZE
TO BE OF BRASS AND FINISHED SMOOTH

Figure 5-2 Nozzle Stream Pitot. *Copyright Insurance Services Office, 1975*

This page includes copyrighted material of Insurance Services Office with its permission. Copyright, Insurance Services Office, 1975.

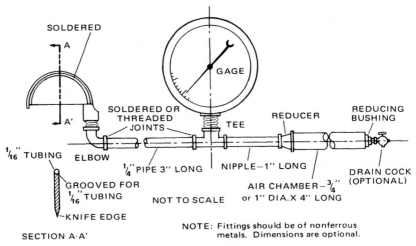

Figure 5-3   Pitot Tube Assembly. *Copyright Insurance Services Office, 1975*

be connected, by brass or other non-ferrous metal pipe fittings, to an air chamber and pressure gage, as shown in Figure 5-3. The pitot tube should be kept free of dirt and the air chamber free of water. Any water that accumulates in the air chamber should be removed after each test. The knife edge, indicated in Figure 5-2, will get battered in service and must be kept sharp to reduce as much as possible the spray that arises when inserting the pitot into the stream. The pitot tube gage should be capable of measuring up to 200 psi, marked for every pound, and labeled every 5 or 10 pounds.

8. *Pressure Gage.* A pressure gage with the necessary fittings for attachment at the pump is required to obtain the discharge pressure. The test gage should be attached to a special tap, a plugged ¼-inch standard pipe thread connection that must be provided on all pumpers. Occasionally, a tapped 2½-inch cap attached to an unused pump outlet is used for a short test. If a gage is used on the suction side of the pump, a gage capable of indicating a vacuum should be provided. Gages should be of good quality and accurate.

This page includes copyrighted material of Insurance Services Office with its permission. Copyright, Insurance Services Offices, 1975.

Apparatus Testing

Those of about 3½-inch diameter are generally most convenient. They should be without a rest pin for the needle, so that needle slippage on the spindle, which would give a false reading, may be readily observed. The pump pressure gage should be capable of measuring up to 400 psi and may be marked for every 5 pounds. Gages should be carefully calibrated (tested) with a weight tester or a standard gage before use. An air chamber or short length of rubber hose in the connection of the pressure gage at the pump will reduce the vibration of the gage needle and permit a more accurate reading.

9. *Revolution Counter or Accurate Tachometer.* Either a revolution counter, with extension if necessary, or an accurate tachometer should be used for obtaining speed readings. Tachometers may be used to check the speed reading obtained with the revolution counter. Some portable commercial tachometers of an electric type are sufficiently accurate.

10. *Watch.* A stopwatch is best; however, any watch with a full sweep second hand that may be clearly read can be used.

11. *Test Data Blanks.* Some test data blanks for recording the test readings and other necessary data should be provided. The use of such forms will help to assure that all needed data are obtained.

Recommended Hose Layouts

The discharge hose layout consists of one or more 2½-inch hose lines running to smooth-bore nozzles of suitable size. The hose performs two functions. The first is to carry the water from the pumper to the nozzle. The second is to provide enough total friction loss to reduce the required pump discharge pressure to the desired nozzle pressure. If only a relatively short length of hose is required to perform the first function, friction loss can be increased by partially closing the discharge gate valve or valves on the pumper or, preferably, by partially closing a gate valve that is inserted in one or more of the discharge lines for that purpose.

Discharge hose lines should be securely fastened to the

This page includes copyrighted material of Insurance Services Office with its permission. Copyright, Insurance Services Office, 1975.

Table 5-2 Hose and Nozzle Layout Suggestions

Discharge in Gallons Per Minute	120 or 150 psi Test
2,000	Two 100' lines, 2" nozzle, in duplicate
1,750	Two 100' lines, 2" or 1⅞" nozzle, in duplicate
1,500	Three 100' lines, 2" nozzle; and one 50' line, 1⅜" or 1½" nozzle
1,250	Three 100' lines and one 50' line; 2¼" nozzle or one 50' line, 1⅝" nozzle, in duplicate
1,000	Three 100' lines; 2" nozzle
750	Two 100' lines; 1¾" nozzle
500	One 50' line; 1⅜" or 1½" nozzle

Note: Where two or more lines are indicated they are to be siamesed into a heavy-stream appliance.

pumper outlets so as to avoid injury to personnel, should the hose come loose from the coupling during the test. In most cases a rope hose tool is adequate for the purpose.

The nozzle size is usually chosen so that the nozzle gives the desired discharge at a nozzle pressure of between 60 and 70 psi. This pressure is not so high that the pitot is difficult to handle in the stream nor so low that the normal inaccuracies of a gage used at low pressure would come into play. The nozzle should always be securely tied. Never run a test in which any person or persons are needed to hold the nozzle. Failure to abide by this precaution has caused serious injuries in several instances.

Table 5-2 shows suggested hose and nozzle layouts, but other combinations of layouts may be used. Longer lines or smaller nozzles may be required if the hose is slightly larger or the lining is unusually smooth. Shorter lines or larger nozzles may be necessary if the hose is slightly smaller or unusually rough.

Safety Precautions

Every precaution must be taken during service tests to protect personnel and equipment from injury or damage. If adequate

This page includes copyrighted material of Insurance Services Office with its permission. Copyright, Insurance Services Office, 1975.

Apparatus Testing

safety measures are not observed, or if personnel get careless, enough dangerous ingredients are present during the test to cause serious problems. The factors that cause problems during service tests are also present during many emergencies or training situations; however, during service tests the hazards are concentrated in a small, relatively confined area. The following are some of the safety precautions that should be observed.

1. Ensure that all test nozzle devices are correctly secured. Failure of the securing device could cause severe injuries or damage.

2. Ensure that guards are in place over all manholes or other openings at the test site.

3. Be sure that sufficient personnel are used to lift heavy objects and that proper lifting techniques are followed.

4. Do not stand in line with the engine fan. If it breaks, a blade will be thrown off from the engine at nearly a right angle.

5. Do not stand directly in front of discharge gates to which the hose lines are attached. The hose at this point is subjected to more pressure than elsewhere in the line. A hose pulling free from the coupling could cause severe injury.

6. Remain clear of charged hose lines, if possible. If it is necessary to work near charged lines, do not stand on or straddle the lines. Stand to one side, where there is less danger of being hit in the event of hose failure.

7. Remember that the engine and the exhaust pipe are hot enough to cause severe burns if touched. The heat remains long after the engine has been shut off.

8. Open and close nozzles, discharge gates, hydrants, and so on, slowly. It is best barely to crack open a nozzle and then let the line fill with water before opening the valve fully.

9. Open and close the throttle valve slowly, thereby preventing rapid pressure changes in hose lines and equipment.

10. Take care not to trip over hose lines or other testing equipment. Remain alert at all times when on the test grounds.

This page includes copyrighted material of Insurance Services Office with its permission. Copyright, Insurance Services Office, 1975.

Recording Results

Readings should be taken on the pitot gage and pressure gage with sufficient frequency to allow one to obtain a good average. If the pressures vary, readings need to be taken with greater frequency than if the pressures hold steady. It is common but faulty practice to read a pressure gage at the highest point in the swing of its needle. The center of the needle swing should always be read, as this is the average pressure. A petcock in the line to the test gage may be throttled to prevent excessive vibration, but if it is throttled too much the gage pointer will no longer indicate the pressure correctly. Never attempt to eliminate the needle vibration entirely. Any leak in the line to the test gage will result in an incorrect gage reading.

Special care should be taken in reading the pitot pressure. The pitot should be held in the center of the stream, with the tip a distance of about one-half the nozzle diameter away from the end of the nozzle. The pitot reading will be erroneously increased if the pitot is brought closer to the nozzle.

The speed should be recorded at frequent intervals, corresponding to the times the pressure readings are taken. Counting the revolutions for one minute generally gives readings of sufficient accuracy.

The best and most accurate method of using a stopwatch is to leave the stopwatch running at all times, engaging the revolution counter at a chosen instant and disengaging it when the hand of the stopwatch makes a passage on the same point on the dial one minute later. Accuracy can be developed and checked by performing the operation on a constant-speed electric motor, a common piece of equipment in the shop or home.

Other data that should be secured are indicated on the test form shown in Figure 5-4. The layout of hose and nozzle and data about the pump and engine should be recorded. The lift is measured in feet. While the theoretical lift of the pump is the vertical distance from the center of the pump to the water level, the lift that should be recorded is the vertical distance from the pressure test

This page includes copyrighted material of Insurance Services Office with its permission. Copyright, Insurance Services Office, 1975.

SERVICE TEST RESULTS

PUMPER NO. ______
MANUFACTURER ______ YEAR BUILT ______
Manufacturer's Model: ______ Manufacturer's Serial No.: ______
ENGINE: Make and Model: ______
PUMP: Make and Model: ______ Rated Capacity: ______ gpm at ______ psi
GEAR RATIO: Engine to Pump: Capacity: ______ 200 psi ______ 250 psi ______
TRANSMISSION GEAR USED: Capacity: ______ 200 psi ______ 250 psi ______
Suction Size: ______ Length: ______ Lift: ______
Hand Counter Speed Check Readings taken from: ______
Ratio to Engine: 1 to ______

CAPACITY TEST:
Layout ______ Nozzle Size ______

Time	Counter	RPM	Tach.	Pump Pressure		Pitot
				Appar. Gage	Test Gage	

200 PSI TEST:
Layout ______ Nozzle Size ______

Time	Counter	RPM	Tach.	Pump Pressure		Pitot
				Appar. Gage	Test Gage	

250 PSI TEST:
Layout ______ Nozzle Size ______

Time	Counter	RPM	Tach.	Pump Pressure		Pitot
				Appar. Gage	Test Gage	

FINAL RESULTS

	Capacity	200 psi	250 psi
Duration			
Ave. Noz. Pressure			
Correction			
Corrected Press.			
Gallons Per Min.			
Displacement			
(Nominal)			
Slip. Per Cent			
Pump Stg. (Ser. Par.)			
Ave. Pump Press.*			
Gage Correction			
Suction Press.			
Net Pump Press.			
RPM–Engine			
RPM–Pump			

* Test gage reading.

REMARKS–

Tested at: ______

Witnessed by: ______

Date: ______

Figure 5-4 Service Test Results Form. ***Copyright Insurance Services Office, 1975***

gage to the water level, as this is what is needed to obtain the actual net pump pressure.

Conducting the Test

The following is an acceptable method of making service tests on pumpers. Tests should be run with the pump drafting water whenever possible. Attention should be paid, at the start of the test, to the ease with which the pump takes suction. Close all discharge, drain, and booster tank valves and petcocks before starting to prime the pump. Make sure the intake gaskets are in place and free from foreign matter. Tighten the intake caps and couplings. Start the priming mechanism, noting the time. After the prime is obtained, operate the pump controls as necessary to develop pressure, and then open one discharge gate to permit the flow of water, also noting the time. The priming devices of pumpers that have a capacity of less than 1500 gpm should be able to take suction in 30 seconds through 20 feet of suction hose, hose of appropriate size and with a 10-foot lift. Those having a capacity of 1500 gpm or greater should be able to take suction in 45 seconds.

In testing a pump there are three variable factors that are so interrelated that a change in one factor will always produce a change in at least one of the others. The three variables are pump speed, net pump pressure, and pump discharge. For example, an increase in the speed of the pump will increase the discharge or the pressure or both. Adjustments of these variables through a change in the position of the engine throttle (this modifies pump speed), a change in the hose layout or gate valve positions (this modifies pump pressure), or a change in the nozzle size (this modifies discharge) are the only ways to reach the standard test condition desired.

The pumper should be operated at reduced capacity and pressure for several minutes, to allow engine and gear transmission to warm up. The pump should then be gradually speeded up until the desired pump pressure is reached. If the pressure will not come up to the desired amount, a length or two of hose may have

This page includes copyrighted material of Insurance Services Office with its permission. Copyright, Insurance Services Office, 1975.

Apparatus Testing

to be added, a smaller nozzle used, or a discharge gate throttled. The pitot should be read to see if the required amount of water is being delivered when the desired pressure is obtained at the pump. If the discharge is not as great as desired and it is believed that the pump will deliver a greater quantity of water, the discharge may be increased by further speeding up the pump. If speeding up the pump increases the pump pressure more than 5 or 10 psi, then a length of hose should be taken out, a discharge gate opened slightly, or a larger nozzle used.

There should be little change in the engine speed once the engine has warmed up. Any change in engine speed must of necessity produce a corresponding change in pump discharge and hence in pitot reading. Thus, other things being equal, any change in pitot reading indicates a change in engine speed. A change in pump speed will also cause a change in pump pressure. Whenever pump speed, pump pressure, and pitot readings do not show corresponding changes, it is safe to say that some reading is in error or some condition has arisen which affects the readings and needs correction. Engine speeds can be changed by working the hand throttle at the operator's position. Automatic relief valves or pressure regulators controlling the speed of the pump should be disengaged during the test. However, it is good practice to check the accuracy of the relief valves or pressure regulators after the test.

The pressure tests should be carried out immediately after the capacity test, with only a few minutes intervening—just long enough to change the hose and nozzle layout. Readings for the pressure tests are taken in the same way as are those for the capacity test.

Short lines of hose are always more convenient for a test layout than long ones. It is generally best to use a single line 100 feet long for the pressure tests and restrict the discharge at the pump outlet valve, or at a valve placed in the line, enough to increase the friction loss so that the desired pump pressure will be obtained with the proper discharge. The valve can be gradually closed as the engine speed is increased, until the pump pressure

This page includes copyrighted material of Insurance Services Office with its permission. Copyright, Insurance Services Office, 1975.

and pitot readings are both as desired. Care should be taken to make sure that the valve does not jar either open or shut; in either case both the discharge and the pump pressure will be affected.

It is important when operating a pumper that the engine temperature be kept within the proper range. Neither a cold engine nor an excessively hot engine will give as good service as one run at the proper temperature.

The oil pressure on the engine should be watched to see that the engine lubrication is properly maintained. Transmission gears should be watched for overheating. Any unusual vibration of the engine or pump should be recorded, and any leak in the pump casing or connections should be noted and taken care of, especially with centrifugal pumps, since they are not self-priming and could therefore lose their supply if there were a leak in the suction line.

Other defects in the performance of the engine or pump should be recorded. Minor defects should be corrected immediately, if possible. Pumpers should be retested after any major repairs.

Troubleshooting

Most tests are conducted without incident. Nevertheless, trouble does develop during some tests and an effort should be made to locate the source of trouble while the pumper remains at the test site. Some common difficulties experienced in testing a pump are outlined in the following paragraphs, along with suggested methods of tracing and correcting these troubles.

Failure to Prime   Failure to prime a centrifugal pump is a frequent source of trouble and is usually caused by an air leak in the assembly. One method of tracing this trouble is to remove all attached hose lines, cap all intake and discharge openings, and operate the priming mechanism in accordance with the manufacturer's recommendations. Study the intake gage to determine the maximum vacuum developed. It should be at least 22 inches of mercury. Stop the primer and attempt to hold vacuum in the

This page includes copyrighted material of Insurance Services Office with its permission. Copyright, Insurance Services Office, 1975.

pump. There is a leak in the pump assembly if the vacuum drops from the maximum to about 12 inches in less than 10 minutes. The leak may be in a valve, drain cock, piping casing, or pump packing. The leak can be located by listening for air movement. Another method is to connect to a convenient hydrant, cap the discharge outlets, open the hydrant, and watch for water leaks. A leak can usually be corrected at the test site.

Failure to Deliver Rated Capacities Two possible causes for failure of the pump to deliver rated capacities of water are insufficient power and restricted intake. Insufficient power results from not running the engine at the proper speed for the desired condition. The operator may be hesitant in advancing the throttle to a fully opened position or may be using the wrong transmission gear position. The engine may be in need of a tune-up, the grade of fuel may be improper for adequate combustion, or there may be vaporization in the fuel line.

Restricted Intake A restricted intake may be the result of any one or more of the following conditions:

1. Suction hose too small
2. High altitude
3. Lift too high
4. Improper type of strainer
5. Clogged strainer at the pump or at the end of the suction hose
6. Aerated water
7. Water too warm (over 85 ° F.)
8. Leaks on the intake side of the pump at the intake or couplings
9. Collapsed or defective suction hose
10. Foreign material in the pump

Insufficient Pressure Insufficient pressure in a centrifugal pump may be the result of pumping too large a volume of water for the power available. In multi-stage pumps, the problem

This page includes copyrighted material of Insurance Services Office with its permission. Copyright, Insurance Services Office, 1975.

may be pumping in the volume position instead of the required pressure position. This can be checked by partially closing off all discharge valves until only a small flow is observed, then opening the throttle until the desired pressure is reached, and then slowly opening all discharge gates and increasing the throttle as necessary to maintain pressure until the desired volume is obtained. An improperly adjusted or inoperative transfer valve may prevent adequate pressure from building up.

Excessive Engine Speed Excessive engine speed may result from use of the wrong transmission gear, a stuck throttle control cable, a restricted intake, or not having the suction hose under a sufficient amount of water.

Slip of the Revolution Counter A slip of the revolution counter or its fitting will show an apparent decrease in speed. Frequent checks should thus be made with the pumper tachometer to verify a change of speed. A clogged pitot tube will cause a drop in the gage reading.

Calculating the Results The quantity of water pumped is obtained from discharge tables, using the average pitot reading—corrected for gage error—and the appropriate nozzle size. If the readings are taken at exact and equal intervals, the proper average is the arithmetical average of the recorded readings. A judgment or estimated average of the readings should be used if the readings are not taken at equal intervals, so that more weight can be given to those readings that are maintained for longer periods of time. This same principle applies to the readings of pump pressure and speed.

The discharge from smaller nozzles may be obtained by taking from the tables the discharge from a nozzle of twice the given size operating at the same nozzle pressure and dividing this discharge by four. The discharge for nozzles larger than those given in the tables may be obtained by taking from the tables the discharge for a nozzle of half the given size and multiplying this by four. This practice is based on the fact that the increase in

This page includes copyrighted material of Insurance Services Office with its permission. Copyright, Insurance Services Office, 1975.

Apparatus Testing

discharge is very nearly proportional to the square of the ratio of the nozzle diameters for the same pitot pressure. A more accurate determination of the discharge for large nozzles may be obtained by using the formula given below:

$$\text{gallons per minute} = 29.83cd^2\sqrt{p}$$

where d = diameter of nozzle in inches
   p = pressure recorded on pitot gage in psi
   c = a constant, varying from 0.990 for a 1-inch nozzle to 0.997 for a 6-inch nozzle

For ordinary use, the formula can be reduced to

$$\text{gallons per minute} = 29.7d^2\sqrt{p}$$

Results of service tests should always be compared with the pumper's original test performance and specified requirements; however, an old pumper cannot be expected to perform as well as a new one. Nevertheless, the pump should meet minimum standards of performance. Necessary corrections should be made immediately in the event that the engine, pump, or any component part fails to meet test requirements. The apparatus should not be returned to service until all corrections have been made. The pumper should be retested after the repairs have been completed.

AERIAL LADDER AND ELEVATING PLATFORM TESTING

Standards for the testing of aerial ladders and elevating platforms have been developed by the Technical Committee on Fire Department Equipment of the National Fire Protection Association. The recommendations of this committee have been incorporated in NFPA Standard 1904, *Testing Fire Department Aerial Ladders and Elevating Platforms*. Fire depart-

This page includes copyrighted material of Insurance Services Office with its permission. Copyright, Insurance Services Office, 1975.

ment officials needing to test this equipment should refer to the standard prior to commencing the testing procedure. In general, the standards provide guidelines for making a visual inspection of the equipment, conducting a load test, and conducting an operation test.

The following information regarding the frequency of testing is reprinted with permission from NFPA 1904–1980, *Testing Fire Department Aerial Ladders and Elevating Platforms* (Copyright © 1980, National Fire Protection Association, Quincy, MA 02269). This reprinted material is not the complete and official position of the NFPA on the referenced subject, which is represented only by the standard in its entirety.

Tests shall be conducted at least annually, after major repairs or overhaul, following the use of the aerial ladder or elevating platform when the ladder or platform may have been subjected to unusual operating conditions of stress or load, or when there is reason to believe that usage has exceeded the manufacturer's recommended aerial ladder or elevating platform operation procedures.

If the aerial ladder or elevating platform is involved in a situation that produces any permanent deformation, the aerial ladder or elevating platform should be placed out of service, and the condition shall be reported in writing to the manufacturer. The aerial ladder or elevating platform shall be repaired in accordance with the manufacturer's written recommendations before it is placed back in service.

The inspections and tests specified in the standard shall be used to supplement, not to replace or modify, any instructions recommended by the manufacturer's maintenance manuals. Since each manufacturer's unit will be somewhat different, specific attention shall be given to the manufacturer's instructions concerning periodic maintenance and inspection checks.

SAFETY TESTING AND EVALUATION

Periodic safety tests and inspections should be made of aerial ladders and elevating platforms. It is recommended that these tests be conducted annually. Fire officials are capable of completing the inspection and testing as outlined in NFPA No.

1904; however, most fire departments have neither the equipment nor the expertise to establish satisfactorily a total safety testing and inspection program and, therefore, must contract with certified inspection and testing companies to perform these functions. The overall objectives of these tests are

1. To detect defective components before they fail in service
2. To minimize the possibility of accidents
3. To reduce equipment failures
4. To determine operational and structural capabilities
5. To determine compliance with regulations

Certified inspection and testing companies have the capability to perform the following types of non-destructive tests and inspections. Non-destructive testing is used to detect defects which are subsurface in nature and not detectable in a visual examination. The types of non-destructive test methods used on a particular apparatus depend upon the metal used in its construction:

1. Magnetic particle inspections to determine defects in ferrous critical weldments that are accessible—The material to be tested is magnetized. Any flaw on the surface will create magnetic poles within the material. Iron powder attracted to the defective area will identify the extent of the flaw. This process is used in such areas as the aerial ladder sections, elevating platform booms, turntable, platform support structure, turret, and basket (see Figure 5-5).

2. Penetrant inspections to determine surface defects in non-ferrous materials such as castings, fiberglass, aluminum, and so forth—The surface to be tested is saturated with a dye or fluorescent penetrant. A developer is then applied to the penetrant. The dye will bleed visibly to the surface. If fluorescents are used, the defective areas can be detected under ultraviolet light. This method is used in the same general areas as those inspected by the magnetic particle process.

3. Dielectric tests to determine the insulating value of non-conductive boom sections, pike poles, hot sticks, and wood and fiberglass ground ladders

4. Ultrasonic inspections for the purpose of making "in place" evaluations of pins, shafts, and bolts for possible

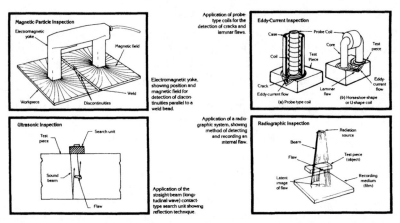

Figure 5-5   Non-Destructive Tests. *Courtesy of Aerial Testing Corporation*

internal defects that would show the item tested to be substandard or weak—In this test, high-frequency vibrations are injected into the surface of the test material. If no flaw exists, the vibrations "bounce back" to the source from the opposite surface. If the vibrations hit a flaw, the signals will return in a different pattern, revealing the location and extent of the flaw (see Figure 5-5).

5. Radiographic inspections to help evaluate the structural integrity of welds, castings, and fiberglass components—Radiation is beamed to the test object. Any flaw will be displayed on the recording film (see Figure 5-5).

6. Eddy currents are used in such areas as ladder support bushings, outrigger support pin areas, turntable support areas, and longitudinal welds in welded tubing to measure or identify seams, laps, cracks, voids, inclusions, hardness, and physical dimensions on non-ferrous material (see Figure 5-5).

7. Engine and hydraulic oil analysis to determine the condition of the internal components of the engine and hydraulic systems—The analysis identifies and measures metallic elements found in the oils. These elements provide a positive means of identifying the critical parts that are wearing.

Apparatus Testing

    8. Steel hardness tests to measure the hardness of steel and steel alloys on pins, bolts, welds, and structural members

    9. Load tests—Using a dynamometer with a test load, proof loads can be accurately applied.

    10. Stability tests on elevating platforms

    11. Visual examination to detect visible defects in main frame, turret, turntable, and boom sections

In addition to timing and operational tests, and proof and stability tests, both aerial ladders and elevating platforms should receive a thorough inspection in order to provide fire officials with the assurance that the apparatus complies with all safety standards and is not likely to fail under operational conditions.

Test Results

Upon completion of the inspection and test, fire officials are given a full report of conditions found. All items noted as deficient on the report should be corrected as soon as possible. Fire officials can use these reports to advantage in the development or enhancement of a preventive maintenance program.

REVIEW QUESTIONS

    1. What are the seven pump capacities that are considered standard?

    2. Where can the standards for pumpers be found?

    3. Pumps manufactured today are designed to meet what discharge requirements?

    4. What are the three different types of tests to which pumpers are subjected?

5. What are certification tests designed to do?
6. Who conducts certification tests?
7. What are the six parts of the certification pump test?
8. What is the basis for acceptability for the governor test?
9. Describe the pumping test.
10. What is the basis for acceptability for the pumping test?
11. What procedure is used for the overload test?
12. What is the basis for acceptability for the overload test?
13. Describe the automatic pressure control test.
14. What is the basis for acceptability for the automatic pressure control test?
15. What is the basis for acceptability for the water tank flow test?
16. What is the basis for acceptability for the vacuum test?
17. What is the minimum number of gated inlets required on a pumper?
18. What is the purpose of a delivery test?
19. How long are delivery tests?
20. What is the difference between the pumping test in the certification program and the pumping test in the delivery test?
21. What three tests should be included in the delivery test procedure in addition to the pumping portion?
22. If possible, who should conduct a service test?
23. How long is the pumping portion of a service test?
24. What is vertical lift?
25. How is the net pump pressure determined when pumping from draft?
26. How is the net pump pressure determined when pumping from a hydrant?
27. What size nozzle tips are desirable for use during service tests?

Apparatus Testing

28. What is a pitot tube used for?
29. What is a revolution counter used for?
30. What size hose is used in service tests?
31. What size nozzle is generally used in service tests?
32. What are some of the safety precautions that should be observed during service tests?
33. How should the pressure on a pressure gage be read when the needle vibrates?
34. Where should a pitot gage be placed in the hose stream?
35. Within what length of time should a 1000-gallon pumper obtain water when drafting through 20 feet of suction hose of appropriate size and with a 10-foot lift?
36. What three variable factors present in testing a pump are so interrelated that a change in one factor will always produce a change in at least one of the others?
37. What should be the position of pressure-controlling devices during service tests?
38. What are some of the common reasons a pumper may fail to prime?
39. What are the two possible causes for failure of a pump to deliver rated capacities of water?
40. What are some of the factors that may cause a restricted intake?
41. What might be the reason for failure to obtain adequate pressure when pumping in the service test?
42. What might be the cause of excessive speed?
43. Who has developed standards for the testing of aerial ladders and elevating platforms?
44. Where are the standards published?
45. How often should tests of aerial ladders be made?
46. How often should tests of elevating platforms be conducted?
47. What are the overall objectives of the safety testing and evaluation tests conducted by private firms?
48. What is the purpose of a magnetic particle inspection?

49. What is the purpose of a penetrant inspection?
50. What is the purpose of a dielectric test?
51. What is the purpose of an ultrasonic inspection?
52. What is the purpose of a radiographic inspection?
53. What is the purpose of an engine and hydraulic oil analysis?

6

Driving Procedures

Driving an apparatus is one of the most important tasks to which a firefighter may be assigned. The mere fact that the success of fire operations depends on the safe arrival of the apparatus and crew on the fireground places a heavy responsibility on the driver. The success of the apparatus operator depends on the combination of attitude, knowledge, judgment, habits, and skills. He must be a mature individual who recognizes the importance of the assignment and accepts without question the need to exercise restraint and good judgment in the operation of apparatus under both emergency and non-emergency conditions.

Good drivers are made, not born. They gain their competence primarily through the development of proper attitudes, acquisition of required knowledge, and department training programs. They must constantly be aware of a number of factors that affect their driving capabilities. The factors which must be considered by a skillful driver are:

1. *Apparatus condition:* The condition and operating limitations of the apparatus are prime factors affecting driving operations. Drivers are responsible for seeing that their apparatus are maintained in good condition at all times. Good condition means that all parts are in operational readiness. The driver's responsibility includes the correction of minor deficiencies, proper preventive maintenance, and the reporting of all needed repairs that require the work of a mechanic. The driver must be aware of the handling characteristics and operating limitations of the apparatus. He must be concerned with the power available, stopping distances, turning radii, weight and balance, and other factors which influence safe operation.

2. *Physical conditions:* Apparatus operators must be aware of the physical features of roads, streets, and terrain in their response districts. They must be concerned with those areas where heavy traffic, drawbridges, train crossings, or obstacles may impede their response. Driving a heavy apparatus in mountainous terrain is very different from driving on flat land. Shifting techniques, braking distances, and other factors that affect operational results change rapidly. Skillful drivers compensate for these differences and make allowances for narrow roads, curves, and other features affecting operations.

3. *Behavior conditions:* People are people, but people are different. Drivers of fire apparatus are generally more skilled and have quicker reactions than the average driver. The drivers of other vehicles vary from the careful to the careless, from the skilled to the unskilled, and from those who consistently do the wrong thing to those who react as expected. A good driver recognizes and compensates for the differences in attitudes, behavior, and reactions of others.

4. *Weather conditions:* Driving at night or in fog is quite different from driving during daylight. Wet streets require different skills and techniques than does dry pavement. The skilled driver compensates for adverse conditions, recognizing the challenge of the hazards.

DRIVER CHARACTERISTICS

The apparatus operator is still the key to traffic safety, in spite of all the mechanical improvements and safety advances that have been made in recent years. He must keep in good physical condition, have sound driving skills and habits, and develop proper attitudes. There are certain characteristics which separate the poor driver from the good. The most noticeable trait is attitude.

Characteristics of a Poor Driver

A driver with a poor attitude usually has an excuse for anything that goes wrong. The other driver was at fault; the street was in poor condition; it was a blind intersection; and so forth. Attitudes are learned, not inborn. Therefore, they can be improved. Unfortunately, there are many men who are considered perfect gentlemen until they climb behind the wheel of a vehicle. Then their personalities change completely. Many assume the role of a "big shot," acquiring all the authority of the red lights, siren, and several tons of heavy apparatus. Tests have shown that these self-styled "expert drivers" usually have an attitude of indifference, which tends to cloud their judgment and often results in a number of accidents that an alert driver would probably have avoided. The following are some of the bad attitude characteristics which prevail in this type of apparatus operator:

1. *Egotism:* This characteristic in a driver causes him to drive as if he owns the road, using the prestige of the apparatus and red lights to bully other drivers. He insists on taking the right of way regardless of the law or of good common sense. His show-off attitude and self-love cause him to take unnecessary risks.

2. *Overconfidence:* Such drivers generally take too much for granted. They have a serene confidence that their apparatus will always perform as they will it to do. They count on other people doing the right thing at all times, and they feel that the red lights and siren of the apparatus provide them with an impenetrable barrier through which nothing can pass to do them harm.

3. *Pride in safety record:* This type of driver swells with pride because of an accident-free record. Safety officers are not as concerned about those who have had an accident recently as they are about those who have not. Many drivers with accident-free records are due for a surprise.

4. *Dependence on experience:* Experience is an excellent teacher, but experience alone is not enough. Experience must be supplemented by comprehensive and continuing training and constant self-analysis of one's capabilities, attitudes, and skills. It must be remembered that experience develops bad habits as well as good ones, and bad habits many times prevail over common sense.

5. *Lack of facts:* Many apparatus operators with poor attitudes depend upon guesses, estimates, legends, and hearsay information instead of relying on facts. A quiz on stopping distances was given to hundreds of drivers. The results showed that 90 percent of those tested guessed short by more than 40 feet—a possibly fatal underestimation. The minimum emergency braking distance with all four wheels locked for a fire apparatus traveling 60 mph is 270 feet. When reaction time and air brake lag are added, the minimum total stopping distance is 367 feet. (See Figure 6-4.)

6. *Impatience:* Impatience is a sign of immaturity that can contribute to accidents. There are many drivers who take needless chances, suppress good judgment, and violate safe driving practices, just to save a little time. This characteristic in a driver causes him to become nervous and upset over minor items. He is likely to lose his temper and, consequently, his judgment very easily. This is the type of driver who blows the horn unnecessarily and expresses anger by driving recklessly.

7. *Indifference:* This type of driver believes it can happen to others but it cannot happen to him. He lives in an illusion, allowing an aura of indifference to replace judgment.

Characteristics of a Good Driver

A good apparatus operator is a defensive driver. A defensive driver is a compassionate and understanding person, the type who knows that all people operating vehicles do not have the knowl-

Driving Procedures

edge, skill, or training which he has, and, consequently, makes allowances for their deficiencies. A good driver is aware that he has no control over the unpredictable actions of other drivers and pedestrians nor over conditions of weather and road. He understands that his alertness at three o'clock in the morning may not be as keen as it was at three in the afternoon.

A defensive apparatus operator drives so as to prevent an accident *in spite of* the actions of others, his temporarily lowered alertness, or adverse traffic, road, or weather conditions. The defensive driver learns to look ahead and watch situations developing, knowing that some drivers will do the unexpected. He anticipates both the predictable and the unpredictable (see Figure 6-1).

Most skillful drivers can become good defensive drivers. Defensive driving requires the efficient management of time and space. Some of the basic elements of defensive driving are:

1. Identifying elements likely to contribute to an accident: pedestrians, vehicles, road conditions, weather, and so on
2. Proper positioning of a vehicle in relation to front, rear, and side traffic

Figure 6-1 Apparatus Drivers Should Always Expect the Unexpected. *Courtesy of the Los Angeles County Fire Department*

3. Selecting a speed that is safe for existing conditions
4. Realizing that signs and signals are no guarantee of safety
5. Predicting potential problems such as vehicles pulling away from the curb, unexpected stopping of vehicles ahead, and radical lane changes by other drivers
6. Allowing for deficiencies of others: reduced skills, deafness, skylarking, and daydreaming
7. Keeping his eyes moving: continuously observing what is happening ahead, to both sides, and to the rear
8. Thinking about what is seen and how it will affect him
9. Having an escape plan: knowing where to go in the event the planned path of travel is suddenly blocked

In addition to being a defensive driver, a good driver has certain further characteristics which contribute to his effectiveness. Some of these are attitude, skill, knowledge, judgment, awareness, good habits, physical fitness, and mental fitness.

Attitude

A good attitude is possibly the most important requirement of being a good driver. A driver's attitude is reflected in his mental or emotional regard for himself, for others, for the apparatus, and for surrounding conditions.

Skill

A good driver is a skillful driver who establishes a good driving performance. Skill in driving does not necessarily mean a safe driving performance. Records show that some drivers of exceptional skill are repeatedly involved in accidents, while others who are less adept have good safety records. Skill is ability plus training. A good driving performance is skill plus a good attitude. It is not how much skill has been developed that is important, but rather the extent to which the skill is applied.

Knowledge

A good driver has knowledge of his apparatus capabilities and limitations, response routes, physical conditions, applicable statutes and ordinances, and department rules. A driver who does not possess this knowledge is apt to be distracted and confused when confronted with unusual situations requiring insight and keen judgment.

Judgment

Good judgment means making the right decision at the right time. Right decisions are not a matter of luck; they are based on proper attitude, mental fitness, knowledge, and experience.

Awareness

Awareness is an important aid to safe driving. A good driver must learn to be aware of engine speed, road speed, and the terrain that affects these speeds. He should also have the "feel" of the throttle, clutch, and brakes. A good driver should never become so engrossed in the operation of the apparatus controls that he neglects to pay attention to traffic or road conditions.

Good Habits

Habits are acquired and learned through training and self-discipline. It has long been recognized that it is much more difficult to break a bad habit than to develop a new and acceptable one. The development of good habits requires an open mind. Good habits are developed through training and the willingness to learn from the experience of others.

Physical Fitness

Illness, fatigue, drowsiness, or the influence of medication will reduce the efficiency of a driver. Drivers should keep themselves in good physical condition. Any driver who is assigned to operate an emergency vehicle should refrain from driving and ask for a relief whenever there is an impairment to his physical well-being, however slight or temporary it may be.

Mental Fitness

Mental fitness means mental alertness. Mental fitness changes from day to day, hour to hour, and minute to minute. A driver who is worried over financial difficulties, domestic problems, department discipline, and the like may not be mentally fit to drive. A driver who cannot clear his mind of such distractions and concentrate on the job of operating his vehicle is not mentally fit to drive and should either remove himself or be removed from driving status.

DRIVING TECHNIQUES

A skillful driver is one who is able properly to coordinate and control the throttle, clutch, transmission, steering mechanism, and brakes. He must be capable of effectively using these devices to move the apparatus smoothly and safely from one location to another. To coordinate these controls properly, a driver must recognize minimum and maximum engine speeds and should understand their application when selecting gears while traveling at various road speeds.

Many people who can drive an automobile effectively with a stick shift seem to have trouble behind the wheel of a heavy fire apparatus. While the basic procedures of driving a car and a fire apparatus are the same, there are certain professional techniques used in driving a fire apparatus that are not as essential in the operation of an automobile. Driving a fire apparatus requires a thorough understanding of the importance of proper starting, throttle, clutch, and transmission coordination and control.

Starting Procedures

When the driver first comes on duty, he should complete the preliminary apparatus check and properly adjust the seat and mirrors.

There are several things that should be done before starting the engine in order to drive the apparatus in a safe manner. The first is to fasten and adjust the seat belt and shoulder harness. Others are:

1. Set the parking brake, if that is the department's regulation.

Driving Procedures

2. Move the gear shift lever to neutral.
3. Make sure the pump control is in ROAD position.

Starting the Engine: Gasoline Engines

Correct starting procedures may vary slightly from one engine to another. The procedures given here for both gasoline and diesel engines are general in nature and should be considered only as a guide.

1. Depress the clutch pedal and place the transmission in neutral; then hold the pedal in the depressed position. This takes much of the load off the battery and starter.
2. Turn the ignition on.
3. Set the choke, if necessary, based on experience with the engine. Use the choke sparingly in warm weather or when the engine is warm.
4. Press the starter button firmly. Only one starter button should be used on apparatus equipped with two, unless a single battery will not start the engine.

Warning: The starter should not be operated continuously for more than 10 seconds (or in accordance with the manufacturer's recommendations or department procedures). If the engine fails to start, wait until the starter motor comes to a complete stop, and then wait an additional thirty seconds before trying again.

5. Release the starter button as soon as the engine starts.

Starting the Engine: Diesel Engine

1. Disconnect the electrical supply from the heating element, if the apparatus is so equipped.
2. Make sure the battery selector switch is on either number 1 or number 2 position, but not both.
3. Make sure the transmission is in neutral.
4. Depress the clutch pedal if the apparatus has a standard transmission.
5. Turn on the battery switch.
6. Press the starter button. Follow the same procedure as that given under gasoline engines if the engine fails to start within ten seconds.
7. Release the starter button as soon as the engine starts.

Idling The Engine

The engine should be allowed to operate through a short warm-up period at 800 to 900 rpm before moving the apparatus, whenever conditions permit. This interval gives the apparatus operator time to check oil pressure, air pressure, and temperature gages. It also gives him the opportunity to observe the engine's performance before moving the apparatus.

After warm-up, the throttle control should be advanced and the engine allowed to run at idling speed. However, idling the engine unnecessarily for long periods of time wastes fuel and fouls the injector nozzles (diesels), while unburned fuel causes carbon formation and oil dilution.

Throttle Control

Getting the feel of the throttle is important. Some throttles have stronger springs than others and therefore require more pressure than might be expected. The best time to get the feel of the throttle is right after the engine has been started. With the transmission still in neutral, the driver should apply a slight amount of pressure in order to determine how much is necessary to obtain the amount of engine speed desired. He should then experiment, trying a little more throttle and a little less to see how the engine speed is affected. Getting the proper feel of the throttle is essential, particularly when it is necessary to coordinate throttle control with clutch control.

Clutch Control

The clutch is the link between the engine and the transmission. There is an interconnection between the engine and the transmission when the clutch is engaged; this link is broken when the clutch is disengaged. A clutch is a rugged device, but it is not indestructible. Proper use will improve its efficiency, prolong its useful life, and reduce strain on other connected components.

Excessive clutch wear is the result of excessive friction caused by improper engine speed at the time the clutch is re-

Driving Procedures

leased. A clutch pedal is released to engage the clutch; consequently, the terms "release" and "engage" in respect to clutch operation are often used interchangeably. The driver must maintain the proper engine speed as the clutch is released.

Even though the clutch is engineered to withstand some friction, excessive wear shortens its life. Every driver should remember that proper clutch care can be achieved by the correct use of the throttle.

Clutch Operation

Most clutch pedals are depressed and released in the same manner; unfortunately, not all of them are adjusted identically. Normally a clutch pedal has a certain amount of free travel before it affects the clutch action; however, some clutches begin to function just as the clutch pedal is moved downward while others do not function until the pedal is depressed two or three inches. As the pedal continues to be pressed downward, the clutch is completely uncoupled; however, the pedal still has more distance to travel before it contacts the floorboard (see Figure 6-2).

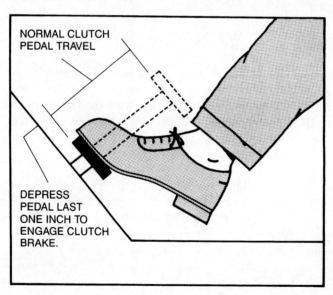

Figure 6-2 Clutch Pedal Travel. *Courtesy of Mack Trucks, Inc.*

The opposite is true when the pedal is released. The pedal travels a short distance from the floorboard before affecting clutch action. Further travel completes the coupling, and then there is some free travel before the pedal is completely released. The distance of pedal travel may not be the same on every vehicle, but the principle is the same.

An apparatus operator should be thoroughly familiar with the action of his clutch. He should know at what points action begins and ends. A driver should always get the feel of the clutch before moving the vehicle. When he is ready to move, he should push the clutch pedal gently downward and feel how far it travels freely before it encounters resistance. This free travel is comparable to the free play in a steering wheel—a sort of built-in safety feature. The resistance felt in the clutch is the start of the actual clutch action.

The driver should then depress the pedal all the way down, put the stick in gear, and slowly release the pedal. He will find that the pedal travels upward a certain distance before the vehicle attempts to move. This distance is important to remember, as the pedal must always come up just this far before the clutch begins to engage.

Misusing the Clutch

There are two common ways of misusing the clutch. One way occurs when the apparatus is in motion with the driver's foot on the pedal, anticipating its use. At such a time the driver generally depresses the pedal slightly, which causes a small, unnoticeable amount of clutch slippage. This is referred to as "riding the clutch" and is damaging to the clutch plate.

The second type of misuse occurs when the apparatus is stopped, the clutch pedal depressed, and the transmission in gear. This situation normally occurs at a stop signal while the driver is waiting for the signal to change. Maintaining the clutch in a disengaged position over a long period of time is damaging to the springs and clutch throw-out bearing. Furthermore, all the ingredients for an accident are present in the event the driver's foot slips off the clutch pedal.

A good general rule for the apparatus operator to follow is to keep his foot off the clutch pedal whenever he is not actually shifting gears.

Gear Shifting

Shifting gears is much more challenging when driving a heavy fire apparatus than it is when driving a car. Most light vehicles use relatively high-speed engines and are equipped with synchromeshed transmissions, which normally offer very little challenge to shifting ability. On the other hand, most heavy-duty fire apparatus utilize slower engines.

Gear shifting techniques require the knowledge and skill of matching engine speed to gear speed without abusing either the engine or the clutch. A driver is required to know the controls, their location, and their peculiarities so well that he can properly coordinate his hands and feet even in total darkness.

It is important to understand that matching engine speed to gear speed demands some change in engine speed. The amount of engine speed should be determined by the driver when the shift is to be made. This amount varies considerably, depending on the road speed of the vehicle while the shift is executed and on which gear is being selected. Shifting to a higher gear generally requires a different change in engine speed than shifting to a lower gear (the latter requires double-clutching).

There are three items which influence the amount of engine speed needed during a shift. These are:

1. The gear to be selected
2. The apparatus speed
3. The engine speed required for that gear at that speed

The driver must determine how much engine speed will be needed in that gear to match the road speed of the apparatus after the clutch has been released at the completion of the shift. He must determine the required amount of throttle, then regulate and hold that amount until the clutch is released. A smooth entry into the selected gear will be accomplished if his judgment has been reasonably correct, but, equally important, the entire shifting process will result in a smooth clutch release.

Double-Clutching

The normal shifting technique for a driver of an automobile is to depress the clutch pedal, use the shift level to move a gear from one intermesh position to another, and then release the clutch pedal. Double-clutching involves an additional step. The driver depresses the clutch pedal, moves the gear from an intermesh position to neutral, releases the clutch pedal, revs up the engine, depresses the clutch pedal a second time, eases the gear to the new intermesh position by feel as engine speed slows, and then releases the clutch pedal.

The objective of double-clutching is to permit the driver to match engine speed with gear speed and, therefore, avoid the clashing of gears. The length of time the clutch pedal is "out" during the double-clutch operation depends upon the driver's ability to regulate the engine speed to match the next gear speed. A driver who can quickly regulate the engine speed has only to let the clutch pedal out and back in again to complete a fast shift. A driver who is unsure of the throttle and has difficulty attaining proper engine speed must keep the pedal out until he does get it regulated properly; then he can depress the pedal to complete the shift. In any case, the pedal should be out while the transmission is in neutral and as the engine speed is regulated for the next gear.

Double-clutching serves a very important function. Making a habit of always double-clutching will make even a downshift smoother and easier.

Using a Clutch Brake

If the apparatus is equipped with a clutch brake, the brake is generally mounted on the countershaft of the transmission. Since the countershaft is always connected to the clutch disc through the constant mesh driving pinion, the action of the brake overcomes the tendency of the clutch disc(s) to continue to rotate at high speed when the clutch is disengaged.

When making an upshift, the driver depresses the clutch pedal the normal amount, separating clutch plates and allowing a shift into neutral. He then presses down slightly farther on the clutch pedal to apply braking force to the freely spinning clutch

Driving Procedures

disc and countershaft. A sense of timing gives just the right amount of braking effort so that the gears can be meshed without the usual time delay.

The following is an example of the proper method of using the clutch brake with a five-speed transmission:

> 1. With the vehicle standing, release the clutch (being careful not to apply the clutch brake, as that may make it difficult to engage first gear), shift to first gear, engage the clutch, and accelerate to full governed speed.
>
> 2. Release the clutch and shift the lever to neutral, depress the clutch the last one inch of travel in order to engage the clutch brake, and quickly shift the lever into second gear. This entire sequence of operations must be done very quickly to avoid road speed drop-off.
>
> 3. Engage the clutch and accelerate to full governed speed.
>
> 4. Repeat step 2 to shift into third gear.

Once the vehicle has attained third speed on an upgrade shift, further clutch brake operation should not be required if the apparatus can attain speed for fourth and fifth gears.

The clutch brake is particularly useful in the lower gears when the apparatus is going uphill, when the road speed drops off more quickly than the engine rpm. The clutch brake should not be used when making a downshift.

Note: A slight but definite resistance to the clutch pedal's downward travel will be felt at the last one inch of travel, when the clutch brake is engaged.

There are no particular recommendations for double-clutching. As with all non-synchromesh transmissions, double-clutching is necessary on a downshift. On an upshift, there is a choice of either double-clutching or using the clutch brake. It will probably be more convenient to use the clutch brake in the lower gears and to double-clutch in the higher gears.

Upshifting

Upshifting is the changing from a lower to a higher gear; for example, from second to third. Upshifting is used to increase momentum or to control engine speed. Controlling engine speed by upshifting means to keep the engine from exceeding the maximum recommended by the manufacturer.

The following procedure is recommended when shifting from a lower to a higher gear:

1. Increase the engine rpm's to near maximum before starting to shift so the engine will not lug after the shift has been completed.
2. Depress the clutch and move the gear shift to neutral while easing off on the throttle.
3. Move the gear shift to the next highest gear; then release the clutch.
4. Regulate the throttle as the situation requires after the shift has been completed.

There will be no gear clash or grind if the throttle has been properly regulated, and there will be no noticeable change of motor or road speed if the proper amount of throttle is applied as the clutch is released.

Downshifting

Downshifting is the changing from a higher gear to a lower gear; for example, from third to second. Downshifting is used:

1. To regain momentum after a reduction in road speed
2. To maintain momentum on upgrades
3. To conserve brakes
4. To control road speed when approaching a downgrade
5. To keep the engine speed within its proper range

It is important to realize when downshifting that the engine must turn considerably faster in a lower gear than in a higher gear at the same road speed. Therefore, the driver must be alert to the minimum and maximum engine speeds (indicated by the tachometer) and understand that unless the engine speed is lowered to near its minimum range in the higher gear before the shift, it will most probably be above the maximum range after the shift to the lower gear has been completed.

A shift to a lower gear should never be attempted if the

Driving Procedures

vehicle is going faster than the engine can move the vehicle in the lower gear. For example, if the vehicle can only go 15 mph in second gear at the maximum engine speed, then a shift into second gear at a road speed of more than 15 mph should never be attempted. Such an attempt may result in a stripped gear, a ruined engine, a twisted driveshaft, or possibly a broken axle. To prevent this from happening, a driver should be familiar with the maximum speed for each gear. If it becomes necessary to downshift when the road speed is too high, then the apparatus must be slowed to the proper road speed for the shift.

The driver who understands the relationship of engine speed to road speed and the variations incurred in changing gears will have little difficulty in executing smooth, safe downshifts.

The following procedure is recommended when downshifting:

1. Make sure the engine is near the minimum of the operating range before the start of the shift so that the rpm will not be above the maximum after the shift is completed.
2. Depress the clutch pedal and shift to neutral, maintaining engine speed.
3. Double-clutch, increasing engine speed to match the gear speed, and engage the next lower gear. Then release the clutch.
4. Regulate the throttle as the situation requires after completing the shift.
5. The gear change will be smooth if the engine speed and gear speed are properly coordinated. Most good drivers learn to recognize when gears are matched by the sound of the engine.

Effects of Terrain on Shifting

When contemplating a shift it is necessary to consider the terrain and the momentum of the vehicle in order to keep the engine speed within its proper range. The momentum of the vehicle will change as soon as the clutch pedal is depressed, depending upon the topography. As soon as the clutch pedal is depressed, the road speed will:

1. Change very little on a level roadway
2. Decrease on an upgrade
3. Increase on a downgrade

The amount of change in the road speed depends primarily upon three factors:

1. The momentum at the start of the shift: The greater the momentum, the less noticeable the change.
2. The steepness of the grade: The steeper the grade, the more the speed is affected.
3. How long a time is needed to make the shift: The longer it takes, the greater the amount of momentum change.

A knowledge of proper throttle regulation is necessary to be able to adjust engine speeds under varying road conditions. A driver should be able to shift gears with little difficulty once he understands how much terrain conditions affect shifting.

Shifting During a Turn

Shifting during a turn should be avoided. A driver who feels that a shift may be necessary when he approaches a turn should complete the shift prior to starting the turn so that both hands are on the wheel throughout the turn. Another advantage of shifting before the turn is that the throttle can be applied once the turn is completed because the apparatus will already be in the proper gear needed to accelerate.

Shifting After a Stop

Quite often a clashing of gears will occur when a driver attempts to shift into the starting gear after a stop has been made. This problem is more pronounced in some apparatus than in others. There are three possible solutions to the problem:

1. Be sure the clutch pedal is fully depressed.
2. Be sure the engine is idling and that adequate time has been allowed before the shift is attempted to permit the gears to stop turning.
3. Shift into a higher gear; then try the starting gear again.

Driving Procedures

Engine Speed Control

Every engine has an rpm range within which it should be operated. This operating range is established by the engine manufacturer. Exceeding the recommended rpm is called, "racing" the engine. Allowing the rpm to drop below the recommended range results in "lugging" the engine. Both racing and lugging an engine will cause extensive damage.

Lugging generally occurs when only a small portion of the total available horsepower is being produced and the engine is operating under excessive load.

Both racing and lugging are the results of operating an engine in an unbalanced condition. When an engine is operating properly, the momentum of the moving parts is in a relatively balanced condition. The pressure of the burning fuel on top of the pistons is sufficient to produce the horsepower required for the conditions. If the engine is attempting to perform an operation that requires more horsepower than the engine is actually producing, the pressure on top of the pistons will exceed the momentum of the moving parts, resulting in an unbalanced condition. This unbalanced condition is called lugging.

An unbalanced condition also exists whenever the momentus of the moving parts exceeds the pressure on top of the pistons. This overbalanced condition is called racing, or overspeeding.

Continued lugging or racing an engine will cause severe damage and greatly shorten the life of the engine. Damage caused by lugging is the result of subjecting the engine to a severe shock, similar to striking the top of the piston with a sledgehammer every time ignition occurs within the cylinder. Damage by racing is the result of a reduced load on the engine, which causes a tremendous inertia in the pistons as they travel upward, which finally results in a severe strain being placed on the pistons, wrist pins, connecting rod bearings, and the lower part of the engine.

From the standpoint of efficiency, economy, and long, trouble-free mileage, it is best to operate a gasoline engine in a range from about two-thirds of peak rpm to about 200 rpm below the peak speed. The accelerator must also be in a position corresponding to about two-thirds to three-quarters torque. For an engine with a peak speed of 2400 rpm, the range as described would be

1600 to 2200 rpm. Within this range, the engine would be more responsive to the accelerator, and its proper torque would be achieved more readily with the opening of the throttle. Use of these engine speeds and selection of the proper transmission gear will achieve the desired results.

Road speed is controlled more by gear selection than by engine speed. The driver selects the proper gear for a given road speed to keep the engine within the recommended operating range. In general, an upshift is made whenever the engine speed approaches the upper limits or that point where it would be considered racing. A downshift is made whenever the engine speed approaches the lower recommended limit or the point where it would be lugging.

An apparatus operator inexperienced in recognizing the minimum and maximum engine speeds may have difficulty in judging the proper time to shift. A tachometer makes the correct control of engine speed comparatively simple. If the apparatus is not equipped with a tachometer, then it is important that the driver learn to recognize the racing point and the lugging point by the sound of the engine.

Steering Control

Many drivers are unaware of the importance of proper hand position when operating heavy-duty apparatus. They believe the apparatus can be kept under the proper control regardless of how their hands are placed on the wheel—this is not true. Improper positioning of the hands may limit the reaction of a driver in emergency situations. Good driving techniques require that the driver keep his hands in such a position that recovery can be made in the event of a tire blowout, wheels hitting holes in the road, and so forth.

The standard recommended position for the hands in straight-ahead driving is referred to as the "ten and two" position—the left hand at the ten o'clock position on the wheel and the right hand at the two o'clock position. The palms should be turned toward the center of the wheel (see Figure 6-3).

When a curve in the road ahead necessitates a turn of the wheel, the good driver prepares for the turn before reaching it. If

Driving Procedures 223

the turn is made to the right, the driver should hold the wheel firmly with his right hand and slide his left hand down to the eight o'clock position, then hold the wheel firmly with his left hand and slide his right hand up to the twelve o'clock position. As the turn is begun, the hands should move clockwise and approach the "ten and two" position as the turn is completed. This procedure permits control of the vehicle while approaching the curve and during the turn. Of course, the hands may have to be in a slightly different position, depending upon the sharpness of the turn. The opposite procedure would be followed when a turn is to be made to the left.

Another point to consider in turning is the position of the thumb. Some drivers grip the wheel so that their thumbs curve around the wheel. A number of drivers have sprained their thumbs when the steering wheel suddenly slipped as a front wheel unexpectedly hit something. To eliminate the danger, the hands should be placed with the thumbs resting on the wheel or the forefingers—*not* around the wheel. The wheel can be held just as firmly, but if it is accidentally spun through the operator's hands, the hands remain on the wheel and in a position to grasp it again without causing injury to the driver.

Both hands should be kept on the steering wheel whenever a turn is being negotiated, when the brakes are being applied, whenever roadway hazards are present, and at all other times that safety precautions dictate. Shifting gears, signaling for turns, and so

Figure 6-3 Safety Dictates the Proper Positioning of the Hands When Driving. The "Ten—Two Position"

forth, which require the removal of one hand from the wheel, should not be performed when any of the conditions just listed exist. When the turn is completed, the driver should allow the steering wheel to slide through his hands as the front wheels straighten out. The driver will be able to stop the turn of the wheel at any desired point by simply tightening the grip of his hands.

Backing Up

A large number of accidents have occurred while apparatus operators were backing the apparatus. A driver should always move forward if he has a choice between moving forward and backing up to reposition the apparatus. If it is necessary to back up, certain precautions should be taken.

A driver should never back apparatus until he is absolutely certain it is safe to do so. The addition of back-up signaling devices to apparatus has helped to alert firefighters and others in the vicinity that the apparatus is backing, but it has not eliminated the problem of hitting objects which might not be apparent to the driver from the operating position.

Whenever possible, other firefighters should guide the driver while he is backing. The guides should stand clear of the apparatus and remain in direct or mirror view of the driver at all times. Communication between the guides and the driver should be maintained by signaling rather than by voice.

If personnel are not available to assist a driver in backing up, then the driver should not move the apparatus until he has gotten out of the seat and made a positive inspection to assure that there is nothing in the intended travel path.

The driver should be aware of the location of the wheels in relation to the ends of the apparatus before backing. The distance that the rear end of the apparatus protrudes beyond the rear wheels (overhang) varies considerably between types of apparatus. An apparatus operator should be aware of this overhang and govern the steering accordingly.

One factor in backing that is often overlooked is that the back wheels are leading and the front end is following. To change the direction of travel when backing, the front end must swing to the

Driving Procedures

side in order to point the rear of the apparatus toward its goal. The sharper the turn, the more the front end must swing.

Brake Control

Too little thought is usually given to brake use. It is assumed that brakes will always slow and stop the apparatus at the driver's discretion. However, some drivers give very little thought to brake conservation. Brakes are one of a driver's best friends. He should give considerable thought to saving them by proper utilization of the gears and proper brake application.

Most newer apparatus are equipped with air brakes. However, some still have hydraulic brakes, often with vacuum boosters. While the general principle of application is the same regardless of the type of brakes, there are some operational differences which exist in comparing each of the systems.

The amount of pressure required on the brake pedal varies from a very light touch to a rather hard application and varies also with the weight of the apparatus. The heavier the apparatus, the more brake pedal pressure is required to stop the apparatus.

In normal stopping more pedal pressure is needed as the brakes are first applied than is needed as the apparatus stops. A driver should apply considerable pressure at the start of the application, ease off as the apparatus slows, release the brakes almost entirely at the end of the stop, and then apply more pressure to hold the apparatus once it has stopped. This method of operation provides a smooth stop and eliminates the lurch or jerk so often noticed as an apparatus ceases its forward motion.

Drivers should also be aware of the effect brake application has on tire slippage. With a mild brake application there is a slight amount of tire slippage, but such slippage increases as more braking force is applied. There is much more braking force or traction with 6 to 8 percent of tire slippage than when the wheels are locked. Locking the wheels can decrease maximum braking by 40 to 50 percent. Not only can a driver brake more efficiently with the wheels rolling, but this method permits control of the apparatus with less chance of entering a skid.

Some situations require that the brake pedal be held down

for a long period of time. The proper application of brakes in these situations can spell the difference between safety and disaster. Although opinions differ as to the proper method of brake application in these situations, it is obvious that the brake linings will heat at any time prolonged brake pressure is applied. Increased heating of brake linings causes them to become ineffective. On the other hand, if a slow pumping action is used on an apparatus equipped with air brakes, the system will quickly use up its air reserve and thus lose its brakes. (This statement is not to suggest that a long, hard application of the brakes should not be used if the situation so demands.) After all, brakes should always be used as necessary to stop the apparatus short of a collision. The point is that drivers should conserve their brakes as much as possible during routine operations so that adequate braking will be available in an emergency if a prolonged application of the brakes is necessary.

Drivers should learn to develop the "feel" of their brakes. They should know approximately how much pressure is required to stop the vehicle at various road speeds and under varying load conditions. They should also be aware of the braking characteristics of their apparatus during wet or other inclement weather conditions.

Air Brakes

Operating the brakes of an apparatus equipped with air brakes varies little from operating the brakes of a passenger car. Proper control of the brakes may be easily accomplished because the operation of the brake pedal requires very little physical effort. The distance the brake pedal is depressed determines the amount of air pressure delivered to the brake chambers, and the brake chamber pressure determines the braking force. The driver of an apparatus equipped with air brakes should keep in mind that he is operating a brake valve that is capable of giving finely graduated brake control.

An apparatus equipped with air brakes should not be moved until the air gage shows the minimum pressure recommended by the manufacturer. Air brakes are not fully effective at pressures below that minimum pressure. The driver should observe the air pressure periodically while operating the apparatus to insure that

Driving Procedures 227

adequate pressure is being maintained. Most systems have a warning buzzer and light for low pressure. The apparatus should be stopped as soon as possible if the buzzer sounds or the light becomes illuminated.

The best way to stop the vehicle is to apply just enough pressure to feel the apparatus slow at the desired rate of deceleration. The brake pedal may be released slightly, as necessary, to assure a smooth stop as the vehicle speed decreases. There should be only sufficient air pressure in the brake chambers to hold the vehicle stationary as the stop is completed. The brakes should never be applied lightly at first and the braking pressure increased as the speed decreases. This type of operation results in a rough stop.

The brake pedal should not be "fanned." Fanning the brakes merely wastes compressed air and does not effect a desirable stop. Neither should the brake pedal be fully depressed; this causes full braking force to be delivered to the wheels. Of course, under emergency conditions this rule would not apply. Normally, the driver should use the engine to assist the brakes. He does this by waiting to disengage the clutch until the last few feet of travel.

Stopping Distance

It is essential that all apparatus drivers thoroughly understand the relationship of speed, perception time, reaction time, air brake lag time, and braking distance to the total stopping distance.

Speed is a critical factor in stopping distances. When the speed is doubled, the minimum emergency stopping distance is approximately quadrupled. This means that the stopping distance is about sixteen times as long for an apparatus traveling at 40 mph as compared with one traveling at 10 mph.

Perception in driving refers to a driver's seeing what he is looking at and recognizing its effect on his movement. Perception varies from driver to driver. The driver who pays close attention to the surroundings will see danger much faster than one who is less observant. Some agencies estimate the average perception time as three fourths of a second. The distance an apparatus travels from the time the driver sees a dangerous situation until he perceives it as such is perception distance.

Reaction time is the time that elapses between a driver's

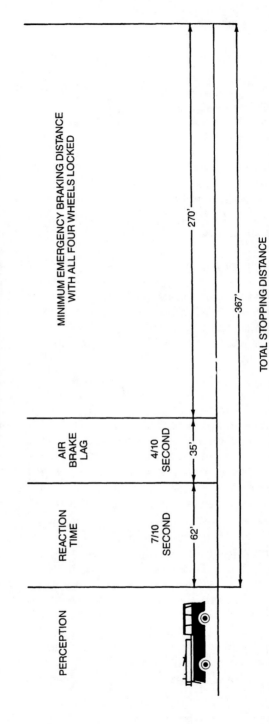

Figure 6-4 Effects of Reaction Time and Air Brake Lag on Total Stopping Distance. *Courtesy of Los Angeles County Fire Department*

Driving Procedures

perceiving a dangerous situation and the time he requires to react to it. Normally, the reaction time of fire department drivers is better than average.

Air brake lag is a time lag between engaging the brake and brake reaction. At least one agency estimates this lag to be about four-tenths of a second.

Perception time, reaction time, and air brake lag time are extremely important to the overall stopping distance of an apparatus. When an apparatus is in motion, it travels about 1½ feet per second for every mile per hour of speed. As an example, an apparatus traveling at 60 mph will cover a distance of approximately 90 feet (1½ × 60) every second.

Figure 6-4 illustrates the impact of reaction time and air brake lag on the total stopping distance of an apparatus traveling at 60 mph. If the average perception time of three-fourths of a second is added to the distances given in the chart, the total stopping distance would be increased to approximately 433 feet, almost one and one-half times the length of a football field. Even at 30 mph the total stopping distance would be 140–150 feet.

Automatic Transmissions

Automatic transmissions in fire apparatus have taken much of the guesswork out of shifting gears. These transmissions provide many advantages for drivers who must stop and go or change speeds frequently. Driving is easier, safer, and more efficient.

The driver of an apparatus with an automatic transmission has two means of partially controlling when the transmission will shift. They are the range selector and the accelerator.

Different transmission models, as well as different vehicle manufacturers, require different designations on the range selectors. The best performance is obtained by using the correct gear range for the particular driving condition. The range selector shown in Figure 6-5 is used for the Allison AT 540, 543, MT(B) 643, 644, HT 740 FS, and 747 automatic transmissions. Several of these transmissions are commonly used on fire apparatus. The selector is used here for the purpose of illustrating the use of the range selector.

230 Introduction to Fire Apparatus and Equipment

Figure 6-5 The Range Selector for the Allison AT 540, 543, MT(B) 643, 644, HT 740, 740FS, and 757 Automatic Transmissions. *Courtesy of Detroit Diesel Allison, Division of General Motors Corporation*

Range Selector Positions

Reverse: Use this position to back the vehicle. Stop the vehicle completely before shifting from a forward gear to reverse or from reverse to forward. The reverse warning signal is activated when the range selector is in this position. Reverse has only one gear. Reverse operation also provides the greatest tractive advantage.

Neutral: Use this position when you start the engine. If the engine starts in any other position, the neutral start switch is malfunctioning. Neutral position is also used during stationary operation of the power takeoff (if the apparatus is so equipped). Use neutral when the apparatus is left unattended while the engine is running; always apply the parking brake.

The D position: The apparatus will start in first gear, and as the driver depresses the accelerator, the transmission will upshift to second gear, third gear, and fourth gear automatically. As the apparatus slows down, the transmission will downshift to the correct gear automatically.

The 3 and 2 positions: Occasionally, road, load, or traffic conditions will make it desirable to restrict the automatic shifting

Driving Procedures

to a lower range. When conditions improve, return the range selector to the normal driving position. These positions also provide progressively greater engine braking power (the lower the gear range, the greater the braking effect).

The 1 position: This position is low gear; use this position when pulling through mud and snow or driving up steep grades. This position also provides maximum engine braking power.

In the lower ranges (1, 2, and 3), the transmission will not upshift above the highest gear selected unless the engine-governed speed is exceeded.

Warning: Do not allow the apparatus to coast in neutral. This practice can result in severe transmission damage. Furthermore no engine braking is available then.

Driving Tips

These driving tips apply to all Allison automatic transmissions and are not restricted to those illustrated in Figure 6-5.

Accelerator Control

The pressure on the accelerator influences automatic shifting. When the pedal is fully depressed, the transmission will automatically upshift near the governed speed of the engine. A partially depressed pedal will cause the upshifts to occur at a lower engine speed. When modulated lockup is provided, closed-throttle operation delays the release of the lockup clutch to provide additional engine braking at lower vehicle speeds.

Warning: Never shift from neutral N to drive D or reverse R at engine speeds above idle: the vehicle will lurch forward or backward and the transmission may be damaged.

Downshift or Reverse Inhibitor Feature

The transmission hydraulic system will not permit a shift into any forward gear at a speed that will cause excessive engine overspeed. Any lower forward range may be selected at any time, but the actual engagement of the gears in that range will not occur until road speed is reduced; downshifting is progressive as road speed decreases. The inhibitor effect will cause downshifts to occur at slightly higher speeds than normal automatic downshifts.

If the shift lever should accidentally be moved to reverse while the apparatus is travelling forward, the transmission is designed not to shift into reverse gear until road speed is very low. To avoid shift shock, always come to a full stop before shifting from forward to reverse or from reverse to forward.

Operating in Cold Weather

Listed below are the minimum fluid temperatures at which the transmission may be safely operated in a forward or reverse range. When ambient temperature is below the minimum fluid temperature limit and the transmission is cold, preheat is required. If auxiliary heating equipment is not available, run the engine to preheat the fluid to the minimum temperature limit before operating in a forward or reverse range. Failure to observe the minimum fluid temperature limit can result in transmission malfunction or reduced transmission life.

Fluid Type	Minimum Fluid Temperature
Dexron*11 or Dexron*	−30°F. (−34°C.)
Type C-3 SAE 10W	10°F. (−12°C.)
Type C-3 SAE 30	32°F. (0°C.)

*Dexron is a registered trademark of General Motors Corporation.

Using the Engine to Slow the Apparatus

To use the engine as a braking force, shift the range selector to the next lower range. If the apparatus is exceeding the maximum speed for a lower gear, use the service brakes to slow the apparatus to a speed at which the transmission may be downshifted safely.

Using the Hydraulic Retarder

Hydraulic retarders are available on the MT 600 and HT 700 series transmissions. The MT 600 retarder is installed on the rear of the transmission in place of the output housing; the HT 700 retarder is installed between the torque converter and the transmission gearing. The function of the retarder is to provide auxiliary braking under all conditions.

An output retarder option is available on MT 600 series transmissions. The retarder is mounted on the rear of the transmission as an integral part of the transmission and provides the apparatus with an auxiliary retardation system. The unit combines both hydraulic and clutch pack retardation capabilities. In most applications, the output retarder is applied in conjunction with the service brakes.

An input retarder option, available on HT 700 series transmissions, provides hydraulic retarder action to slow the apparatus. The retarder is located between the torque converter and transmission gearing and is an integral part of the transmission. Maximum retarder effect in this series occurs at high retarder rotor speed. Selecting a lower hold range position when using this type of retarder is recommended for maximum effect.

Partial retarder application is available when maximum application is not needed.

Observe the following cautions when driving an apparatus equipped with an input or output retarder.

Apply and operate the retarder with engine at closed throttle only.

Do not use the retarder when road surfaces are slippery.

Do not apply retarder control or de-energize the system at the master control switch.

Observe transmission and engine temperature limits at all times. Select the lowest possible transmission range to increase the cooling system capacity and total retardation available.

In the event of overheating, decrease the apparatus speed to reduce retardation power requirements.

Observe the retarder "alert light" to ensure that the apparatus control system is functioning properly.

Do not operate the input or the output retarder simultaneously with an engine exhaust brake; extreme torque loads can be produced in the range section, damaging the transmission.

Auxiliary Transmission

Select the desired auxiliary gear ratio while the apparatus is stopped. Do not shift the auxiliary while the apparatus is travelling.

Two-Speed Axle

The two-speed axle may be shifted from low to high or high to low while the apparatus is moving without damaging the transmission. The axle or vehicle manufactuer's recommendations should, however, be followed for shifting the axle. Axle shifts should be made with the transmission in the highest gear to prevent a transmission shift from coinciding with an axle shift.

Towing or Pushing

Before towing or pushing a disabled apparatus, the driveline should be disconnected or the drive wheels lifted off the road. The engine cannot be started by pushing or towing.

Caution: Failure to disconnect the driveline or lift the driving wheels before pushing or towing the apparatus can cause serious transmission damage.

Parking Brake

There is no "park" position in the transmission shift pattern. Therefore, always put the selector in neutral and apply the parking brake to hold the apparatus when it is unattended.

Driving on Ice or Snow

The Allison automatic transmission continually provides a proper balance between required power and good traction. The driver can have better control of his apparatus because of this smooth, constant flow through the drive train. When an apparatus is being driven on ice or snow, any acceleration or deceleration should be made gradually.

Driving Procedures

Rocking Out

If the apparatus is stuck in deep sand, snow, or mud, it may be possible to rock it out. Shift to D and apply a steady, light throttle (approximately 800–900 rpm—never full throttle). Then, by moving the range selector between drive and reverse, rock the apparatus free. Time these shifts to take advantage of the forward and reverse momentum. If the driving wheels spin, apply less throttle.

Temperatures

The transmission oil temperature is indicated in some apparatus by a gage specifically designed for this purpose and in other apparatus by the engine coolant temperature indicator. Extended operations at low vehicle speeds with the engine at full throttle can cause excessively high oil temperatures in the transmission. These temperatures may tend to overheat the engine cooling system as well as to cause possible transmission damage.

If excessive temperature is indicated by the engine coolant temperature gage, stop the apparatus and determine the cause. If the cooling system appears to be functioning properly, the transmission is probably overheated. Shift to neutral and accelerate the engine to 1200–1500 rpm. This should reduce the sump temperature to operating level within a short time.

Normal transmission operating temperature is 160–220° F. (71–105° C.). Maximum transmission-to-cooler oil temperature for AT, MT, and HT (non-retarder models) is 300° F. (149° C.). Retarder models allow a maximum of 330° F. (165° C.). If excessive temperature is indicated by the transmission oil temperature gage, stop the apparatus and shift to neutral. Accelerate the engine to 1200–1500 rpm and allow the temperature to return to normal (two or three minutes) before resuming operation.

Caution: The engine should never be operated for more than 30 seconds at full throttle with the transmission in gear and the output stalled. Prolonged operation of this type will cause the transmission oil temperature to become excessively high and will result in severe overheating damage to the transmission.

If the transmission overheats during normal operation, check the transmission oil level.

If high temperature in either the engine or the transmission persists, stop the engine and have the overheating condition investigated by maintenance personnel.

DRIVING CONDITIONS

The complicated task of driving a heavy-duty fire apparatus is further aggravated by changing road and weather conditions. While the basic principles of safety apply under all conditions, certain variations are applicable to meet changing conditions. Some of the most common driving conditions with which a fire apparatus operator must cope are: freeways, curves, intersections, terrain, night, fog, rain or wet weather, snow, and winter driving in general. The objective of this section is to introduce some guidelines for coping with these variables.

Freeway Driving

Freeways crisscross most of the major cities in the United States. Traffic flow on these arteries varies from a snail's pace to the fastest found in the city. A mistake by one driver on a freeway could instantaneously result in a multi-vehicle incident, possibly involving numerous deaths. Apparatus drivers must be particularly alert, putting into practice some safety rules which have proved to be worthwhile.

1. If at all possible, adjust the speed of the apparatus when entering a freeway to most nearly match that of the traffic in the lane to be entered. Drivers who stop or slow to a near-stop when entering a freeway generally create a traffic hazard.

2. Move from the right-hand lane once on the freeway, signaling before making the adjustment. The right-hand lane is a freeway trap. It presents the constant hazard of having to give way to entering traffic and passing close to vehicles in trouble which have stopped off the roadway on the right

shoulder. Those operators driving in the right-hand lane may also find vehicles cutting in front of them or braking unexpectedly as exits are approached, as these drivers make a last-minute decision to depart from the freeway.

3. Do not change lanes unnecessarily, once established. When lane changing is required, check all traffic conditions and signal properly before making the move.

4. Do not follow the vehicle ahead too closely. Rear-end collisions are the most frequent type of freeway accidents. At least one-third of the freeway incidents have been contributed to by following too closely. A general rule is to allow at least one apparatus-length separation for every 10 mph of speed.

5. Extend your vision to several cars ahead. Be on the alert for indications that these vehicles are slowing or might have to slow—stop lights, driver signals, or congested traffic ahead.

6. Maintain the speed of the other vehicles in your lane. If the lane is moving faster than you like, check traffic conditions, signal, and move to the right into another, slower lane.

7. Check your side and rear view mirrors frequently. Keep abreast of conditions behind you. How close are vehicles following? How fast are vehicles approaching?

8. Do not make any unexpected moves except in an emergency. If possible, keep other vehicle operators informed of your intentions by means of proper signaling.

9. Have an escape plan. Know what you would do if the driver ahead makes a sudden stop or something else unusual develops (see Figure 6-6).

10. When driving under emergency conditions on a freeway, it is normally best to curtail the use of the siren and depend only on the use of the light to request the right of way. (State law, however, may require both emergency light and siren to maintain emergency vehicle status.)

Skids

Skids cannot be classified as a driving condition; however, they are the possible result of improper techniques when certain

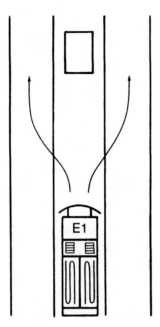

Figure 6-6 Have an Escape Plan

conditions are encountered. They are presented here as an adjunct to driving conditions.

Skids are caused; they do not merely happen. They are the result of some improper action on the part of the driver combined with road contributions. The most likely actions on the part of the driver which cause a skid are excessive speed, improper braking technique, and sudden steering movements. The most likely factors contributed by the road are wet weather, ice, or spillage; however, it is possible to go into a skid on a perfectly dry roadway.

Research has shown that vehicles actually hydroplane when travelling at high speeds during heavy rains. Hydroplaning occurs when the tires ride on the water and are no longer in contact with the road surface. It has been estimated that the front tires will begin to lose contact with the roadway at approximately 30 mph, and that they may lose all contact near 55 mph. The problem is more acute with underinflated tires. An apparatus is operating on the verge of a skid once hydroplaning occurs.

There are some positive steps a driver can take to avoid hydroplaning. The first is to slow down. If the situation is one in which slowing is difficult, such as on a freeway where it is safest

Driving Procedures 239

to keep pace with the traffic flow, then drive in the tracks left by the vehicles ahead. These tire tracks are, usually, relatively free of water. Even in heavy rains the tracks will remain for several hundred feet.

Slamming on the brakes on a slippery roadway will most likely cause a skid. The tires lose all traction once the wheels are locked. Of course, the best procedure is not to become trapped so that a quick stop must be made. If the occasion arises, however, it is best to pump the brakes and keep the wheels from becoming locked.

Curves are always skid traps, but they are more so during wet weather. Slow down well ahead of a curve and avoid braking during the turn.

Do not make the mistake of thinking that a dry roadway is skidproof. A quick stop with locked wheels will cause the tires to heat and melt the rubber. The vehicle then slides on the molten rubber, just as it would on wet pavement.

Despite all precautions, a driver may become involved in a skid. His greatest help during the skid will be his free-rolling wheels on all axles. The turning wheels provide a considerable amount of side force to assist in straightening the apparatus. The driver should be able to correct the skid with the assistance of the turning wheels if the proper steps are taken (see Figure 6-7). Here are some suggested rules to follow in the event of a skid:

1. Act quickly, but do not overreact.
2. Do not slam on the brakes.
3. Take your foot off the accelerator pedal.

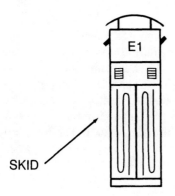

Figure 6-7 Turn the Front Wheels in the Direction of the Skid

4. Steer in the direction the rear end of the apparatus is sliding, but do not oversteer.

5. Straighten the wheels when the grip on the road is regained.

6. Apply throttle and resume driving as the apparatus moves in the inline direction.

Curves

Centrifugal force tends to throw an apparatus toward the outside when it enters a curve. This force increases rapidly with speed. If the speed is doubled, the centrifugal force is multiplied by four. It goes without saying that speed should be controlled in the curve.

Any shifting that should be done for a turn should be done prior to entering the turn. This precaution permits the driver to keep both hands on the wheel during the turn and provides him with the power necessary to accelerate.

The apparatus should be slowed prior to entering the curve. Apply a slight amount of throttle while in the curve. This provides better traction and control of the apparatus and assists in preventing skids.

Do not brake in a curve except when absolutely necessary,. If braking is required, then apply intermediate taps to the pedal, to avoid locking the wheels. Improper handling of the brakes can result in either a skid or possibly in rolling over the apparatus.

Intersections

More accidents involving fire apparatus responding to an emergency take place at intersections than at any other location. Never enter an intersection without having complete control of the apparatus. The apparatus should be in a gear which can provide quick acceleration if necessary. Have your foot on the brake pedal ready to apply the brakes. As you approach the intersection, you should look first to the left, then to the right, back to the left, and then proceed if it is safe to do so.

Be on the alert for hazards. Watch out for the driver who

Driving Procedures

insists on taking the right of way, expecting all others to do the stopping. Do not argue with him. Let him go first. Watch for pedestrians who step unexpectedly from the curb and start across the street.

Intersections controlled by signal lights present a special hazard when the red light is against a responding emergency vehicle. The driver should slow down, stop if necessary (or if required by law or department regulations), and not proceed until he is certain that all traffic having the green light is aware that he is requesting the right of way.

Downhill Driving

Only general situations can be examined when considering this phase of driving. Exceptional circumstances or a combination of specific conditions may necessitate actions other than those outlined below.

Downshifting while traveling downgrade is not recommended. The best procedure is to gear down prior to starting down the grade. In the event an error in judgment results in selecting a gear too high for the grade, the driver should use the brakes to slow the vehicle, stopping if necessary, and engage the proper gear.

The accepted procedure is to drive downgrade in a gear low enough to permit the engine speed to be in the medium range and still hold the vehicle at a safe, steady road speed. The gear required to travel up a grade is generally the best for downgrade level. The apparatus will slow down when the driver eases off the throttle if the apparatus is in the proper gear; it will gain speed when additional throttle is applied. That the apparatus does this is a good indication that the driver is in the proper gear. This technique gives the driver more control over the apparatus and reduces excessive braking.

Uphill Driving

The problem of many drivers when they attempt to climb a steep grade is that they shift from a lower gear to a higher gear only to find that the apparatus immediately starts to lug in the higher gear. It then becomes necessary to shift back to the lower gear, which is sometimes difficult. Downshifting while climbing a grade is not always easy.

If possible, the best procedure to follow when climbing a hill is to get into the gear that is needed to climb the grade and then to stay in it. Operate near the middle of the operating speed range, taking care neither to lug nor to race the engine. It is best to shift into the climbing gear while the apparatus still has adequate momentum. With experience, a driver will be able to judge the required gear while approaching the hill and downshift before or soon after starting up.

One-Way Streets

Most vehicle codes permit apparatus responding to an emergency to proceed against traffic on a one-way street. However, the decision as to whether or not to respond against traffic should be left to the judgment of the company officer and the apparatus operator. Conditions are too variable to set any hard-and-fast rules. Normally, the time of day, the traffic conditions, the type of emergency, and the like should be analyzed prior to considering travelling longer than half a city block against traffic. A good general rule to follow is that extreme caution should be exercised during all travel on one-way streets, and that any extended travel against traffic should be avoided whenever possible.

Night Driving

Less driving is done at night; therefore, statistically, fewer traffic deaths should occur at night. That, however, is not the way it actually happens. Records indicate that the vehicle mileage death rate at night is three times higher than it is during the day. Yet the only factor that has really changed is the lessened visibility.

Drivers can not see as well at night because of two things—not enough light and too much light. Not enough light is the result

Driving Procedures

of night darkness. Too much light is caused by man's attempt to compensate for the darkness.

The multitude of lights reflecting on an apparatus windshield cause glare and, consequently, reduced vision. The bright lights of an oncoming vehicle can temporarily blind a driver, resulting in the apparatus travelling some distance before driver vision returns. It is estimated that vision recovery after being blinded by the headlights of an oncoming vehicle can take as long as seven seconds. An apparatus traveling at 50 mph will cover a distance of approximately 500 feet in this time.

Driving an apparatus at night is a matter of being seen, seeing, and operating within the driver's visibility stopping distance.

Being seen involves the proper use of vehicle lights and turn indicators. Seeing involves the acceptance of the fact that night vision is not as good as day vision, and learning to keep the multitude of lights from further reducing that vision. Some aids to assist in this task are:

1. Keep the windshield of the apparatus clean. Dust on the glass tends to increase and distort reflections.
2. Use the side mirrors.
3. Do not look directly at the headlights of oncoming vehicles. Keep your eyes slightly to the right of approaching traffic.
4. Avoid looking at bright lights on buildings, signs, and the like.

Operating within the visibility stopping distance of the apparatus requires knowledge of the apparatus stopping distance at various speeds and understanding of the distance illuminated by the headlights. A prime example of overdriving the vision operating capability of the driver is driving at 50 mph when the headlights can illuminate objects for a distance of only 200 feet. Most apparatus cannot be stopped in that distance, considering perception and reaction times.

One last warning on night driving: Some pedestrians think a driver can see them because they can see the apparatus. This is not true. The combination of darkness and dark clothing makes some pedestrians almost impossible to see. Three out of every eight people killed in city traffic accidents are pedestrians. Take extra care at night so as not to contribute to this statistic.

Fog

A few years ago a fire occurred during the night in a fairly large industrial building in southern California. The incident was not reported to the fire department until the next morning when shocked employees reporting for work found the building and its contents in ashes. The building was isolated, but was not too far from a normally well-travelled road. The problem? During the night the entire area had been blanketed by a heavy fog. This was not a driving incident, but it certainly illustrates the problem. Visibility is at its worst in dense fog.

Fire apparatus should remain in quarters during heavy fog conditions except to respond to an emergency. Even then, remember that it is better to drive safely and reach the emergency than to become involved in an accident. Movement should be slow, with all warning lights on and headlights on low beam. This improves the chances of the apparatus being seen and throws the headlights down on the road, where the light is needed, rather than out into the fog, where it will be reflected back. Speed should be limited so that an emergency stop can be made within the visibility limit, regardless of how short this distance might be.

Rain or Wet Weather

Wet weather is particularly hazardous, especially the first rain after a dry spell. Roadways are coated with oil, dust, and dirt, making them extremely slick when wet. Stopping distance is at least doubled—and in some instances tripled. In addition, visibility may be reduced by half. Fortunately, most people are aware of the hazards and use more caution during wet weather. The following are some guidelines which may prove beneficial:

1. Slow down to compensate for the reduced braking efficiency and reduced visibility.
2. Have headlights on low beam. Compensate for the other driver's reduced visibility.

Driving Procedures 245

3. Keep windows from fogging. Open a window part way; open a vent; use defrosters.

4. Do not drive through deep water if it is unnecessary. A gasoline engine can easily be shorted out by the fan throwing water on the ignition wires.

5. Slow down when going through standing water. Remember that the brake linings will become wet if the water is of any depth at all; dry them out, once through the water, if reduced brake efficiency is noticed. This can best be done while driving slowly. Depress both the brake pedal and accelerator pedal at the same time; the heat and friction will dry out the brakes.

6. Keep to the crown of the road if possible. The water is not as deep at this point.

7. Avoid streetcar tracks or other objects which might be unusually slippery in wet weather.

8. Avoid applying brakes in a puddle; this could initiate a skid.

Winter Driving

Winter adds the hazards of snow, ice, and freezing conditions to the multitude of challenges already facing the fire apparatus driver. Driving during winter is at best demanding; at its worst, extremely dangerous. It requires caution, skill, and patience to prevent accidents. The following are some tips to aid in combating winter driving problems:

1. Road conditions change fast. Be on the alert for these changes.

2. Think ahead. Anticipate turns and stops; make them slowly and without any quick steering movements or braking actions.

3. Keep headlights on low beam when visibility is restricted because of sleet or swirling snow.

4. Be extremely careful when the outside temperature is warming or when the sun is out. This extra heat may begin to melt the ice, and wet ice is twice as slippery as dry ice.

5. Be especially cautious near intersections. The road-

way can become unexpectedly slippery as a result of the polishing effect of stopping and starting traffic.

6. Be on the alert for shady areas such as underpasses. These areas may be icy when other parts of the road are dry. Reduce speed and continue straight ahead if an unexpected ice patch is encountered.

7. If you are involved in a skid, follow the directions outlined under skid procedures.

8. Maintain more distance between your apparatus and the vehicle ahead than you normally would during dry weather. Stopping distances become multiplied several times on slippery streets.

9. Chains provide the most traction for severe snow and ice conditions. Use them as conditions warrant.

10. Use the rocking technique for getting out if the apparatus becomes stuck in snow. It is best to remove the snow from the tires first by turning the wheels back and forth. Then rock the apparatus in a rhythmic motion by shifting from low to reverse while using a slight pressure on the accelerator pedal. Avoid racing the engine or spinning the wheels. If sand or salt is available, sprinkle some under the drive wheels if you encounter any spinning at all.

EMERGENCY RESPONSE

The response to an emergency begins when the driver gets behind the wheel. He must start the engine and prepare to respond. He should fasten and adjust his seat belt and shoulder harness, turn on the radio, and turn on all necessary lights. A check to see that all members of the company are safely aboard should be made before the transmission is engaged. Then the driver is ready to drop the transmission in gear and move out at the officer's command.

Response Routes

The apparatus operator should, upon leaving quarters, follow as closely as possible response routes of travel that have been established by pre-fire planning sessions. These routes should be known by the officers and drivers of companies from other stations who will also be responding to the alarm. Following predetermined response routes reduces the possibility of unexpectedly meeting other responding equipment at intersections. In selecting these response routes, particular attention should be paid to major traffic arteries which will safely allow more rapid movement of emergency equipment.

Normally, all fire department vehicles responding from a multiple-company station should use the same response pattern. Apparatus should proceed in single file when more than one vehicle is responding. A reasonable distance should be maintained between the apparatus. If too much distance is allowed, however, a motorist who yielded to the first apparatus may pull out in front of the second. No apparatus should pass another unless the latter is disabled.

Officers and drivers should also recognize the need to develop alternate routes of response which might be necessary because of traffic condition, street blockage, or other situations that require companies to respond from locations other than regularly assigned quarters.

Apparatus Speed

The use of excessive speed while responding to an alarm can seldom, if ever, be justified. Speed, per se, is not of prime importance. Excessive speed may cause serious and unnecessary accidents which not only prevent the apparatus from reaching a fire but may cause death or injury to firefighters and civilians.

Of even greater importance is the psychological effect which speed has on personnel. The natural result of speeding to a fire is that it induces a lack of logical judgment once the fireground is reached. The most effective approach to the control of a fire may then go unrecognized. The driver who makes certain of an alarm location and responds with sane and safe driving will arrive with a company prepared for action. Of major importance during a response is that the driver of the apparatus keep the vehicle under control at all times, and especially at street intersections.

A good rule to follow when responding to an alarm is the basic speed law. Very simply, this rule states, "Never exceed a speed which is reasonable and proper for existing conditions, even where the law permits a speed higher than that at which you are driving." Safe, prudent speeds vary with such factors as driver reaction time, driver condition, brake efficiency, condition of pavement, weather, and traffic congestion. A speed which is reasonable when there are few persons or vehicles on the street may be excessive in heavy traffic or at hours when school children are on streets. Over the same stretch of roadway, speeds that are safe at certain times are unsafe at others.

Exemptions From the Vehicle Code

Most vehicle codes exempt drivers of emergency vehicles from complying with some specific sections of the code under certain circumstances. At least three conditions must exist under most codes before these exemptions apply:

1. The vehicle must be driven in response to an emergency.
2. The vehicle must be displaying an emergency light.
3. The driver must be sounding a siren.

Although all of the above conditions exist, the apparatus operator still must drive with due regard for the safety of people using the highway.

Use of the Red Light and Siren

When responding to an emergency, fire apparatus must have the siren and flashing emergency lights operating in order to have emergency vehicle status. The driver should, however, remember that these devices are used to request the right of way—not to demand it. He must still proceed with caution and use all means at his command to prevent an accident.

To guard against accidents, an apparatus operator can also sound an air horn as reasonably necessary to request the right of way. Some people can hear an air horn but have difficulty hearing the high pitch of a siren. Care must be taken that the air horn does not drown out the siren. To do so would require the driver to obey all rules of the road, as the apparatus would no longer be exempt from the vehicle code.

The apparatus operator should be particularly prudent in his use of the siren during cold or rainy weather, because car windows are usually rolled up tightly at such times, making it difficult for other drivers to hear the approach of emergency vehicles. As a result of the increasing use of air conditioning systems by motorists, it is probable that windows will also be rolled up during extremely warm weather. With the windows rolled up, together with the noise from a radio and surrounding traffic, it is likely that some motorists will not hear the siren at all.

The siren is particularly ineffective when approaching intersections. The sounds cannot bend around buildings; consequently, drivers approaching from side streets may not hear the siren until the last minute. This situation can result in an accident if the driver of the emergency vehicle does not slow down or stop before proceeding through an intersection. As an example, a driver approaching from a side street at 30 mph will be travelling at approximately 45 feet per second. It will take approximately three-fourths of a second to perceive the danger if he hears the siren when he is 50 feet from the intersection. If the driver then reacts immediately, he will travel approximately 66 feet before he hits the brakes (perception time plus reaction time). This amount of time will put the vehicle in the intersection. The chances are that a collision could not be avoided if the apparatus had also entered the intersection.

Controlled intersections where the red light is against the responding emergency driver are generally more dangerous than non-controlled intersections. Drivers moving with the green light have a feeling of security that they are free to continue through the intersection. Those who hear a siren as they approach a signal with the green light in their factor have difficulty in adjusting themselves to the fact that they should yield the right of way. They must hear the siren sufficiently far in advance to make the adjustment and react to the situation.

When approaching intersections, the apparatus operator should be particularly alert for other apparatus responding to the emergency. Remember that police cars, ambulances, and other emergency vehicles will be responding to the same incident and might be arriving from *any* direction. These drivers will be requesting the right of way just as the fire apparatus driver is. It is possible that the siren of one apparatus will drown out that of another, particularly if the sirens are not being used effectively in a fluctuating manner. More than one serious accident has occurred between two responding emergency vehicles.

Approaching the Emergency

The driver should slow down as he approaches the location of the emergency. This will give the company officer adequate time to assess the situation and make a decision as to the proper action to take. Second and later apparatus follow standard operating procedure—which may be to wait at a distance for orders. The first-in officer gives assignments to following companies by radio if the situation is not covered by standard operating procedure.

Positioning Apparatus at Emergencies

The proper position to spot the first-arriving company is one where the mission can be most satisfactorily accomplished without interfering with other incoming companies. It is important that a pumper be spotted sufficiently close to the emergency so that hose reel lines and preconnected lines can be utilized, if desired. Ideally, the pumper should stop just beyond the fire building. The officer then has a view of three sides of the building (when separated) and room is left for the aerial ladder truck. Remember that the best spot for the aerial ladder truck is normally directly in front of the fire building. A pumper arriving first should, if possible, save this location for the truck. Lines can be laid over, under, or around the truck, but if the truck cannot be properly spotted, it may be impossible to use the aerial ladder.

Apparatus arriving after the aerial ladder truck should be careful not to park so as to block the removal of ladders from the rear of the truck. Many times it is difficult to move apparatus, once spotted, because of the arrival of additional units and other traffic.

It is usually better for the driver of a pumper approaching a vehicle fire from the rear to pass the vehicle and park in front. Firefighters have been seriously injured while positioning lines to

Driving Procedures 251

the rear of a vehicle fire when the gasoline tank ruptured and quickly spread the fire.

Apparatus Not in Use

When only personnel—not the apparatus—are needed at the emergency, the apparatus should be parked in a safe location that permits later mobility for possible use. Operation of the engine and use of lights are determined by department rules and regulations.

Pumper Operators

The operator of a pumper should remain with his apparatus at the emergency until he is ordered to do otherwise.

Aerial Truck and Elevating Platform Operators

These operators are responsible for the raising and lowering of aerial ladders and elevating platforms at emergencies. They should, under normal circumstances, remain with the apparatus as long as the aerial ladder or elevating platform is in use. Of course, if it becomes obvious that the aerial ladder has been raised to a position in which it will be kept for an extended period of time, then the operator should report to his commanding officer for assignment. The operator of an elevating platform must stay with the apparatus as long as the platform is out of its bed. The department rules will, however, guide the operator; he may be the second roof man.

Idling Motors

Apparatus engines should be kept running until such time that the driver is assured that the apparatus will not be used. While waiting, the driver should maintain the engine rpm at a sufficiently high speed to keep the battery from being discharged.

The drain on the battery with all lights, radio, and other electrical equipment working is extensive if engine speed is maintained below the alternator cut-in speed.

Backing Up at Emergencies

There are many conditions at any emergency which require that an apparatus driver back up the vehicle. All unnecessary backing should be kept to a minimum, however. If it becomes necessary to back up, then the driver should secure a guide, if one is available. If a guide cannot be found, the driver should climb out of the vehicle and make sure the space behind the apparatus is clear before it is moved.

Many drivers of pumpers back to the hydrant whenever a line has been laid going in. Before backing, the driver should consider the feasibility of going around the block instead. In many cases this will be much quicker and safer. It also gives the operator a better opportunity to spot the apparatus properly at the hydrant.

Note: Much of the information contained in the first part of this chapter was provided through the courtesy of the California State Fire Marshal's Office.

REVIEW QUESTIONS

1. Upon what factors does the success of a driver depend?
2. Are good drivers made or born?
3. What are some of the factors which must be considered by a good driver?
4. What are some of the characteristics of a poor driver?
5. What are the characteristics of a defensive driver?
6. What is a defensive driver?

Driving Procedures 253

7. What are some of the basic elements of defensive driving?

8. What is probably the most important requirement for being a good driver?

9. What are some of the things that should be done before starting the engine?

10. Why should the clutch pedal be depressed when starting the engine?

11. What is the maximum length of time that the starter should be operated continuously?

12. What is probably the most significant sign of a really good driver?

13. What function does the clutch play?

14. About how much clutch pedal travel does it take to uncouple the clutch?

15. What are the two common ways of misusing the clutch?

16. What are the three items which influence the amount of engine speed needed during a shift?

17. Describe the method for double-clutching.

18. What is the objective of double-clutching?

19. Describe the proper method of using a clutch brake.

20. When is the most effective use made of a clutch brake?

21. What is the purpose of upshifting?

22. Describe the proper procedure for upshifting.

23. What is the purpose of downshifting?

24. What might be the result if a downshift is made at too high a road speed?

25. Describe the proper procedure for downshifting.

26. What is the general rule about shifting during a turn?

27. If a driver finds that there is a clashing of gears as he attempts to shift into the starting gear after a stop has been made, what three steps can he take to try to correct the problem?

28. Define racing an engine.

29. Define lugging an engine.

30. When does racing an engine most often occur?

31. When does lugging an engine generally occur?

32. From the standpoint of efficiency, economy, and long, trouble-free mileage, at what range of engine speeds is it best to operate a gasoline engine?

33. Is road speed controlled more by gear selection or by engine speed?

34. In general, when should upshifts and downshifts be made?

35. What is the standard recommended position for the hands in straight-ahead driving?

36. What is the proper procedure that should be followed before backing an apparatus?

37. What are some of the factors that determine the amount of pressure required for brake application?

38. Describe the proper use of the brakes in stopping.

39. What is the proper method of stopping a vehicle with air brakes?

40. What is the effect on stopping distance of doubling the speed?

41. What is the average perception time of a driver?

42. What is the average reaction time of a driver?

43. What two means does an operator of an apparatus with an automatic transmission have to partially control when the transmission will shift?

44. With an Allison MT 640 automatic transmission, which range selector position provides the greatest tractive advantage?

45. What might be the result of coasting an apparatus in neutral when it is equipped with an automatic transmission?

46. What effect will a partially depressed position of the accelerator pedal have on the shifting of an automatic transmission?

47. What is the purpose of the downshift or reverse inhibitor feature of an Allison automatic transmission?

48. What would happen in an automatic transmission if

Driving Procedures

the shift lever is accidentally moved to reverse while the apparatus is travelling forward?

49. What is the function of a retarder on an automatic transmission?

50. What are some of the precautions that should be taken when driving an apparatus equipped with an input or output retarder?

51. What should be done before towing or pushing a disabled apparatus equipped with an automatic transmission?

52. What is the procedure for rocking out an apparatus which is stuck in deep sand, snow, or mud if the apparatus is equipped with an automatic transmission?

53. If the coolant temperature gage indicates an excessive temperature and the cooling system appears to be functioning properly, the transmission is probably overheated. What should be done to reduce the temperature?

54. What is the normal operating temperature of an automatic transmission?

55. What is the maximum transmission-to-cooler oil temperature for a non-retarder automatic transmission? For a retarder model?

56. What are some of the safety rules for freeway driving?

57. About what portion of freeway incidents have been attributed to following too closely?

58. What are some of the most likely actions by a driver that can cause a skid?

59. What is meant by hydroplaning?

60. What are some of the positive steps a driver can take to avoid hydroplaning?

61. What steps should be taken by a driver involved in a skid?

62. What effect does speed have on centrifugal force as an apparatus negotiates a curve?

63. Where is the most likely place for an apparatus to have an accident when responding to an emergency?

64. What is the proper way for an apparatus driver to enter an intersection?

65. What is the proper procedure to follow in regard to shifting when driving downhill?

66. What is generally the proper gear to use when traveling downhill?

67. What is the best procedure to follow in regard to shifting gears when traveling uphill?

68. What is a good rule to follow in regard to one-way streets?

69. What are some of the aids a driver can use when driving at night to keep lights from reducing his vision?

70. What is the general rule regarding speed that should be followed when responding in fog?

71. What are some of the guidelines that should be followed when driving in rain or wet weather?

72. What are some of the tips that can be used to assist in winter driving?

73. When does the response to an emergency begin?

74. What is meant by "predetermined" response routes?

75. What is meant by the "basic speed law"?

76. What three conditions must exist before a driver of an emergency vehicle can be exempt from certain sections of the vehicle code?

77. What are some of the problems a driver of an emergency vehicle may have in regard to other drivers hearing his siren?

78. What is the proper procedure for an apparatus driver to take as he approaches the emergency?

79. What are some of the general considerations for positioning apparatus at emergencies?

80. What care should be taken in regard to idling motors at an emergency?

7

Fire Pumps

There are three general classifications of pumps used on fire apparatus: main pumps (midship or front mount), booster pumps, and priming pumps.

A *main pump* has a capacity of 500 gpm or more. These pumps are used primarily for supplying hose lines 2½ inches or larger, but are also used for supplying smaller lines on those apparatus not equipped with booster pumps. Main pumps are of the centrifugal type. Main pumps receive their supply of water from the hydrant distribution system or from an auxiliary water supply source when drafting operations become necessary.

Booster pumps are permanently mounted on an apparatus and have a rated capacity of less than 500 gpm. These pumps are primarily used for supplying hose lines 1½ inches and smaller. Both centrifugal and positive displacement pumps are used as booster pumps. These pumps generally take water from the tank carried on the apparatus; however, the piping configuration on some apparatus is so designed that the booster pump can either take water directly from a hydrant or draft.

Priming pumps are used to prime the main or booster pump when it is necessary to operate from draft. These pumps are of the positive displacement type.

THEORY OF POSITIVE DISPLACEMENT PUMPS

A *positive displacement pump* will theoretically discharge a given amount of material with every revolution of the pump shaft—hence the term positive displacement. The amount discharged depends on the total space within the pump casing from the inlet to the discharge side of the pump. Positive displacement pumps are capable of moving gases as well as liquids; thus they make ideal priming pumps on fire apparatus, for they fill the need to move air as well as water.

The amount of water that can theoretically be discharged from a positive displacement pump is determined by the total space within the pump casing. This space is referred to as the *theoretical displacement* of the pump. The theoretical displacement is usually stamped on a tag attached to the pump casing. Discharge is given in gallons per revolution (gpr).

Perhaps the best way to visualize the theoretical displacement of a positive displacement pump is to imagine a single cylinder of a piston pump that is constructed in the same way as a cylinder in a two-stroke-cycle internal combustion engine. Imagine that the cylinder is taking in and discharging water rather than taking in a combustible mixture and discharging exhaust gases, and that it is discharging its contents on each revolution of the crankshaft.

Figure 7-1A shows the piston at TDC with the intake valve open. Water is drawn into the cylinder as the piston starts its downward movement, filling the space vacated by the piston together with that in the combustion chamber.

The cylinder is filled with water when the piston is at BDC (Figure 7-1B). The intake valve is closed and the exhaust valve is open.

The upward movement of the piston forces the water out the exhaust port (Figure 7-1C). The amount discharged equals the

Fire Pumps

amount taken in. While piston pumps as illustrated are not currently being installed on fire apparatus, the general principles involved do apply to those positive displacement pumps that are used.

It is not difficult for engineers to calculate the amount of water the pump will theoretically discharge in one cycle. In the case of our piston pump, the amount would equal the volume of water that the cylinder would hold from TDC to BDC. The amount in the combustion chamber cannot be counted, as the back pressure on the discharge side of the pump keeps this water in the cylinder until the exhaust valve closes. For purposes of illustration, assume that the cylinder holds one gallon. It takes one revolution of the crankshaft to complete a cycle; therefore, one gallon of water will theoretically be discharged on each revolution of the crankshaft.

Carry the principle of a positive displacement pump one step further, in order to obtain an understanding of how pressure is developed by the pump. Imagine that the water discharged from

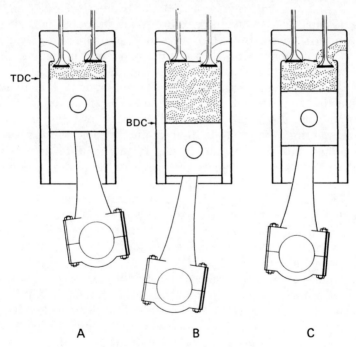

Figure 7-1 A Piston Pump

the cylinder enters a closed compartment, where it is stored until its use is required. A given amount of water will be discharged into the storage chamber every time the crankshaft turns over. This will continue until the chamber is full. Any attempt to add more water will result in a pressure buildup within the chamber. This pressure buildup will cause a back pressure within the cylinder, and will also exert a force on the storage chamber walls. The pressure buildup will continue until something is done to relieve it, or something breaks. Either water must be drawn from the storage chamber, or a relief valve must be provided to relieve some of the pressure and take out some of the water. To be fully effective, the relief valve must be capable of taking out the water as fast as it is being discharged into the storage chamber; otherwise the pressure will continue to increase.

Let's say that the pump is discharging 60 gpm at 60 rpm, and that the pressure in the storage chamber has built up to 150 psi before the relief valve starts working. If a discharge gate is now opened and water is discharged from a nozzle at 60 gpm, the pump pressure will remain at 150 psi. If the discharge from the nozzle is reduced to 50 gpm without any corresponding change in the pump's speed, the relief valve will start operating again. A rapid drop in pressure will occur if any attempt is made to draw more water from the storage chamber than the pump is supplying.

To carry the illustration a step further, imagine that an additional line from the pump is required to discharge an additional 60 gpm from the nozzle, for a total discharge of 120 gpm. As the discharge from the pump is directly proportional to the speed of the pump, and twice as much water is now required as before, it will be necessary to increase the speed of the pump by 100 percent, or from 60 rpm to 120 rpm.

For illustrative purposes, the actual discharge of the pump equaled the theoretical discharge. However, the theoretical discharge is never reached during actual pumping operations. This is due partly to the back pressure that forces some of the water from the discharge side of the pump back to the suction side of the pump and partly to the leakage of water through pump clearances as the pump rotates. The actual amount of water a pump can discharge at various rpm's is determined by flow tests. The difference between the amount of water that a pump should theoretically deliver and the actual amount the pump does deliver is referred to as *slippage*. The slippage of a pump is usually ex-

Fire Pumps

pressed as a percentage. The percentage of slippage of a pump depends upon the condition of the pump and the pressure buildup. In general, the higher the pressure, the greater the slippage. While the slippage of each pump is different, new pumps as a group will deliver 90 to 95 percent of their theoretical capacity at 120 psi, 80 to 90 percent at 200 psi, and 75 to 85 percent at 250 psi. The percentage of the theoretical capacity discharged is referred to as the *volumetric efficiency*. The percentage of slip can be determined by subtracting the volumetric efficiency from 100, or, expressed as a formula:

Perceentage of slip = 100 − volumetric efficiency

The volumetric efficiency is the measurement of the condition of the pump. At a given pump speed, the higher the volumetric efficiency, the better the condition of the pump. The volumetric efficiency is the ratio between the amount of water a pump actually discharges and the theoretical amount of water it should discharge. Expressed as a formula:

$$\text{Volumetric efficiency} = \frac{\text{discharge} \times 100}{\text{theoretical discharge}}$$

QUESTION: A rotary gear, positive displacement pump is discharging 290 gpm at a pressure of 125 psi. The pump is turning at a speed of 380 rpm. What is the volumetric efficiency and the amount of slippage if the plate on the pump reads 0.85 gpr?

ANSWER: Given that

theoretical discharge at 380 rpm
= (380) (.85)
= 323 gpm

then

$$\text{volumetric efficiency} = \frac{(290)(100)}{323}$$

= 89.78%

and

percentage of slip = 100 − 89.78
= 10.22%

TYPES OF POSITIVE DISPLACEMENT PUMPS

Two general types of positive displacement pumps are used in the fire service: the piston type and the rotary type. The majority, by far, are of the rotary type.

Piston Pumps

Piston pumps have been used in the fire service longer than any other type of pump; however, over a period of years their use on fire apparatus has been diminishing. The few remaining in service have been designed primarily for situations requiring extra-high pressure.

The principle of operation is similar to that of the piston pump described earlier. The difference is in the type of valves, the valve arrangements, and the piston construction. The high pressures that are demanded are obtained by utilizing several cylinders in series, all operating from a single crankshaft. The employment of multiple cylinders helps reduce pulsations, in addition to providing increased pressures, and therefore contributes to a smoother water flow.

Rotary Pumps

There are three basic types of *rotary pumps* in current use on fire apparatus: the *gear*, the *lobe (cam)*, and the *vane*. These pumps are employed primarily as priming pumps and booster pumps. To the author's knowledge, no rotary pumps are currently being installed as main pumps on new apparatus; however, there are a few still in service as main pumps on older apparatus.

Rotary Gear Pumps

Figure 7-2 shows a schematic of a typical rotary gear pump. These pumps vary slightly from one manufacturer to another, primarily in the shape of the gears, the number of teeth, and the

Fire Pumps

method of transmission of power to turn the pump. Regardless of the manufacturer, the principle of operation is the same.

Rotary gear pumps consist of two intermeshing gears which revolve in opposite directions within a closely fitted casing. One gear is designated the driving gear, while the other is referred to as the driven gear.

Each gear of the pump used for illustration in Figure 7-2 has six teeth. The teeth intermesh in turning, much as they would in any gear train. Gear G1 rotates in a clockwise direction, while gear G2 turns in a counterclockwise direction. The gears turn away from each other on the intake side and toward each other on the discharge side.

Rotary gear pumps are positive displacement pumps and so are capable of pumping either air or water. The air or water enters the pump through the intake manifold, where it is trapped in the pockets formed by two adjacent teeth. As the gear turns, the trapped water is carried along the pump casing walls toward the discharge side of the pump, where it is released. The close meshing of the teeth on the discharge side keeps the water from returning to the intake side of the pump and also contributes to the pressure buildup as the gears try to squeeze the water between them.

It should be noted that one tooth on gear G1 is juste departing from the discharge opening as a tooth on gear G2 approaches the

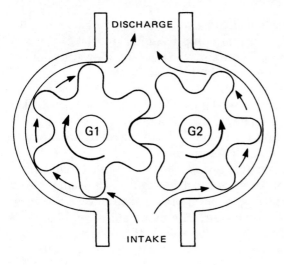

Figure 7-2 A Typical Rotary Gear Pump

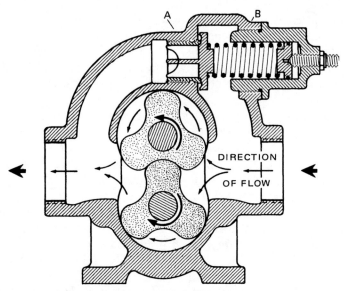

Figure 7-3 A Waterous Rotary Lobe Pump. *Courtesy of Waterous Company*

opening. This arrangement of the gears tends to produce a nearly continious discharge which assists in the reduction of pulsations.

Rotary Lobe (or Cam)

Figure 7-3 is a schematic of a Waterous rotary lobe pump. The principle of operation of this pump is similar to that of the gear pump. Water is trapped between the lobes as they turn, and is released on the discharge side of the pump.

One difference between this type of pump and the gear pump is in the method of driving the lobes. The shape of the lobes makes it impossible for one lobe to drive the other; consequently, each lobe is independently driven. The shafts of the rotors are properly synchronized to the meshing of the lobes.

As explained earlier, either the water discharged from positive displacement pumps must be used at the same rate of discharge, or some means must be provided to release the water and restrict the pressure buildup. The pump shown in Figure 7-3 incorporates a built-in relief valve. When the pressure acting on the piston at point A exceeds the spring tension pressure (B)

Fire Pumps   265

holding the piston closed, the piston will move to the right, allowing the water to be returned to the suction side of the pump.

Rotary Vane Pumps

Figure 7-4 is a schematic of a rotary vane pump. This pump consists of a single rotor within, and eccentric to, the casing. The rotor is considered eccentric as its center (C1) is offset from the center of the casing (C2).

Within the rotor are four identical vanes (V1, V2, V3, V4) that are free to move in and out of the slots (S1, S2, S3, S4) as the rotor turns. These vanes slide in and out, partially because of cam-shaped guides and partially because of the centrifugal action of the rotor. The vanes maintain contact with the surface of the casing as they move.

The constant in-and-out movement of the vanes as the rotor turns increases and decreases the volume of the space between two adjacent vanes. For example, the volume of the space between vanes V-3 and V-4 is greater than the volume of the space between vanes V-1 and V-4.

The space between two adjacent vanes is at a minimum when the first vane passes the discharge port. As the first vane passes the port, the space between it and the following vane will begin to increase. The space between vanes V-4 and V-1 is in the expanding position. The space between these two vanes will be

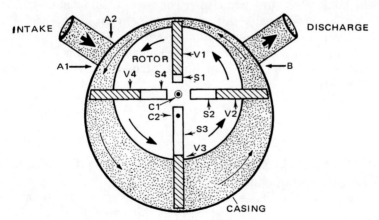

Figure 7-4 A Schema of a Rotary Vane Pump

filled to its maximum with water or air when vane V-1 reaches point A-1. The air or water is trapped between the vanes and moves along the casing wall toward the discharge outlet as the rotor turns.

The space between two vanes starts diminishing in size as the first vane reaches point B, forcing the air or water from the pump. As they are larger than the intake and discharge openings, the vanes are kept from moving into these spaces as they pass. The space between vanes V-1 and V-2 in Figure 7-4 is well into the diminishing phase.

CENTRIFUGAL PUMPS

The principle of a *centrifugal pump* can be demonstrated with a bucket partly filled with water. Swinging the bucket in a circle will hold the water in the bucket. The faster the bucket is swung, the greater the weight of the water in the bucket will seem to be. This effect is produced by the tendency of the water in the bucket to try to escape but to be held in place by the bottom of the bucket.

This can be demonstrated further by poking a hole in the bottom of the bucket. Water will now be discharged from the hole as the bucket is swung in a circle. The faster the bucket is rotated, the greater the intensity and reach of the stream discharged. The principle is that a revolving motion will create a force that will impel an object outward from the center of rotation. The force produced is referred to as centrifugal force—hence the name centrifugal pump.

The Impeller

The heart of centrifugal pump is a revolving disc known as an *impeller* (see Figures 7-5, 7-6, and 7-20). The impeller is a circular blade that has a hole in the center and is closed at the sides. The center hole is referred to as the eye (Figure 7-5). Water enters the eye of the impeller through the intake manifold in a

Fire Pumps   267

direction parallel to the shaft, and is hurled to the outer edge of the turning impeller by centrifugal force. The direction of movement of the water is controlled by the impeller blades (or vanes), while pumps using additional diffusion vanes help straighten out the flow. The speed of the water increases as it moves from the eye toward the outer edge. It also increases as the impeller is turned at a faster rate.

There is always a continuous path for the water to flow through a centrifugal pump, whether the pump is turning or in a static position. This feature provides an advantage in that it allows the pump to use the incoming hydrant pressure fully. This advantage will be more completely explored during the discussion on series-parallel operations.

Impellers are generally made of bronze or a bronze alloy. They are both mechanically and hydraulically balanced to provide

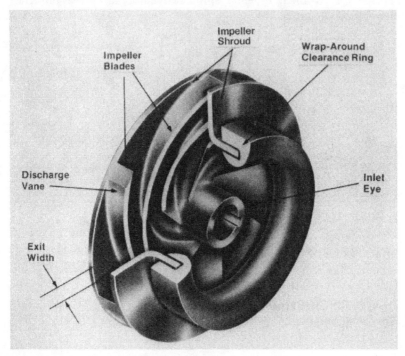

Figure 7-5 The Parts of an Impeller. *Courtesy of Hale Fire Pump Company*

Figure 7-6A A Single Impeller Splined to the Shaft. *Courtesy of Hale Fire Pump Company*

Figure 7-6B Two Impellers Splined to the Same Shaft. *Courtesy of Hale Fire Pump Company*

Fire Pumps

for vibration-free operation. The impellers are splined or keyed to the shaft. The shaft is usually made of stainless steel.

The Volute

The second main part of the centrifugal pump is the *volute*. The volute is a spiral-shaped section of the body enclosing the impeller, and it carries the water from the impeller out to the discharge manifold. It can best be visualized by referring to Figures 7-7 and 7-9. The volute is the area between the impeller and the wall of the pump casing. It should be noted that the cross-sectional area (distance from the impeller to the casing) is constantly increasing at a uniform rate as it approaches the discharge outlet. The volute enables the pump to handle an increasing amount of water but at the same time allows the water to remain at or near the same velocity throughout its entire movement within the pump casing. This is necessary in order to achieve a continuity of flow.

Figure 7-7 demonstrates the path a single drop of water would take from the time it enters the pump until it is discharged. The drop enters the eye as the result of an outside force. Hydrant pressure causes the movement when one is pumping from a hydrant. The movement is the result of atmospheric pressure when one is working from draft. Once in the eye, a drop of water moves through the volute by centrifugal force. The number of drops moving through the pump at the same time can be regulated by adjusting pump speed, transfer valve setting, discharge valve opening, or nozzle size.

Displacement Principles

Three factors involved in a centrifugal pump operation are so closely related that a change in one will automatically result in a change in another. These factors are pump speed, discharge pressure, and quantity of discharge. Within limits governed by the design of the pump, if the pump speed is held constant, an increase in discharge pressure will result in a decrease in the amount of water discharged, and vice versa. If the discharge pres-

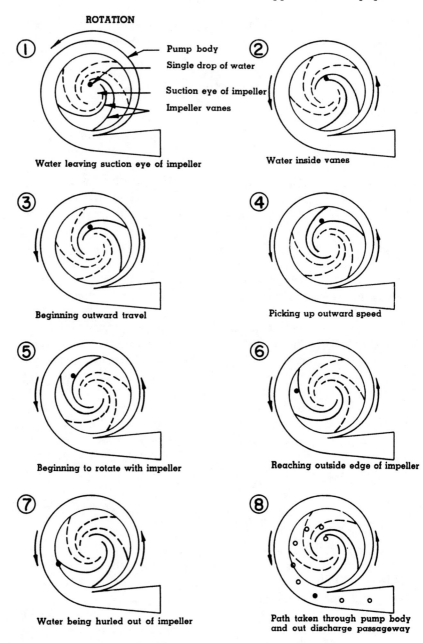

Figure 7-7 How a Centrifugal Pump Operates. *Courtesy of Waterous Company*

Fire Pumps

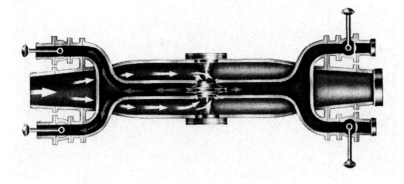

Figure 7-8 The Flow of Water in a Single-Stage, Dual-Suction Pump. *Courtesy of Hale Fire Pump Company*

sure is held constant, an increase in pump speed will result in an increase in the discharge, and vice versa. If the discharge is held constant, an increase in pump speed will result in an increase in discharge pressure, and vice versa.

The degree of relationship between these factors is usually expressed by two general principles. One principle is that *when pumping at a constant pressure, the quantity of water discharged is directly proportional to the speed of the pump*. This basically means that as long as the pressure remains the same, the amount of water discharged will change at the same rate as the pump speed. If the speed is doubled, the discharge will be twice as great. Decrease the speed by a fourth, and the discharge will be reduced by a fourth. This principle can be expressed by the formula:

$$\text{new quantity} = \frac{\text{new pump speed}}{\text{old pump speed}} \times \text{old quantity}$$

or

$$\text{new pump speed} = \frac{\text{new quantity}}{\text{old quantity}} \times \text{old speed}$$

QUESTION: A pump is discharging 300 gpm at a pressure of 100 psi while turning at a speed of 500 rpm. How much water will be discharged if the pressure remains at 100 psi but the pump speed is increased to 750 rpm?

ANSWER: Since

$$\text{new quantity} = \frac{\text{new pump speed}}{\text{old pump speed}} \times \text{old quantity}$$

where new pump speed = 750 rpm
 old pump speed = 500 rpm
 old quantity = 300 gpm

$$\text{new quantity} = \frac{750}{500} \times 300$$
$$= (1.5)(300)$$
$$= 450 \text{ gpm}$$

Therefore, a pump discharging 300 gpm at a pressure of 100 psi while turning at 500 rpm will discharge 450 gpm if the speed is increased to 750 rpm and the pressure remains at 100 psi.

QUESTION: A pump is discharging 500 gpm at a pressure of 120 psi while turning at a speed of 750 rpm. What pump speed would maintain a discharge pressure of 120 psi but increase the flow to 700 gpm?

ANSWER: Since

$$\text{new pump speed} = \frac{\text{new quantity}}{\text{old quantity}} \times \text{old speed}$$

where new quantity = 700 gpm
 old quantity = 500 gpm
 old speed = 750 rpm

$$\text{new pump speed} = \frac{700}{500} \times 750$$
$$= (1.4)(750)$$
$$= 1050$$

Therefore, in order to increase the discharge of a pump from 500 gpm to 700 gpm when the pump is turning at 750 rpm at a pressure of 120 psi, it would be necessary to increase the speed of the pump to 1050 rpm if the pressure remained the same.

Both of these problems illustrate situations where one or more lines are being used from a pump and it becomes necessary to take off additional lines. To maintain the same discharge pres-

Fire Pumps

sure while adding another line, the pump operator would have to open the discharge gate slowly, while simultaneously increasing the speed of the pump. Close coordination must be maintained between the opening of the discharge gate and the increasing of the pump speed to ensure that there is not a temporary pressure drop.

A second principle involving the three interrelated factors is that *when pumping at a constant quantity, pressure is directly proportional to the square of the pump speed.* This basically means that, as long as the flow remains the same, it is not necessary to increase the pump speed by 50 percent in order to increase the discharge pressure by 50 percent, or to double the pump speed in order to double the discharge pressure. The principle is expressed by the formula:

$$\text{new pressure} = \left(\frac{\text{new pump speed}}{\text{old pump speed}}\right)^2 (\text{old pressure})$$

QUESTION: One wants to increase the nozzle pressure on a constant flow tip that is discharging 300 gpm. The pump discharge pressure is 150 psi while the pump is turning at 500 rpm. What would be the discharge pressure if the pump speed were increased to 600 rpm?

ANSWER: Since

$$\text{new pressure} = \left(\frac{\text{new pump speed}}{\text{old pump speed}}\right)^2 (\text{old pressure})$$

where new pump speed = 600 rpm
old pump speed = 500 rpm
old pressure = 150 psi

$$\text{new pressure} = \left(\frac{600}{500}\right)^2 (150)$$
$$= (1.2)^2 (150)$$
$$= (1.44)(150)$$
$$= 216 \text{ psi}$$

Therefore, while discharging 300 gpm, it was only necessary to increase the speed of the pump from 500 rpm to 600 rpm in order to increase the pump discharge pressure from 150 psi to 216 psi.

Pump Slippage

Pump slippage is not a factor in centrifugal pumps because the continuous path allows water to flow from the intake side of the pump to the discharge side whether or not the pump is turning. Water can also flow through the pump in a reverse direction when the pump is standing still. When connected to a positive pressure source, such as a hydrant, the continuous pathway results in both the intake and discharge manifolds filling, once water is let into the pump.

When more than one line is taken from a pump, the shutting down of a single line will reduce the load on the engine and cause a slight amount of slippage within the pump. When all lines are shut down the slippage automatically increases to 100 percent and the pump simply churns. This takes place without any damage to either the pump or the engine. It is necessary under this condition to reduce the throttle setting, as the pump will have a tendency to run away. The continuous churning of the water will cause an increased water temperature within the pump due to friction, and, therefore, some corrective action should be taken. A slight opening of a pump discharge gate (5 to 20 gpm) will usually alleviate the churning and temperature buildup.

Single-Stage Pumps

A *single-stage* centrifugal pump consists of a single impeller mounted on a shaft and enclosed in a pump housing (see Figure 7-8). The pressure is controlled by the pump speed, while the volume is controlled by the impeller design (inlet size and exit width).

The single-stage pump has a minimum of controls and is relatively simple to operate. Water enters the intake side of the pump, where it is directed into the eye of the impeller. Some single-stage pumps are equipped with double intakes (Figure 7-8); this particular pump design allows water to enter the impeller simultaneously from both sides. The discharge of the water is directed to the discharge manifold, where it is available for use at the discharge gates.

Single-stage centrifugal pumps have an advantage in their simplicity of operation but are handicapped in their ability to obtain pressures of 300 psi and above. Larger impellers would be

Fire Pumps

more effective; however, limited physical space on pumps and the high capacity rating requirements restrict the practical size.

Multi-Stage Pumps

In *multi-stage* pumps, the discharge from one impeller is connected directly into the intake side of another impeller. The number of stages in a multi-stage pump refers to the number of impellers connected in series. Multi-stage pumps are used whenever the discharge pressure requirements exceed the capability of a single-stage pump.

Figure 7-9 is a schematic of a two-stage pump. Figure 7-10 is a schematic of a four-stage pump. As the discharge from the first stage is delivered directly into the intake of the second stage, and the discharge from the second stage is delivered directly into the third stage, and so on, the total capacity of a multi-stage pump is limited to the capacity of the first stage. However, each stage is capable of taking full advantage of the pressure received from the previous stage, and then adding its own pressure capability prior to transferring water to the next stage. Figure 7-10 illustrates this principle.

Consider that each of the impellers in Figure 7-10 is capable

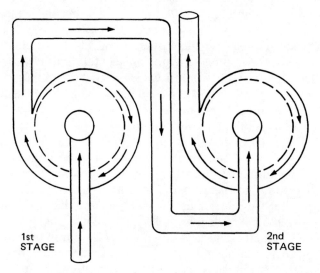

Figure 7-9 A Schema of a Two-Stage Pump

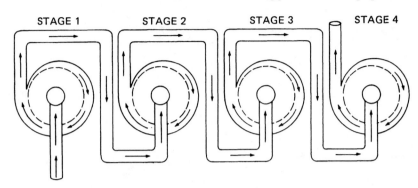

Figure 7-10 A Four-Stage Centrifugal Pump

of a capacity of 150 gpm at 150 psi, and that all the impellers are turning at the same speed. Also, for illustrative purposes, consider that the water enters stage 1 at zero pressure. The impeller of stage 1 would receive the water at zero pressure and discharge 150 gpm at 150 psi into the eye of the stage 2 impeller. The stage 2 impeller would take the 150 gpm and discharge it into the impeller of stage 3 at 300 psi, adding 150 psi of its own to the 150 psi it received. The 150 gpm would be picked up by the third-stage impeller and sent on to the fourth stage at 450 psi, the 300 psi it received plus 150 psi of its own. The fourth-stage impeller would add 150 psi of its own and discharge the 150 gpm into the discharge manifold, where it would be available for use at 600 psi.

The ability of a centrifugal pump to take full advantage of the pressure received applies to a single-stage pump as well as to the first stage of a multi-stage pump. Consider that the impellers in the two-stage pump in Figure 7-9 are each rated at 250 gpm at 150 psi, and that the pump is connected to a hydrant that is delivering water into the pump at a flow pressure of 50 psi.

The first-stage impeller would pick up the water and, if working at capacity, deliver 250 gpm at 200 psi into the eye of the second-stage impeller. The 200 psi results from the pump's 150 psi capability added to the 50 psi of incoming pressure. The second-stage impeller would pick up the 250 gpm and discharge it at 350 psi, adding 150 psi of its own to the 200 psi it received.

The inherent advantage of a multi-stage pump over a single-stage pump is the ability to deliver pressures at relatively low speeds. This advantage is particularly valuable with larger capacity pumps.

Multi-stage centrifugal pumps are usually built as a single unit, with each impeller mounted on a common shaft. The pump housing consists of a divided case, with each impeller housed within its own casing. Water moves from one impeller to another automatically, with no action required by the pump operator. Four-impeller, four-stage pumps are high-pressure pumps.

Series-Parallel Pumps

A series-parallel pump can be operated as either a series pump or a parallel pump. When it is operating as a series pump, the water from one impeller is discharged directly into the eye of the second impeller; the pump is thus functioning as a multi-stage pump. When it is operating in parallel, each impeller discharges water independently into the discharge manifold; the pump is thus acting as two independent single-stage pumps that are discharging their contents into a common manifold.

The two-stage, series-parallel pump has long been the standard of the fire service. It provides a wide range of capacities over a range of pressures roughly twice as wide as that possible with a single-stage pump. A 1000-gpm series-parallel pump will deliver 500 gpm at 300 psi net pump pressure at essentially the same speed and power demand as is required for 1000 gpm at 150 psi, while a 1000-gpm single-stage pump must be driven at a speed 30 to 35 percent higher to deliver 500 gpm at 300 psi and will require 80 to 85 percent more power. More important, in special cases, is the series-parallel pump's ability to develop net pump pressures of up to 600 psi; the single-stage pump is usually limited to a net pump pressure of about 300 psi.

Another advantage of a series-parallel pump over a single-stage pump or multi-stage pump is its ability to adapt to a wide variety of operational demands without resorting to excessive pump speeds. It can meet the demand for increased pressure when long lays have been made but volume is not an important factor.

Series-parallel pumps have two impellers, with each impeller enclosed within its own volute. The impellers are keyed to a common shaft and housed within a common casing. The impellers are of identical construction and, being keyed to the same shaft, turn at identical speeds. A transfer valve at the outlet of the

278 Introduction to Fire Apparatus and Equipment

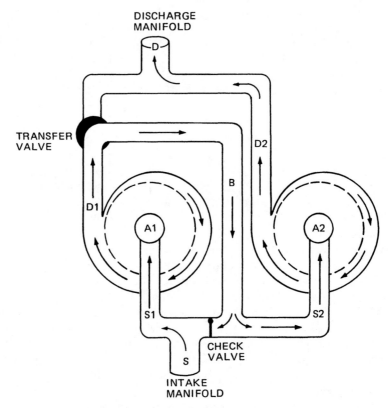

Figure 7-11 Series (Pressure) Operation of a Series-Parallel Pump

first-stage volute directs water either to the discharge manifold for parallel or volume operation or into the eye of the second-stage impeller for series or pressure operation.

Figures 7-11 and 7-12 are schematics of typical piping arrangements of a series-parallel pump with two impellers. The intake and discharge manifolds are common to both impellers. For purposes of explanation consider that each impeller is capable of discharging 500 gpm at 150 psi, and that each impeller is operating at capacity.

Series Operation

When the hydrant is first opened, and before the pump is engaged, water will enter the common intake manifold (Figure

Fire Pumps

7-11) and flow into impeller one (A1) through piping S1 and into impeller two (A2) through piping S2. The water will continue through both impellers and into all piping until the entire system is filled. At this point water will be available at the discharge outlets at hydrant pressures. For purposes of explanation, assume that the hydrant pressure is zero. When the pump is engaged, the water will leave impeller one (A1) by way of piping D1, pass through the transfer valve that is set for series operation, and enter piping B. The pressure from the pump, as it is greater than the incoming pressure, will close the check valve when water from impeller one reaches it through piping B. The water will then pass through piping S2 into the eye of impeller two (A2). Assume that impeller A1 is discharging 500 gpm at 150 psi. Impeller two will receive 500 gpm at 150 psi from impeller one, add 150 psi of its

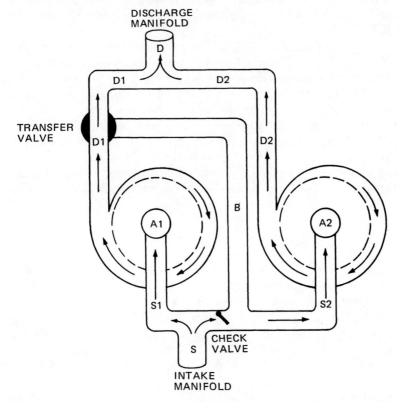

Figure 7-12   Parallel (Volume) Operation of a Series-Parallel Pump

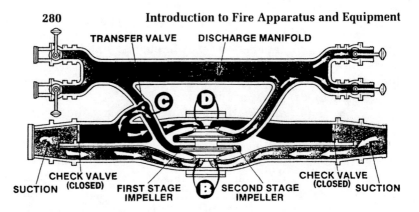

Figure 7-13 The Flow Through a Series-Parallel Pump When Pumping in Series. *Courtesy of Hale Fire Pump Company*

own, and then discharge the 500 gpm at 300 psi through piping D2 into the common discharge manifold (D), where it will be available for use.

This quantity of water at 300 psi is sufficient to supply two extra-long lays of 2½-inch hose equipped with 1-inch tips and still provide adequate nozzle pressures.

Figure 7-13 shows the flow through a series-parallel pump when pumping in series (pressure). The incoming water is directed to the suction side (eye) of the first-stage impeller at point B. Flow from the first-stage impeller at an increased pressure is directed through the transfer valve C to the suction side (eye) of the second-stage impeller at point D. Flow from the second-stage impeller is directed to the discharge manifold. This position achieves maximum pressure from the pump.

Parallel Operation

The transfer valve (Figure 7-12) is set so that each impeller will operate independently. Before the pump is engaged, water will enter the common intake manifold (S) and flow through S1 and S2 into each of the impellers. The water will flow from the impellers through piping D1 and D2, and on into the common discharge manifold (D). It will become static when the entire system of piping is filled. The incoming flow pressure will hold the check valve open. When the pump is engaged, each of the impellers will turn, discharging its 500 gpm into the common

Fire Pumps

discharge manifold at 150 psi. A total of 1000 gpm at 150 psi will be available for use at the discharge outlets. This quantity of water at 150 psi is sufficient to supply five short 2½-inch lines equipped with 1-inch tips, or four short 2½-inch lines equipped with 1⅛-inch tips, or three short 2½-inch lines equipped with 1¼-inch tips, or a portable monitor equipped with a 1000 gpm spray nozzle (as long as the monitor is placed reasonably close to the pumper and an adequate selection of hose is made to keep the friction loss to a minimum).

Figure 7-14 shows the flow through a series-parallel pump when pumping in parallel (volume). The incoming water flow is split at point A to direct approximately one-half the total flow to each impeller. Check valves on both sides are open. The centrifugal force of the impellers increases the pressure which is directed to the discharge manifold. This position achieves maximum flow from the pump.

Determining Series or Parallel Operation.

A series-parallel pump offers a pump operator the choice as to whether to operate the pump as a multi-stage pump or as two single-stage pumps working independently but discharging water into a single discharge manifold. In order to obtain the maximum efficiency from the pump, the operator must know when to use the series position and when to use the parallel position. The choice is made through the manipulation of the transfer (or changeover)

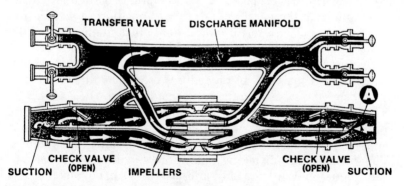

Figure 7-14 The Flow Through a Series-Parallel Pump When Pumping in Parallel. *Courtesy of Hale Fire Pump Company*

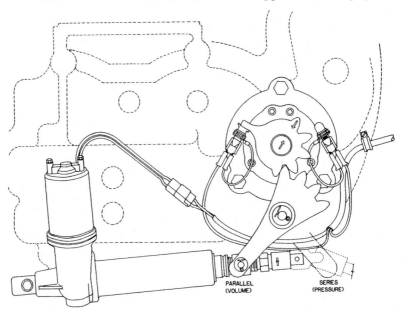

Figure 7-15   A Transfer Valve Actuator. *Courtesy of Waterous Company*

valve (Figure 7-15). Both manual and power-operated transfer valves are installed on fire apparatus.

Power actuators are generally capable of moving the transfer valve when the pump pressure is relatively high, but manually operated valves are difficult to change unless the pump pressure is below 100 psi. Regardless of the type of system one is working with, care should be taken in its operation, particularly when switching from the parallel to the series position. A switch from parallel to series will result in a virtually instantaneous doubling of the pressure, which could cause hose lines to rupture or place firefighters in a very dangerous situation. Pump operators should always follow the rules of procedures established by the department, safety being the paramount consideration.

The best transfer valve position for any particular pumping condition will depend on the characteristics of the particular pump-engine combination used and on whether the pump is being used at draft, is taking water from a hydrant, or is part of a relay operation. While no hard-and-fast rules can be established as to when to pump in series and when to pump in parallel, there are two general guidelines which might prove useful.

One guideline is to use the position (series or parallel) that will give the desired result at the lowest engine speed.

A second guideline is to use the series position when pumping through long single lines, and the parallel position when a number of large lines have been taken from the pump and the total discharge requirement will be a considerable volume of water.

When the apparatus is parked in quarters awaiting the next response, the transfer valve should be kept in the position most likely to be used. In general, the transfer valve should be kept in the series position in those districts where most fires require the use of a single line. It might be best to keep the valve in the parallel position in industrial areas or those where fires requiring large quantities of water are frequent. Pump operators should be aware, however, that it is possible for a transfer valve to freeze up if left in one position for too long a period of time. It is therefore essential that the valve be exercised periodically to assure operability.

Cavitation

One of the menaces with which a pump operator must learn to cope is that of *cavitation*. The operator must be familiar with the causes of cavitation, be able to recognize when a pump is cavitating, and know how to correct the situation.

Cavitation can occur in almost any pump, regardless of size, type, or design. It happens whenever the pressure at any point inside a pump drops below the pressure of the water being pumped. The point within a centrifugal pump where the lowest pressure normally occurs is at the impeller eye in back of the leading edge of the impeller. What happens is that the vapor pressure of the water causes thousands of small bubbles or vapor cavities to form when the water reaches the low pressure (vacuum) area. These bubbles mix with the water, causing a condition similar to that below the surface of the water in a pot on a stove when the water is near the boiling point. The action within the pump is the opposite of that of a boiling pot of water—the bubbles are compressed and implode, whereas a bubble in boiling water explodes when it reaches the surface. The water-bubble mixture leaves the low pressure point within the pump and starts its flow through the impeller, where it is subjected to a substantial increase in pressure. The pressure collapses the bubbles, producing an intense shock. This shock strikes the walls and impellers, causing severe damage to the metal. The shock may also cause pitting near the impeller vane tips and rapid corrosion of some pump materials, due to the action of the free oxygen which may be released.

Cavitation damage is cumulative and progressive. A centrifugal pump that is allowed to cavitate for too long a period of time will eventually suffer impeller failure.

While cavitation damage to a centrifugal pump may be severe, the damage to piston pumps and rotary pumps is even more severe. This is due to the fact that the pressure rise from the vacuum condition causing the problem is much faster in a positive displacement pump than in a centrifugal pump. The main damage to a piston pump occurs in the head of the piston, which becomes eroded. Usually the rotors show the most damage in rotary pumps.

A pump operator may have several indications that a pump is cavitating. One is a rattling sound resembling the sound of gravel going through the pump. This sound may be too faint to be heard, due to the noise of the engine and of the pump turning; however, if it is heard, the pump operator should take immediate steps to stop the cavitation.

Another indication is an increase in rpm without an increase in pump pressure. This is usually a positive sign that cavitation is taking place. Pump operators can use this indicator to check whether or not the pump is cavitating if they think they heard the sounds of stones traveling through the pump. The operator should open the throttle gradually, while watching the engine tachometer and pump discharge pressure gage. Cavitation is occurring if there is an increase in rpm without a corresponding increase in pressure. If cavitation is occurring, the throttle should be reduced immediately, as any further increase will result in additional damage.

Cavitation occurs when a pump operator is attempting to "run away" from the water. When pumping from a hydrant, operating in a relay, or receiving water from some other positive pressure source, the pump will never cavitate as long as the operator keeps a positive pressure on the intake gage. Unfortunately, all intake

Fire Pumps

gages are not 100 percent accurate, and cavitation may be occurring with the intake gage still indicating a few pounds of positive incoming pressure. Maintaining at least 10 psi incoming pressure on the gage will usually be sufficient to compensate for any gage errors. Of course, if the soft suction hose begins to flatten, it is a pretty clear indication that the pump is near the cavitation point.

When one is operating from a hydrant, the following conditions can contribute to runaway problems:

1. The hydrant outlet capacity may not be sufficient for the quantity of water required.
2. There may be a partial collapse of the soft suction hose lining which results in a restricted water flow.
3. The suction hose may be too small. This is most likely to occur when unusual situations are encountered and the pump operator has to connect to the hydrant through a section of 2½-inch hose rather than the regular suction hose.

Common sense dictates the corrective action that should be taken in these situations. Steps should be taken to either increase the flow of water into the pump or reduce the flow leaving the pump. With a double hydrant, it may be possible to lay another supply line from the other hydrant outlet into the pumper. Or it may also be possible to stretch a line from another source into the pump, possibly another hydrant, if one is located within a reasonable distance. It would be preferable, in this case, to use a secondary hydrant that is connected to a different water source from that of the hydrant being used. If there is absolutely no means of increasing the flow into the pump, then it is necessary to reduce the throttle to the point where cavitation ceases.

It is also possible to run away from the water when operating from draft. The intake (suction) gage will indicate cavitation while drafting by showing an abnormally large number of inches of mercury. The most reliable indicator when drafting is the increase in rpm without a corresponding increase in discharge pressure. Several conditions may be contributing factors:

1. The suction hose is too small.
2. The lift is too high.
3. The strainer is partially blocked.
4. The water temperature is too high.

It is obvious that if the suction hose is too small, it should be replaced with a larger hose or an additional suction line extended from the pumper to the water source; if the lift is too high, another source of water must be used; if the strainer is partially blocked, it must be cleared; or if the water temperature is too high, another source of water should be sought; however, a complete shutdown is needed to correct most of these factors. This may not be possible if firefighting operations have commenced, and the hosehandlers working the lines are in a critical position. The best that can be done then is to reduce the flow until cavitation stops. This emphasizes the importance of adequately sizing up a drafting situation prior to making an operational commitment. Consideration should be given to lift conditions, suction hose requirements, and the volume of water that is needed or might be needed before the emergency is abated.

Theoretical Pump Capacity

A pump is much like an engine in that it can do so much work and no more. The amount of work an engine can do is determined by its rated horsepower. The amount of work a pump can do is determined by the pump rating. A pump is rated at a given discharge at a given pressure—for example, at 1000 gpm at 150 psi. The pressure rating refers to net pump pressure or the pressure actually produced by the pump.

The work capability of a pump is expressed in pounds-gallons. The pounds-gallons are determined by multiplying the rated discharge by the rated pressure. For example, a pump rated at 150 gpm at 150 psi would have a work capability of 22,500 pounds-gallons (150 × 150), while a pump rated at 1250 gpm at 150 psi would have a work capability of 187,500 pounds-gallons (1250 × 150).

It should be emphasized that nationally known pumps can pump more than their rated capacity and rated pressure. In fact, the spurt test requires pumps to supply their rated capacity at 165 psi—15 psi above the capacity test pressure.

The amount of water a pump can *theoretically* discharge once the rated pressure has been exceeded is determined by the formula:

Fire Pumps

$$\text{discharge} = \frac{\text{pounds-gallons}}{\text{net pump pressure}}$$

QUESTION: A pump is rated at 1250 gpm at 150 psi. What is the theoretical maximum amount of water it can deliver when pumping at a net pump pressure of 140 psi?

ANSWER: The theoretical maximum amount of water the pump can deliver is 1250 gpm (its rated capacity). The maximum at 140 psi will be less than 1250 gpm. The formula cannot be used until the rated pressure has been exceeded.

QUESTION: A pump is rated at 1500 gpm at 150 psi. What is the theoretical maximum amount of water it can deliver at a net pump pressure of 200 psi?

ANSWER: Since

$$\text{discharge} = \frac{\text{pounds-gallons}}{\text{net pump pressure}}$$

where pounds-gallons = (1500)(150)
                    = 225,000

net pressure = 200 psi

$$\text{discharge} = \frac{225{,}000}{200}$$
$$= 1125 \text{ gpm}$$

In actual practice, nearly all pumps in good condition are capable of delivering more than their rated discharge when working from a strong hydrant. In these conditions, the hydrant is contributing to the total discharge.

Additionally, in actual practice, most pumps are not capable of delivering the total theoretical quantity of water once the rated pressure of the pump has been exceeded. This is because the volumetric efficiency of a pump decreases at higher pressures. This decrease of volumetric efficiency is reflected in *pump ratings*.

Pump Ratings

The pump ratings that are most often quoted are those at 150 psi, such as 750 gpm at 150 psi or 1500 gpm at 150 psi. Actually, pumps are given three separate ratings:

capacity at 150 psi, net pump pressure
70 percent of capacity at 200 psi net pump pressure
50 percent of capacity at 250 psi net pump pressure

The ratings take into consideration the reduction of volumetric efficiency at increased pressures. The effect of this reduction is apparent when the theoretical discharge of a pump is compared with the rated discharge. The comparison of a 1250-gpm pump at 200 psi and 250 psi is as follows:

200 psi

$$\text{rated discharge} = 70\% \text{ of capacity}$$
$$= (.70)(1250)$$
$$= 875 \text{ gpm}$$

$$\text{theoretical discharge} = \frac{(1250)(150)}{200}$$
$$= \frac{187{,}500}{200}$$
$$= 937.5 \text{ gpm}$$

This reflects a reduction of 62.5 gpm, or 6.67 percent.

250 psi

$$\text{rated capacity} = 50\% \text{ of capacity}$$
$$= (.50)(1250)$$
$$= 625 \text{ gpm}$$

$$\text{theoretical discharge} = \frac{(1250)(150)}{250}$$
$$= \frac{187{,}500}{250}$$
$$= 750 \text{ gpm}$$

This reflects a reduction of 125 gpm, or 16.67 percent.

It should be noted that pumps are rated at net pump pressure. Net pump pressure is the actual pressure produced by the pump. When one is working from a hydrant, the incoming pressure aids the pump; therefore, the net pump pressure is less than the discharge pressure. Net pump pressure can be determined by subtracting the incoming pressure from the discharge pressure. For example, a pump discharging water at a pressure of 180 psi when the intake gage reads 80 psi would be operating at a net pump pressure of 100 psi (180 − 80).

When one is working from draft, the net pump pressure is always greater than the discharge pressure. This happens because the pump must do the work necessary to raise the water from its source, in addition to discharging it once it is received. For example, if the pump does the equivalent of 20 psi of work while moving the water from the source to the pump, and then discharges the water at a pressure of 125 psi, the net pump pressure would be 145 psi (20 + 125). The overall result is that a pump will reach its rated discharge capabilities at a much lower discharge pressure when working from draft than when working from a hydrant.

Series-Parallel Variations

The majority of the series-parallel pumps in service are of the type that house two impellers within a single pump body attached to a single shaft, with each impeller having its own volute. When series operation is desired, water is routed from one impeller to another by a transfer valve. This type of arrangement satisfies the operational needs of most fire departments. However, some communities have special problems that require special pumping configurations. Pumps are available to meet these special needs.

One variety of the series-parallel type pump is a series-parallel *three-stage* model. This pump is used in those situations where extra-high pressures are required. It is basically a series-parallel pump to which a third stage has been added. One manufacturer accomplishes this by installing a longer impeller shaft in the main pump and adding a high pressure impeller, an extra volute, and a control valve with a drain/vent linkage.

Figure 7-16 An Extra-Pressure-Stage Fire Pump. *Courtesy of Waterous Company*

Figure 7-16 illustrates a Waterous pump that utilizes the extra stage principle. The third-stage impeller is mounted on the same shaft as the other two impellers, and therefore rotates whenever the pump is in operation. However, it pumps water only when the third-stage control valve is open. The third stage is engaged by opening a quarter-turn control valve, which allows water from the main pump to enter the extra-stage intake inlet. Pressures of up to 800 psi can be produced with this arrangement, with the discharge capacity sufficient to supply several booster lines.

Another variety is the series-parallel *four-stage* pump. This pump was designed primarily to meet the pressure demands of supplying water to the upper floors of a high-rise building through the standpipe system, should the building's pumps be out of service or should the building not be equipped with a pumping system. The pump is capable of delivering relatively large volumes of water at pressures as high as 600 psi.

The pump has four impellers and four volutes. The series-parallel option provides for a wide range of operations. Rated capacity is obtained when all stages are operating in parallel (Figure 7-17). Maximum pressure is obtained when the pump is operated as a four-stage series pump (Figure 7-18). A third option provides intermediate discharges at intermediate pressures (Figure 7-19).

Another variety of the series-parallel pump is referred to as a

Fire Pumps

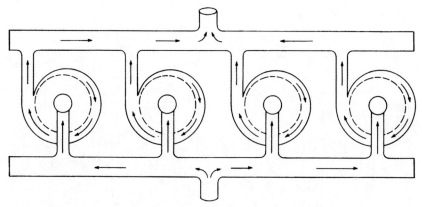

Figure 7-17 A Four-Stage Series-Parallel Pump Operating in the Capacity Position

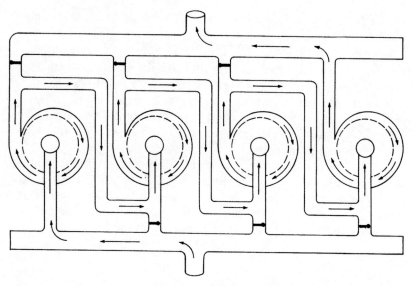

Figure 7-18 A Four-Stage Series-Parallel Pump Operating in the Series Position

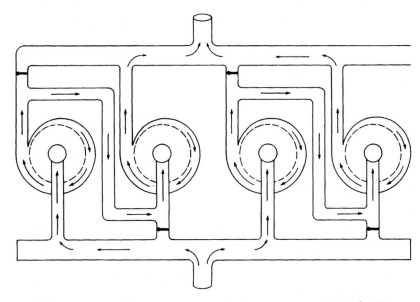

Figure 7-19 A Four-Stage Series-Parallel Pump Operating in the Intermediate Position

duplex multi-stage pump. This pump is in reality two separate pumps capable of working together in series when high pressures are required. The pump consists of two impellers of different sizes, each mounted on a separate impeller shaft. One impeller is referred to as the volume impeller, while the other is referred to as the pressure impeller.

The volume impeller is engaged whenever situations require a large volume of water. The pressure impeller is engaged whenever less water is required but pressures up to 250 psi are desirable. The two impellers are used in series whenever higher pressure demands (up to 350 psi) are encountered.

MAIN PUMPS

A main pump should have a capacity of at least 500 gpm so that it can supply two standard hose streams of 250 gpm each. Pumps larger than 500 gpm are manufactured in 250-gpm incre-

Fire Pumps

ments. The largest made for general use has a capacity of 2000 gpm. With few exceptions, main pumps are of the centrifugal type, with the single-stage and two-stage series-parallel being the most common. The capacities of these standard pumps at 150 psi, 200 psi, and 250 psi are shown in Table 7-1.

While all of the above-listed pumps are available, the large majority of those installed on new apparatus have a rated capacity of 1000 gpm, 1250 gpm, or 1500 gpm. The use of the 500-gpm pump is restricted to those areas where hazards are minimal, where special problems are not encountered, and where large fires are not expected. Large-capacity pumps, such as the 2000-gpm pump, are purchased primarily for the high-value area of larger cities or in extensive industrial areas where fires requiring large volumes of water and the use of heavy streams are anticipated.

Although main pumps may be either front-mount or midship, the vast majority are midship. Front-mount pumps are preferred in certain communities to meet the problems of the particular area. They are usually installed on commercial chassis. The pumps are driven from the front end of the engine, utilizing a clutch and pump transmission (see Figures 7-20 and 7-21). This type of power arrangement permits operation of the pump while the apparatus is moving, a particular advantage on brush and grass fires. Proponents of front-mount pumps also claim that this type of mounting provides better access to some streams and ponds. Front-mount pumps are available in 500-gpm, 750-gpm, 1000-gpm, and 1250-gpm capacity, with the 750-gpm pump the most popular. A 750-gpm Hale front-mount pump is shown in Figure 7-22.

Midship pumps are mounted approximately halfway between the front and rear axles. Discharge gates are on both sides of

Table 7-1 Rated Capacities of Standard Pumps

Rated Capacity at 150 psi	Rated Capacity at 200 psi	Rated Capacity at 250 psi
500	350	250
750	525	375
1000	700	500
1250	875	625
1500	1050	750
1750	1225	875
2000	1400	1000

294　　Introduction to Fire Apparatus and Equipment

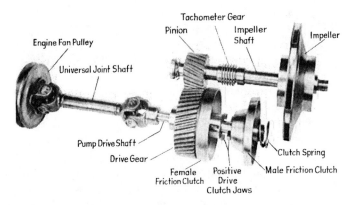

Figure 7-20 The Power Train for a Front-Mount Pump. *Courtesy of American Fire Pump Company*

the apparatus. Power is generally supplied from the engine to the pump by one of two methods. The most common is to install the pump transmission behind the road transmission and in line between the road transmission and rear axle. (See Figure 7-21.) This type of transmission or gearbox is called a split-shaft drive. With this type of installation, the road transmission is generally placed in direct drive for pumping. A separate transmission is used to provide the required engine-to-pump gear ratio. Figure 7-23 shows a midship pump. Figure 7-24, a cross-sectional view of a midship pump, shows the pump power arrangement. This type of mounting does not permit the pump to be used when the apparatus is moving.

Another method is to take power from the engine ahead of the engine transmission. (See Figure 7-21.) This method uses a flywheel drive power take-off. The system has the same advantage as a front-mount pump: one can use the pump while the apparatus is moving. This is particularly valuable at aircraft crashes where it is necessary to use foam or light water while advancing on the fire.

Booster Pumps

Booster pumps have a capacity of less than 500 gpm. They are rated at capacity at 150 psi, and generally at half capacity at

Fire Pumps

200 psi. Pumps in current use are of two basic types: rotary and centrifugal. The rotary type is a positive displacement pump and therefore does not require priming.

The primary source of water for the booster pump is the water tank carried on the apparatus. However, some apparatus are designed and plumbed to permit the booster pump to take water directly from a hydrant and to draft if necessary.

Booster pumps are used to supply small lines. Some apparatus use only hose reels for booster pump operations, while others have 1-inch, 1½-inch, or 1¾-inch outlets from which cotton-jacketed (or equivalent) hose may be used.

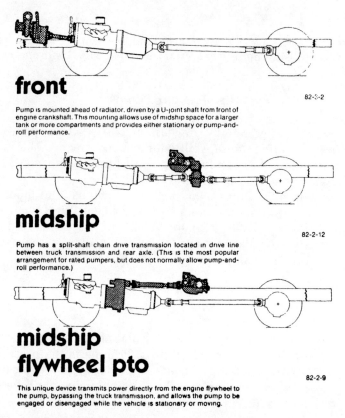

front

Pump is mounted ahead of radiator, driven by a U-joint shaft from front of engine crankshaft. This mounting allows use of midship space for a larger tank or more compartments and provides either stationary or pump-and-roll performance.

midship

Pump has a split-shaft chain drive transmission located in drive line between truck transmission and rear axle. (This is the most popular arrangement for rated pumpers, but does not normally allow pump-and-roll performance.)

midship flywheel pto

This unique device transmits power directly from the engine flywheel to the pump, bypassing the truck transmission, and allows the pump to be engaged or disengaged while the vehicle is stationary or moving.

Figure 7-21 Basic Mountings of Pump Transmissions. *Courtesy of Waterous Company*

Introduction to Fire Apparatus and Equipment

Figure 7-22 A 750-gpm Hale Front-Mount Pump Installed on a Chevrolet Chassis. *Courtesy of FMC Corporation, Fire Apparatus Operation*

High-Pressure Pumps

Some booster pumps are designed to produce extra-high pressure, providing the capability for true fog streams. The pump shown in Figure 7-25 is a four-stage centrifugal pump, capable of discharging 60 gpm at pressure of up to 1500 psi.

As with main pumps, booster pumps can be front-mount or midship. The front-mount pumps are of the centrifugal type, with capacity ratings of 300 gpm to 400 gpm. The power-to-pump arrangement is similar to that used on main pumps, including a clutch and transmission.

Priming Pumps

Priming pumps are positive displacement pumps used to prime the main pump when drafting. Priming is the process of replacing air in the intake lines and passageways of the main pump with water. Priming and the priming process will be discussed in more detail in the following chapter.

Fire Pumps

Pump Maintenance

Pumps should be expected to perform satisfactorily without any major repairs for a minimum of twenty years; however, the actual life span will depend upon a number of variables. While the frequency of use and type of service performed are important, more important are operator techniques and preventive maintenance.

Good operator techniques include the avoidance of a number of situations which cause pump damage. The most harmful of these cannot always be avoided; however, pump operators should do so if possible. This condition is pumping water containing sand or other abrasives. The situation most often arises while drafting and is generally the result of allowing the suction strainer to come in contact with the bottom of the drafting source. Operators should take care to ensure that the strainer is kept at least a foot from the bottom with a greater distance maintained, if possible. Of course, it is also possible to take in abrasives from positive

Figure 7-23 The Four Major Components of a Two-State Midship Pump—Body, Impeller Shaft, Cover, and Pump Transmission. *Courtesy of Waterous Company*

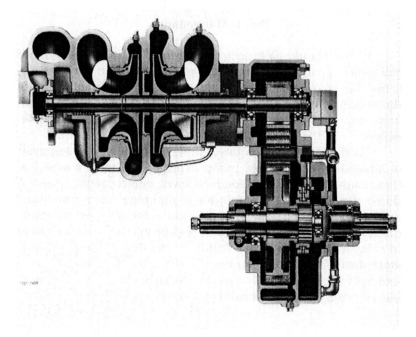

Figure 7-24 A Cross-Sectional View of the Transmission-to-Pump Arrangement. *Courtesy of Waterous Company*

pressure sources such as hydrants. Under emergency conditions when time is of the essence, it is not possible to flush out these systems; however, when drilling or in non-emergency situations where a few minutes are not important, pump operators should take the precaution of flushing out the system if there is any doubt at all that sand, dirt, or other abrasives may be discharged when the flow first commences.

Another operational situation to avoid is pump cavitation. Cavitation was discussed in more detail earlier in this chapter; however, it will be mentioned here to refer to a situation where the pump operator attempts to run away from the water. Pump cavitation can generally be recognized to be taking place whenever the discharge pressure fails to increase with an increase in pump speed.

A third operational condition which causes pump damage is the development of excessive heat within the pump. This is generally caused by the pump operator allowing the pump to churn for

Fire Pumps 299

an excessive period. Whenever all lines are shut down for any length of time, a pump operator should either take the pump out of gear or move water through the pump. This can be accomplished by partially opening an unused discharge gate or opening the tank fill line (if the pump is so equipped), which will discharge the water to the tank and out the tank overflow. If this method is used, it is generally best to remove the tank cap in order to reduce the possibility of a pressure buildup within the tank.

An operational technique which should increase pump life is flushing the pump after use whenever there is any suspicion that abrasives were encountered or the pumping situation involved the use of salt water. Backflush the pump by connecting a line from a hydrant to a discharge gate and opening an intake gate. Eventually open all gates to flush all piping. Flushing should

Figure 7-25   A Four-Stage Centrifugal Pump with a Capacity of 60 gpm at 1500 psi. *Courtesy of W.S. Darley & Company*

HALE PUMPS — MAINTENANCE CHECK LIST

YEAR _____ UNIT # _____

TRUCK MFG. _____

PUMP MODEL & SERIAL NO. _____

WEEKLY

Recommended test on the relief valve system or governor; test transfer valve (if applicable). Test the priming system and check oil level in priming oil tank—lubricate all valves, discharge, suction, hose drain and multi drain.

MONTHLY

MONTHLY CHECK LIST

	JAN.	FEB.	MAR.	APR.	MAY	JUN.	JUL.	AUG.	SEP.	OCT.	NOV.	DEC.
Complete weekly checks												
Lubricate remote controls												
Check controlled packing leakage (Adjust if necessary - 8 to 10 drops per min.)												
Perform Dry Vacuum Test - Per NFPA-1901 Para. 11-2.2.4 22" Minimum Vacuum - Loss not to Exceed 10" Vacuum in 10 minutes												
Check drive line bolts												
Lubricate suction tube threads and clean strainer, inspect gaskets.												
Check oil level in pump gear box, if contaminated replace with SAE EP 90 oil in midship pumps; SAE 10W-30 in front mount pumps												

ANNUALLY

Complete all previous checks ✓ on-all-questions
Check gauge calibration
Check oil level in auto lube assembly (SAE-EP 90 oil)
Lubricate power transfer cylinder and power shift cylinder with vacuum cylinder oil.
Change pump gear box oil and refill (SAE-EP oil midship pumps; SAE 10W-30 front mounts).
Check individual drain lines from pump to multi drain to ensure proper drainage and protection from freezing.
Lubricate transfer mechanism on two stage pumps.
Run yearly standard pump test (Underwriters) to check pump performance levels—chart provided below.
Repacking of pump is recommended every two or three years.

NOTE: The above general recommendations are provided for normal use and conditions. Extreme conditions or variables may indicate a need for increased maintenance. Good preventative maintenance lengthens pump life and ensures greater dependability. Consult service chart in operator's manual for detailed information.

	Capacity @ 150PSI	70% Capacity @ 200PSI	50% Capacity @ 250PSI
Hose Layout			
Nozzle Size			
Nozzle Pressure			
Gallons Per Minute			
Pump Pressure			
Engine Speed			
Original Engine Speed			
Lift & Suction Hose			

Hale Fire Pump Company • Conshohocken, Pa. 19428 • 215-825-6300 • TWX 510-660-8931   a [ ] company

Chart 7-1   Hale Pumps Maintenance Check List, *Courtesy of Hale Fire Pump Company*

Fire Pumps

continue until all traces of abrasives or salt water have been removed.

Equal to the importance of good operational techniques to increase pump life is the need to develop and adhere to a good pump maintenance program. Chart 7-1 contains a weekly, monthly, and annual maintenance check list for Hale pumps. Pump operators assigned to apparatus having other than Hale pumps should utilize the check list from the manufacturer of the pump installed on the apparatus. However, in the absence of such a list or a preventive maintenance program developed by the operator's department, the applicable segments from the Hale check list can be used for the development of a good pump preventive maintenance program. Experience has proven that a good operational technique program together with a good pump maintenance program will increase pump life by a substantial number of years and reduce the need for major pump repairs to an absolute minimum.

Pump Troubleshooting

Despite the development of good preventive maintenance programs and the attempt by pump operators to use effective operational techniques, mistakes will be made, operational conditions less than desirable will be encountered, and mechanical parts or systems will become defective. Following is a diagnostic chart (Table 7-2) for midship Hale pumps which has been provided through the courtesy of the Hale Fire Pump Company. This chart will help pump operators determine the causes of problems and take corrective action.

Table 7-2 Midship Service Diagnostic Chart
Courtesy of Hale Fire Pump Company

Condition	Possible Cause	Suggested Correction
PUMP WILL NOT ENGAGE		
Standard Transmission with Manual Pump Shift	Clutch not fully disengaged or malfunction in shift linkage	Check clutch disengagement. Drive shaft must come to a complete stop before attempting pump shift. Check pump shift linkage. Lubricate and adjust if necessary.
Automatic Transmission with Manual Pump Shift	Automatic transmission not in neutral position	Repeat recommended shift procedures with transmission in neutral position.
Standard Transmission with Power Shift system	Insufficient air or vacuum supply in shift system	Repeat recommended shift procedures. Check system for loss of vacuum or air supply. Check for leaks in system. Employ shift override procedures as follows: Hole is provided in shifting shaft to accomplish emergency shifting. Complete shift of control in cab and proceed to complete shift of lower control manually.
Automatic Transmission with Power Shift system	Automatic transmission not in neutral position	Repeat recommended shift procedures with transmission in neutral position.
	Pump shift attempted before apparatus has been brought to a complete stop	Release braking system momentarily; then reset and repeat recommended shift procedures.
	Premature application of	Release braking system momentarily. Then reset and repeat recommended

	parking brake system (before apparatus comes to a complete stop)	shift procedures.
	Insufficient air or vacuum in shift system	Repeat recommended shift procedures. Check system for loss of air or vacuum. Check for leak in system. Employ manual override procedures if necessary. (See Standard Transmission with Power Shift.)
	Air or vacuum leaks in system	Attempt to locate leak and take necessary steps to repair. Leakage, if external, may be detected audibly. Leakage could be internal and not as easily detected. *Note: Do not leave cab after completing pump shift unless shift indicator light is illuminated or a speedometer reading is noted.*
PUMP WILL NOT PRIME OR LOSES PRIME	No oil in priming oil tank	Refill priming oil tank with SAE30 motor oil.
	Engine speed too low (Rotary gear primer)	Increase engine rpm to recommended priming range 1000 to 1200 rpm.
	Electric priming system	No recommended engine speed is required to operate the electric primer; however, 1000 engine rpm will maintain apparatus electrical system while providing enough speed for initial pumping operation.
	Defective priming system	Check priming system by performing dry vacuum test per NFPA 1901, Paragraph 11-2.2.4. If pump is tight but primer pulls less than 22 inches of vacuum, it could indicate excessive wear in primer.

Table 7-2 Midship Service Diagnostic Chart
Courtesy of Hale Fire Pump Company
(Continued)

Condition	Possible Cause	Suggested Correction
	Defective priming valve (Electric)	Defective sealing rings—replace if necessary. Lubricate rings. Priming valve stuck open will allow loss of prime, also will permit unnecessary running of electric priming motor—assure complete closure of priming valve, dismantle, and lubricate if necessary.
		Note: Weekly use is recommended to keep priming system in good operating condition.
	Defective priming valve (Air or vacuum)	Priming valve not opening—employ manual override to open valve. Inactivity may cause the above. Remove and lubricate when practical. Defective diaphragm in priming valve—replace.
	Suction lifts too high	Do not attempt lifts exceeding 22 inches except at low elevations.
	Blocked suction strainer	Suction strainer must be at least two feet below water surface to prevent whirlpooling. Remove obstruction from suction hose strainer; do not allow suction hose and strainer to rest on bottom of water supply.
	Suction connections	Clean and tighten all suction connections. Check suction hose and suction hose gaskets for possible defects.
	Primer not operated long enough	Proper priming procedures should be followed. Do not release primer control before assurance of complete prime. Open discharge valve slowly during completion of prime to assure same.

	Note: Do not run primer over 30 seconds in attempting prime. If prime is not attained in 30 seconds, stop and look for possible cause, i.e., air leaks, blocked suction, and so on.
Air trap in suction line	Avoid placing any part of the suction hose higher than the suction intake. Suction hose should be laid with continuous decline to water supply. If trap in hose is unavoidable, repeated priming may be necessary to eliminate air pocket in suction hose.
Pump pressure too low when nozzle is opened	Reprime pump and maintain higher pump pressure while opening discharge valve slowly.
Air leaks	Attempt to locate and correct air leaks.
	Use the following procedures to locate air leaks:
	Perform dry vacuum test on pump per NFPA 1901, paragraph 11-2.2.4, with 22-inch minimum vacuum required with loss not to exceed 10-inch vacuum in 10 minutes. If a minimum of 22-inch vacuum cannot be attained, priming device or system may be defective or leak is too large for primer to overcome (i.e., valve open). Loss of vacuum indicates leakage and could prevent priming or cause loss of prime.
	Attempt above dry prime and shut engine off; audible detection of leak is often possible.
	Connect suction hose from hydrant or the discharge of another pumper to pressurize pump with water and look for visible leakage and correct. A pressure of 100 psi should be sufficient. Do not exceed pressure limitations of pump or pump accessories or piping connections.
	Check pump packing during attempt to locate leakage. If leakage is in excess of recommendations, adjust accordingly, following instructions outlined in pump manual.

Table 7-2 Midship Service Diagnostic Chart
Courtesy of *Hale Fire Pump Company*
(Continued)

Condition	Possible Cause	Suggested Correction
INSUFFICIENT PUMP CAPACITY	Insufficient engine power	Engine power check or tune-up may be required for peak engine and pump performance.
		Engine linkage not opening throttle fully.
	Transfer valve not in proper VOLUME position	Place transfer valve in VOLUME position (parallel) when pumping more than ⅔ rated capacity (does not apply to single-stage pumps). For pressure above 200 psi, pump should be placed in PRESSURE (series) position.
	Relief valve set improperly	If relief valve control is set at too low a pressure, it will allow relief valve to open and bypass water. Reset relief valve control, per recommended procedures, to correct pressure requirements.
		Other bypass lines (i.e., foam system, inline valves) may reduce pump capacity or pressure.
	Engine governor set incorrectly	Engine governor, if set for too low a pressure setting when on automatic, will decelerate engine speed before desired pressure is attained. Reset governor according to recommended procedures.
	Apparatus transmission in wrong gear or clutch is slipping	Recheck pumping procedures for recommended transmission gear or range. Utilize mechanical speed counter on pump panel to check actual speed against possible clutch or transmission slippage or inaccurate tachometer. (Check manual for proper speed counter ratio.)
INSUFFICIENT PRESSURE	Check similar causes for insufficient capacity	Recheck pumping procedures for recommended transmission gear or range. Utilize mechanical speed counter on pump panel to check actual speed

Cause	Remedy
Transfer valve not in PRESSURE position	against possible clutch or transmission slippage or inaccurate tachometer (check manual for proper speed counter ratio). For desired pump pressures above 200 psi transfer valve should be in PRESSURE position. Does not apply to single-stage pumps.
Impeller blockage	Blockage in the impeller can prevent loss of both *capacity and pressure.* Backflushing of pump from discharge to suction may free blockage. Removal of one-half of pump body may be required to remove blockage.
Worn pump impellers and clearance ring	Worn impellers and clearance (wear) rings will reduce both pump capacity and pressure. Installation of new parts required, considered major repair.
Impeller blockage	Same as impeller blockage above.

ENGINE SPEEDS TOO HIGH FOR REQUIRED CAPACITY OR PRESSURE

Cause	Remedy
Worn pump impellers and clearance rings	Installation of new parts required, same as above Worn pump impellers and clearance ring.
Blockage of suction hose entry	Clean suction hose strainer of obstruction and follow recommended practices for laying suction hose—keep off the bottom of the water supply but at least 2 feet below the surface of water.
Defective suction hose	Inner liner of suction hose may collapse when drafting and is usually undetectable. Try a different suction hose on sample pump test mode for comparison against original hose and results.

Table 7-2 Midship Service Diagnostic Chart
Courtesy of Hale Fire Pump Company
(Continued)

Condition	Possible Cause	Suggested Correction
	Lift too high, suction hose too small	Higher than normal lift (10 feet) will cause higher engine speeds, high vacuum, and rough operation. Larger suction hose will assist above condition.
	Apparatus transmission in wrong range or gear	Check recommended procedures for correct transmission selection.
RELIEF VALVE DOES NOT RELIEVE PRESSURE WHEN VALVES ARE CLOSED	Incorrect setting of control valve (Pilot valve)	Check and repeat proper procedures for setting relief valve system.
	Relief valve inoperative	Possibly in need of lubrication. Remove relief valve from pump, dismantle, clean, and lubricate. Weekly use of the relief valve system is recommended.
RELIEF VALVE DOES NOT RECOVER AND RETURN TO ORIGINAL PRESSURE SETTING AFTER OPENING VALVES	Dirt in system causing sticky or slow reaction	Relief valve dirty or sticky. Follow above instructions for disassembling, cleaning, and lubrication.
		Blocked; clean with small wire or straightened paper clip.

RELIEF VALVE OPENS WHEN CONTROL VALVE IS LOCKED OUT	Drain hole in housing or piston blocked	Clean hole, same as above.
UNABLE TO ATTAIN PROPER SETTING ON RELIEF VALVE	Wrong procedures	Check instructions for setting relief valve control and reset.
	Blocked strainer	Check and clean strainer in supply line from pump discharge to control valve. Check schematic in pump manual for exact location. Check and clean tubing lines related to the relief valve and control valve.
	Foreign matter in control valve	Remove and clean.
	Hunting condition	Insufficient water supply coming from pump to control valve—check strainer in relief valve system.
		Foreign matter in control valve—remove and clean.
EM GOVERNOR THROTTLE CONTROL KNOB DIFFICULT TO TURN	Dampening needle blocked	Remove dampening needle and clean capillary tube per "governor service" in pump manual.
	Governor unit dirty and in need of lubrication	Remove, clean, and lubricate per manual.
		Clean panel strainer and replace filter. Weekly use of governor system is recommended.
GOVERNOR SHAFT DOES NOT RETURN TO SHUT DOWN POSITION (IDLE)	Governor unit dirty and in need of lubrication	Remove, clean, and lubricate per manual.
		Recommend weekly use of governor system for best operation.

Table 7-2 Midship Service Diagnostic Chart
Courtesy of Hale Fire Pump Company
(Continued)

Condition	Possible Cause	Suggested Correction
UNABLE TO SET GOVERNOR ON AUTOMATIC AND RETAIN REFERENCE PRESSURE	Leak in system	Check for leaks; external leaks will be visible, internal leaks. ("O" rings) will cause pressure loss in reference system; replace sealing rings.
	Reference tank filled with water	Drain reference tank. Reference tank should have independent drain and must be drained with same. Multi-drain valve will not drain reference tank.
	Improper procedure used	Reset governor on automatic. Caution: Wait a full 3 seconds before pulling actuator; longer if engine is slow responding to throttle setting.
GOVERNOR SLOW RESPONDING ON AUTOMATIC SETTING	Dampening needle out of adjustment	Check manual for proper adjustment of dampening needle.
	Faulty throttle cable or linkage	Check throttle cable for proper clamping—see manual for recommendation.
GOVERNOR WILL NOT STOP HUNTING	Improper adjustment of dampening needle	See governor operating and instruction in pump manual for proper adjusting.
	Check valve ball missing	Check unit for check valve ball and replace if missing.
LEAK AT PUMP PACKING	Adjust pump packing	Follow procedures in manual "packing adjustment" (8 to 10 drops per minute of leakage preferred).

	Replace pump packing	Follow pump manual for replacement of packing. Packing replacement is recommended every 2 or 3 years.
WATER IN PUMP GEAR BOX (Midship only)	Leak coming from above pump	Check all piping connections and tank overflow for the possibility of spillage falling directly on pump gear box.
	Excessive leakage at pump packing	Follow above procedures for adjustment or replacement of packing. Excess packing leakage permits the flushing of water over the gear box casing to the input shaft area. Induction of this excessive water may occur through the oil seal or speedometer connection. Inspect and replace oil seal if necessary. Check speedometer connection cap and tighten if necessary. Install modification of additional slinger and front bearing cap if desired; available from factory (not required on models after 1972). Drain contaminated oil from gear box, flush with lighter oil (SAE30), drain again, and replace with SAE-EP-90 gear oil.
DISCHARGE VALVES DIFFICULT TO OPERATE	Lack of lubrication	Recommend weekly lubrication of discharge and suction valve. Use good grade petroleum base grease or silicon grease.
	Valve in need of more clearance	Add gasket to valve cover (per manual). Multi-gasket design allows additional gaskets for more clearance and free operation.
		Note: Addition of too many gaskets to valve will permit leakage.
REMOTE CONTROL DIFFICULT TO OPERATE	Lack of lubrication	Lubricate remote control linkages and collar with oil.

REVIEW QUESTIONS

1. What are the three general classifications of fire pumps used on fire apparatus?
2. What are main pumps?
3. What are booster pumps?
4. What are priming pumps?
5. What is a positive displacement pump?
6. How is the amount of water that can theoretically be discharged from a positive displacement pump determined?
7. What is slippage?
8. What is the relationship between slippage and pressure?
9. What is meant by the volumetric efficiency of a pump?
10. How is the percentage of slip determined?
11. What formula is used to determine the volumetric efficiency of a pump?
12. What are the two types of positive displacement pumps used in the fire service?
13. What are the three basic types of rotary pumps in current use on fire apparatus?
14. What is the heart of a centrifugal pump?
15. What is an impeller?
16. What is the volute?
17. What are the three factors involved in a centrifugal pump operation that have such a close interrelationship that a change in one will automatically result in a change in another?
18. If the speed of a centrifugal pump is doubled, what effect will this have on the discharge?
19. What is the relationship between pressure and pump speed when one is pumping at a constant quantity?
20. What part does slippage play in centrifugal pumps?
21. What does a single-stage centrifugal pump consist of?
22. What are multi-stage pumps?

Fire Pumps

23. What does the number of stages in a multi-stage centrifugal pump refer to?

24. What is the inherent advantage of a multi-stage centrifugal pump over a single-stage centrifugal pump?

25. What is a series-parallel pump?

26. How much pressure will a series-parallel pump develop when operating in the series position as compared to the pressure in the parallel position, when the pump is turning at the same speed in both positions?

27. When a series-parallel pump is pumping in parallel, what portion of the total quantity of water discharged does each impeller produce?

28. What piece of equipment does a pump operator manipulate to change a series-parallel pump from the series to the parallel position?

29. Which would be the most hazardous when pumping at high pressure—to change from series to parallel, or from parallel to series? Why?

30. What are the two general rules regarding when to pump in series and when to pump in parallel?

31. In what position should the transfer valve be kept when the apparatus is parked in quarters?

32. What causes cavitation?

33. At what point in a centrifugal pump does the lowest pressure normally occur?

34. What happens when a pump is cavitating?

35. In which pump will cavitation cause damage more quickly—a centrifugal or a rotary gear?

36. What are some of the indications a pump operator has that a pump is cavitating?

37. What is a good method for a pump operator to use to keep a pump from cavitating when working from a hydrant?

38. What are some of the conditions that will contribute to running away from the water when one is operating from a hydrant?

39. When one is operating from draft, what is the most reliable indicator of water runaway?

40. What are several conditions that will contribute to a pump's running away from its water when it is operated from draft?

41. How is the work capability of a pump expressed?

42. How is the pounds-gallons capability of a pump determined?

43. What formula is used to determine the amount of water a pump can theoretically discharge, once its rated pressure has been exceeded?

44. In actual practice, once the rated pressure of the pump has been exceeded, most pumps are not capable of delivering their total theoretical quantity of water. Why?

45. What are the three separate ratings that pumps are given?

46. What is net pump pressure?

47. How is the net pump pressure determined when one is working from a hydrant?

48. How does a series-parallel three-stage centrifugal pump work?

49. For what purpose was a series-parallel four-stage pump developed?

50. What is a duplex multi-stage pump?

51. What is the minimum size of pump suitable for a main pump?

52. What size are the majority of the main pumps installed on new apparatus?

53. Where are the two locations for mounting main pumps?

54. Where are midship pumps mounted?

55. On an aircraft crash apparatus would it be better to install the main pump behind or ahead of the engine transmission? Why?

56. What is the maximum capacity for a booster pump?

57. How long should pumps be expected to last without any major repairs?

58. What are some of the variables that affect a pump's life span?

Fire Pumps

59. When is a pump operator most likely to encounter abrasives in the water?

60. How can a pump operator avoid heat buildup within a pump?

61. When should a pump be flushed after use?

8

Pump Accessories

For the purposes of this text, pump accessories will be considered those pieces of equipment or devices, other than the pump, that:

1. A pump operator must manipulate during pumping operations
2. Assist the pump operator in determining the operational readiness of the engine and the pumping equipment
3. Contribute to the overall efficiency of pumping procedures

Pump accessories includes priming devices, pressure control devices, pumping devices, monitoring devices, and automation devices. Most of the pump accessories or controls are located on, or adjacent to, the pump operator's panel.

THE PUMP OPERATOR'S PANEL

The *pump panel* is the pump operator's operating base during pumping situations. From here, the pump operator should be able to set pressure control devices, operate priming devices, open and close all gated intake and discharge lines, and keep track of pump intake pressure, pump discharge pressure, oil pressure, engine temperature, and information from other monitoring devices. It is also desirable that pump drain controls be located on, or in close proximity to, the pump operator's position and that the pump operator be able to receive and transmit radio messages from this location. All pump accessories should be designed and installed so that they are adequately protected against mechanical injury and weather, Proper illumination should be provided for night operations.

The number and types of pump accessories will vary from apparatus to apparatus. Generally, front-mount pumps use fewest accessories, sometimes confining the number to a few pump controls and an intake and discharge gage. Most apparatus with midship pumps have well-designed, well-lighted, and well-equipped pump panels (see Figure 8-1).

Pump panels are generally located on the driver's side of the apparatus, immediately behind the door to the cab. The advantage of this location is that the operator can step from the driver's seat to the ground and be in position to commence pumping operations. This is particularly advantageous when water is pumped directly from the tank into small lines.

Two disadvantages of this particular location are the poor visibility of lines taken off discharge gates on the opposite side of the apparatus and the safety of the operator on traveled streets. Several manufacturers have tried to compensate by locating the pump panel in an elevated position, midship on the apparatus (see Figure 8-2). From this position, the pump operator has a fairly good view of the entire fireground situation.

Pump Accessories

Figure 8-1 A Pump Operator's Panel. *Courtesy of Mack Trucks, Inc.*

Figure 8-2   A Top-Mounted Pump Operator's Panel. *Courtesy of W.S. Darley & Company*

PRIMING DEVICES

Pump operators take water from two completely different types of sources. One type of source provides water to the pump under pressure and is commonly referred to as a *pressure source*. The most common pressure source is the fire hydrant. The apparatus pump also can take water from the booster tank, which provides pressure as a result of its elevation of a foot or so above the pump.

The other type of source requires a pump operator to draft. Drafting sources include ponds, streams, swimming pools, tidewaters, and so on. In drafting operations, water moves from the source into the pump because of a difference between the atmospheric pressure and the vacuum created within the pump by a priming device. A more thorough explanation of drafting operations will be given in the chapter on engine company operations; at this point, it is only necessary to know that priming refers to the process of replacing air with water in the intake lines and passageways of the pump.

Pump Accessories

Positive Displacement Priming Pumps

Rotary positive displacement pumps are used on fire apparatus as priming pumps. Both *rotary vane* and *rotary gear* priming pumps are currently in service. These pumps are capable of moving air as well as water; therefore, they can cause a vacuum within the main pump.

The vacuum in the main pump is measured in inches of mercury (Hg). One Hg is equivalent to a negative pressure of .49 psi, or about one-half of one pound per square inch. In round figures, a vacuum of 10 inches Hg would be equivalent to a negative pressure of about 5 psi. The general requirement is that priming systems be capable of providing a vacuum of at least 22 inches Hg at sea level. Most rotary-type priming pumps in good condition are capable of producing a vacuum of at least 26 inches Hg.

A vacuum is said to exist whenever the pressure within the main pump is below that of the surrounding atmosphere. The amount of vacuum is expressed in inches of mercury. A pump whose suction gage reads 20 inches Hg would be said to be pulling 20 inches of vacuum.

Figure 8-3 shows a schematic of a Waterous priming system. The Waterous system is a typical system and is used to illustrate the components and principles involved. The system consists of a priming pump, a priming valve, and a priming tank.

The Priming Pump

Figure 8-4 shows an exploded view of a rotary gear priming pump. Figure 8-5 shows a schematic of the gears within the pump casing. The pump contains two seven-toothed rotors that mesh in the same manner as a pair of spur gears. During operation, oil from the auxiliary priming tank (see Figure 8-3) provides lubrication and an airtight seal between the rotors and the head plates. The rotor teeth unmesh at the intake side of the pump, trapping air between the teeth of the rotors and the pump casing. As the rotors turn, the air trapped between the rotor teeth and the casing is moved from the intake side of the pump to the discharge side. The rotor teeth mesh at the discharge side of the pump, forcing the air from between them to the pump discharge opening. As the prim-

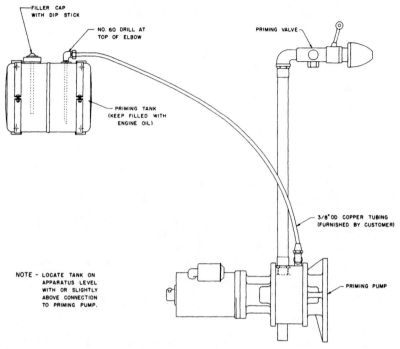

Figure 8-3 A Schema of a Waterous Priming System. *Courtesy of Waterous Company*

ing pump evacuates air from the intake lines and the main pump, atmospheric pressure forces water into the pump.

Priming pumps are driven by one of three methods—by an electric motor, by gears, or by a V belt. The particular drive used depends on the type of pump, the type of pump transmission, the type of priming valve, and the needs of the user. Some of the priming pumps employ a back-up drive system. For example, an electric pump may have a drive gear available in the event of electrical failure.

The Priming Valve

Figure 8-6 is a schematic of a *priming valve*. The valve is installed in the line that connects the main pump with the priming pump. It is normally carried in the closed position. When the

Pump Accessories

valve is opened during priming operations, it allows air to flow from the various main pump passages to the priming pump, which discharges it into the atmosphere. After the main pump is primed, the priming valve is closed to prevent air from leaking into the main pump while it is in operation. If for some reason the priming valve does not fully close or the valve is faulty, water will probably work its way through the priming pump eventually and leak out the priming pump discharge outlet after the pump has been disengaged.

The Waterous valve is mounted on top of the main pump, and consists essentially of a stainless steel sliding plunger (A) operating in a bronze valve body (B) containing three inlet ports (C) and an outlet port (D). A rubber insert on the plunger (E) is forced against the seat in the body when the valve is closed, assuring a tight seal between the inlets and the outlet.

The purpose of the three inlet ports is to permit the pump operator to tap into the main pump at three different locations concurrently. Two of the ports are used for exhausting air, while

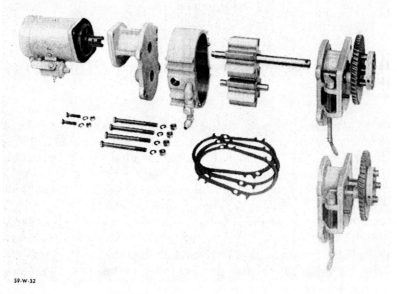

Figure 8-4 An Exploded View of a Waterous Rotary Gear Priming Pump. *Courtesy of Waterous Company*

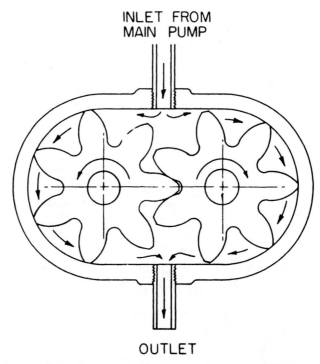

Figure 8-5 A Rotary Gear Priming Pump. *Courtesy of Waterous Company*

the third is used as a safety feature on the priming valve. A relief groove in the valve body connects one inlet port to the area behind the plunger. The pressure at the rear of the plunger is at least equal to that on the opposite end of the plunger when the main pump is operating. This prevents the priming valve from opening accidentally, provided the valve has been properly installed.

Any leakage in the priming valve would result in an air leak, which could cause the main pump to lose its prime once the priming pump is disengaged. Generally the first sign of leakage is water running out of the priming pump when the pumper is connected to a positive pressure source.

Three different systems are currently being used to activate priming valves—solenoid-actuating systems; manual actuating systems; and combination solenoid-manual actuating systems.

The Priming Tank

The *priming tank* is normally mounted on the apparatus in a readily accessible location in order to facilitate checking and refilling. The tank should be mounted level with, or slightly above, the intake connection on the priming pump. During operation, oil from the priming tank lubricates the pump rotors and other moving parts and provides an airtight seal in the pump chamber. A vent hole in the outlet elbow of the oil reservoir prevents oil from being siphoned off when the priming pump has stopped running. It is important that this vent hole be kept open at all times. If oil continues to be discharged from the priming pump discharge outlet after the priming pump has been disengaged, it is probably due to the vent hole becoming plugged.

It is recommended that the main pump be capped and that all discharge and drain valves be closed after each operation. The priming pump should then be run until several drops of oil come out of the discharge pipe. The priming tank should be refilled after the priming pump has been shut down.

Priming Operations

Several operations must be performed to properly prime the main pump.

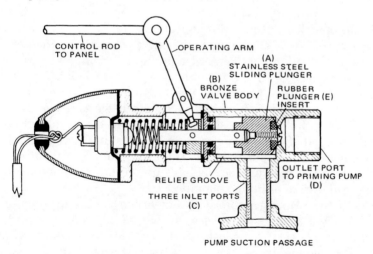

Figure 8-6 Priming Valve Diagram. *Courtesy of Waterous Company*

1. All discharge valves, drain valves, and all other openings must be closed.
2. Care must be taken to ensure that all intake connections are tight.
3. The priming valve in the line between the main pump and the priming pump must be opened.
4. The priming pump must be engaged.

Items 3 and 4 can be performed either as a two-step process or simultaneously, depending on the particular pumping configuration on the apparatus. Some pumps are arranged so that the operator must open the priming valve and engage the priming pump separately. Other pumps are designed so that the priming valve is opened and closed with the engagement and disengagement of the priming pump.

On most pumpers that use a solenoid-actuated priming valve, depressing the priming switch button on the control panel simultaneously opens the solenoid-operated valve and starts the priming pump motor. The priming valve can be opened manually if the solenoid should, for any reason, fail to open it.

A rod leading out of the control panel operates the priming valve on those pumpers equipped with a manual-electric priming valve. As the rod is pulled (or pushed, if so installed), it closes a switch mounted on the valve to start the priming pump motor.

The priming valve should be kept open and the priming pump operated until a solid stream of water is flowing from the priming pump discharge pipe. The first discharge from the pipe is a mixture of air and water. A false prime will result if the priming pump is disengaged at this point. It would be possible to engage the main pump and build up pressure on the discharge gage, but the pump would lose its prime once a discharge gate was opened.

The priming pump should be disengaged and the priming valve closed after a solid stream of water is discharged from the priming pump discharge pipe. The main pump can then be engaged, or the pressure can be built up on those pumps that are engaged during priming operations.

If the priming pump does not prime the main pump and discharge a steady stream of water within 30 seconds, stop the pump and check for air leaks. Forty-five seconds are allowed for pumps having a capacity of 1500 gpm or greater. Care should be taken to make sure that oil from the priming tank lubricates the priming pump.

PRESSURE CONTROL DEVICES

The conditions under which a pump operator functions are many and varied. He may pump at times through small lines and at other times into heavy-stream appliances that will place a maximum demand on his pump and skills.

A pump operator pumping a single 2½-inch line will determine the required engine pressure and then adjust the throttle to turn the pump at the speed required to produce the necessary pressure. If the line is suddenly shut down, the load on the engine will be instantly relieved, causing a sudden increase in the pump's speed.

The increase in pump speed would result in a pressure increase in the second line if the pump operator happened to be pumping through two 2½-inch lines. This increase in pressure could burst a hose line, placing people at the end of the line in extreme danger if they happened to be in a critical spot. If the line did not burst, the rapid change in operating conditions could result in mechanical damage—or, worse still, the increased pressure could actually cause the workers at the end of the line to lose control of the nozzle, in which case it could whip around with killer force. The primary concern of fire officials in these situations is to ensure that personnel are not injured. A secondary concern is that the equipment not be damaged. Pressure control devices are employed to ensure that neither occurs.

Two types of pressure control devices are used in the fire service—*relief valves* and *pressure governors*. Relief valves control pressure by opening a bypass between the discharge and suction sides of the pump. Pressure governors control the pressure by controlling the engine throttle setting. Most relief valves and pressure governors are designed to operate through a pressure range of approximately 75 psi to 300 psi, and to hold the pressure within 30 psi of the pressure setting if all lines are shut down; however, a Hale relief valve operates up to 600 psi. The relief valve can be used on either positive displacement pumps or centrifugal pumps. Pressure governors can be used only on centrifugal pumps. Personal preferences by fire officials for one system or another usually depend upon the conditions of operation and upon how well their present system works under these conditions.

Relief Valves

Relief valves are used in most pumping systems to prevent excessive pressures caused by changes in the conditions under which the system is operating. There are many different kinds of relief valves, and each functions in a slightly different manner, depending upon the type of pump, the type of power turning the pump, and the operating conditions. Regardless of the type, each relief valve works to prevent pressures from going above the desired setting by bypassing water from the discharge side of the pump back to the suction side of the pump. This method of controlling the pressure keeps both the pump and the engine under a constant load.

Relief valves are an absolute necessity on positive displacement pumps. Without a relief valve, the resultant pressure buildup would stall the engine any time one or more discharge valves were closed while the pump was operating. If the pump was being driven at a high speed or by a large engine, the pump and related equipment might be seriously damaged before the engine stalled out.

Relief valves control the pump pressure on centrifugal pumps by preventing the engine from racing if one or more discharge valves are closed. In addition, the relief valve prevents high-pressure surges which might burst hoses on open lines or injure the people manning the hoses. A relief valve thus serves as both a pressure regulator and a safety valve.

Fire department relief valves designed for use with main pumps employ a minimum of two components. One component bypasses the water and is referred to as the *bypass valve*. The other component is mounted on the pump control panel so that the pump operator can set the desired pressure; this is a *pilot control*.

Pump Accessories

Some relief valves combine these two components into one unit; others separate the components.

There are probably as many different types of relief valves in service as there are pump manufacturers. Consequently, it is extremely difficult to make comparisons. The Waterous system is one that employs separate components. This system will be used to illustrate the principles involved in the design and operation of a relief valve. The illustrations and explanations are provided by courtesy of the Waterous Company.

Most fire department personnel are familiar with the way a simple spring-loaded relief valve works. Everything depends on the force of the spring holding the valve closed. Figure 8-7 is a drawing of an ordinary pipe with a simple spring-loaded valve installed in a hole in the side. The spring tension can be varied by turning a handle in and out. The spring can be set so that 100 psi will compress it enough to open the valve. When water pressure in the pipe reaches 100 psi, the valve opens, bypassing water and relieving the pressure. In order to bypass more water, it will take more pressure inside the pipe to open the valve farther, because the farther the spring is compressed, the more resistance it offers.

Sometimes all that is needed is a simple device like this. For more exacting control, especially at higher pressures and capacities, a device must be added that can control pressures using

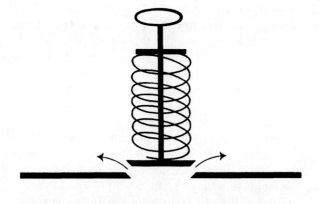

Figure 8-7 A Simple Spring-Loaded Relief Valve

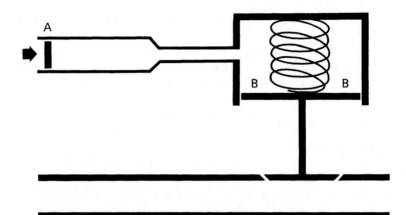

Figure 8-8 The Addition of Two Hydraulic Cylinders to the Simple Spring-Loaded Relief Valve

hydraulic force as well. In Figure 8-8, two hydraulic cylinders (A and B) have been added to the system. Through use of a small piston (A), the same pressure is applied to the much larger area on top of the cylinder in the big relief valve (B). The small cylinder provides a big hydraulic advantage in controlling the main valve—just as the hydraulic brakes in a car do. This means that the valve can be set to bypass large volumes of water with only a slight pressure rise; one need only greatly reduce the pressure behind the piston in the main valve (B). But this control is still strictly manual. An operator must push in the piston on the small cylinder to increase the force against the top of the big cylinder, and must pull out the small piston to decrease the force. At this point, the system demands constant attention to the pilot control.

It is necessary to design into the system some means of eliminating the need for constant attention. The system must be made to operate automatically once it is set.

One method uses the water pressure or hydraulic force from the pipe to operate the whole system. Such a unit also needs a system that will, at the same time, bypass large quantities of water with only a small pressure rise in the pipe.

Figure 8-9 shows the addition of a tube (A) from the main pipe, with a constriction in it that is connected to the back (top) of the large relief valve cylinder. If left as shown, the valve would

Pump Accessories

always stay closed, because water at the same pressure as that in the pipe is also pushing against the much larger area on the back or top side of the big relief valve. However, another tube (B) is added to permit water to flow out of the system, or "dump." Water is metered through the constriction (A) into the pressure chamber of the relief valve and is let out through the other pipe (B). This second pipe has a needle valve (C) on it to control the rate of escape. In fact, this valve can be set so that water can escape from the cylinder more easily than it can enter through the constriction. Now suppose the needle valve is set so that water is flowing through the constricted pipe to the back side of the piston at a rapid rate. The faster the water flows through the constriction, the greater the pressure drop there, and the lower the pressure behind the piston in the large relief valve.

Merely by adjusting the needle valve setting (and thus the rate at which water is dumped), the hydraulic force behind the piston can be varied, so that the relief valve will open part or all the way at any predetermined pressure in the pipe. However, at this point the system is still manual. Someone must keep a sharp eye on the gages and keep adjusting the needle valve to compensate for pressure changes in the pipe. In the pilot, or control, valve of the Waterous relief valve system, the modulating of the needle

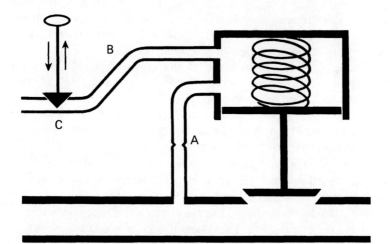

Figure 8-9   The Addition of a Central Mechanism

332 Introduction to Fire Apparatus and Equipment

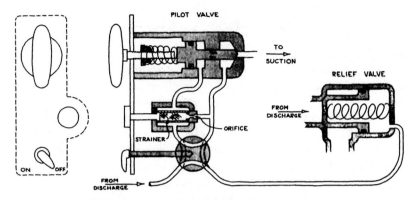

Figure 8-10 A Schema of the Entire System

valve which dumps water to the pump intake is done automatically through changes in discharge pressure. The principle of operation, however, is roughly the same.

The spring in the main relief valve plays only a small part now. The hydraulic force behind the main valve does most of the job, and this force is controlled by the amount of water dumped.

The relief valve proper is mounted on the pump between the intake and discharge, and the pilot or control valve is mounted on the operator's panel of the apparatus. This separation of the two valves makes the whole system more flexible and easily adaptable to most mounting situations. The pilot valve, with its four-way valve, controls the operation of the relief valve.

Figure 8-10 is a schematic of the entire system. The pilot valve on the left is shown in two views in order to provide a better look at the four-way ON-OFF valve. The relief valve is shown on the right. It should be noted that the needle valve can move away from its seat (to the left) in response to an increase in discharge pressure, to allow water to flow back to the suction.

Figure 8-11 shows how the relief valve fits onto the pump between the intake and the discharge. The arrows show the direction of water flow when the valve is open and bypassing.

Figure 8-12 shows what happens when the four-way valve is turned to the OFF position. This takes the entire relief valve system out of operation, which a pump operator would want to do if he intended to pump at pressures at above 300 psi. Water from the

Pump Accessories

pump discharge goes directly through the four-way valve to the main valve chamber, which is on the right-hand side of the main valve in this illustration. The pressure is the same on both sides of the main valve, but the area is greater on the main valve chamber end (right), so the force is greater from that end. This greater force, plus the force of the spring, keeps the valve closed, so there is no flow back to the pump intake.

Figure 8-13 shows what happens when the pump discharge pressure is *lower* than the pilot valve setting. In this illustration the four-way valve is ON, so the entire system is in operation. This time, the water flows from the pump discharge to the pressure chamber of the pilot valve on the left. The piston end of the needle valve has discharge pressure pushing it to the left, opposed by pressure exerted by a spring pushing it to the right. The pump operator can regulate the spring force by turning the handle of the pilot valve. As long as the hydraulic force on the piston is less than the spring force, the needle valve will remain closed, and

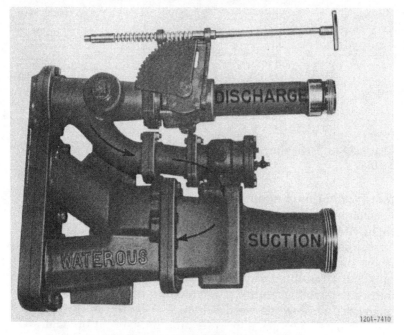

Figure 8-11 Relief Valve Water Flow

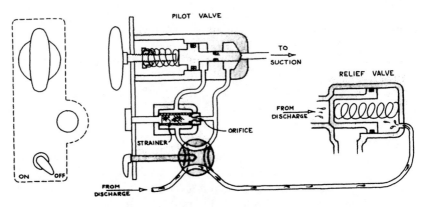

Figure 8-12 System Set in OFF Position

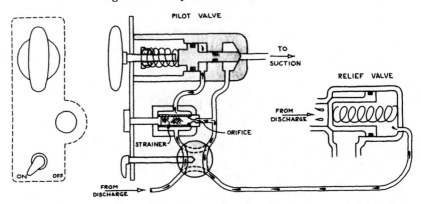

Figure 8-13 System in ON Position—Discharge Pressure Less than Pilot Setting

water cannot flow rapidly through the orifice; however, water at discharge pressure can reach the main relief valve chamber (on the right) and keep the main valve closed.

Figure 8-14 shows what happens when the pressure rises above the "setting." When hydraulic force on the piston of the pilot valve becomes greater than the spring force, the needle valve moves to the left and allows water to flow back to the suction. This increases the flow rate so that there is a significant drop in pressure across the orifice and lowers the pressure in the main relief valve chamber. The force on the small end of the main valve (left) is now greater than the force at the large end so the main valve

Pump Accessories

opens to allow part of the water being pumped to bypass back to the suction, reducing the discharge pressure. The main valve opens just enough to reduce the discharge pressure to the pilot valve setting. If the discharge pressure drops below the set pressure, the pilot valve reseats, and pressure builds up behind the large end of the main valve and closes it.

Suction Relief Valves

Suction relief valves are designed to protect the pump from excessive incoming pressures. Some pumpers are equipped with built-in valves, in which case the relieving pressure is generally set at the factory. Portable valves which can be attached to the suction inlet are also available. Portable valves are normally set in the field to relieve at the pressure deemed desirable. The principle of operation of suction relief valves is the same as that for pressure relief valves; however, with pressure relief valves the water is bypassed from the discharge side of the pump back to the intake side of the pump, while with suction relief valves the water is dumped to the ground.

Pressure Governors

Relief valves and pressure governors perform basically the same function but in two different ways. While a relief valve

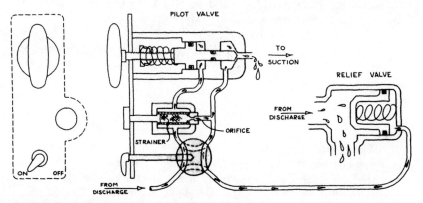

Figure 8-14 System in ON Position—Discharge Pressure Greater than Pilot Setting

bypasses water from the discharge side of the pump to the intake side of the pump, a pressure governor adjusts the discharge pressure by manipulation of the throttle control. For example, suppose a pumper is pumping through two 2½-inch lines, each discharging 250 gpm. If one line was suddenly shut down and the apparatus was not equipped with some type of pressure control device, there would be a fairly good pressure increase on the other line. There would not be a surge on the second line if the apparatus was equipped with a relief valve (and it was properly set), as water would immediately be bypassed back to the intake side of the pump. The pump and engine would continue to run at the same speed, and the pump would continue to discharge 500 gpm; however, only 250 gpm would be going to the open line, while the other 250 gpm would be circulated back to the intake side of the pump.

A pressure governor works differently. Suppose it is set at 150 psi, and the conditions are the same as they were with the relief valve. When the one line is suddenly shut down, a tube from the discharge side of the pump transmits the resulting pressure rise to some sort of device (there are at least a half dozen different types in service) which in turn cuts back the throttle. The engine and pump then run more slowly, just fast enough to supply the remaining line with 250 gpm at 150 psi.

There are, as mentioned, many different types of pressure governors in service, each working in a different manner. The Waterous governor will be used to illustrate the principles involved; the illustrations and explanations are provided through the courtesy of the Waterous Company.

The Waterous pressure controller consists of a cylinder assembly, a directional flow valve, and an accumulator. These are described in detail below.

Cylinder Assembly

The *cylinder assembly* consists of a cylinder and two heads, enclosing a hydraulically balanced piston. This entire assembly is mounted in a bracket, in which it slides when the throttle is manually adjusted. The heads have one-fourth NPT ports for connection to the directional flow valve.

Pump Accessories

Directional Flow Valve

The *directional flow valve* (DFV) is a pressure-actuated, four-way spool valve. Pump discharge pressure acts on one end of the spool and is opposed by a spring on the other. When the pump discharge pressure reaches 75 pounds per square inch, gage (psig), the spring is compressed, and the valve is actuated. When the pump pressure drops below 30 psig, the spring deactivates the valve.

A needle valve is mounted on the directional flow valve to control the system's response rate and dampen excessive pressure fluctuations.

A quarter-turn ball valve is mounted on the panel plate with the directional flow valve to place the system in or out of operation.

Accumulator Assembly

The *accumulator* is a one-gallon, two-chamber tank, with a bladder separating a gas chamber from a water chamber. The gas side of the accumulator is precharged at the factory to 75 psig.

Principles of Operation

After the pump has been primed and is operating, water flows from the pump through the line (6) and the strainer (16) to the bottom port of the direction flow valve (1), through the lower cross holes in the spool (15), and out lines (7) and (8). Water flows through line (7) to the throttle end of the cylinder, through line (8) and the ON-OFF valve (5) to the side port of the DFV. Through internal porting in the DFV, the water then flows to the needle valve (4), and through line (11) to the tube end of the cylinder. With the ON-OFF valve (5) in the OFF position (open), all parts of the controller are equally pressurized, and the piston in the cylinder remains stationary relative to the cylinder (see Figure 8-15).

When pump pressure reaches 75 psig, it moves the spool in the DFV upward against the spring force, changing the flow pattern. Because the accumulator is precharged to 75 psi, water flows into it when pump pressure exceeds that figure.

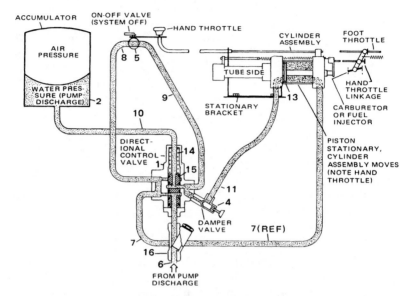

Figure 8-15 System Off (Manual Operation)—Manual Control of Pump by Hand Throttle

After the spool (15) shifts upward, water flows from the pump through line (6) to the bottom port of the DFV, through the lower cross holes in the spool, out line (9), and through the ON-OFF valve (5). The water then flows through line (8) back into the top side port of the spool, out the top of the spool, through line (10), and into the accumulator (see Figure 8-16).

When the desired pressure is reached and the pressure in the accumulator equals discharge pressure, the ON-OFF valve (5) is turned on, activating the system for automatic pressure control. Turning the ON-OFF valve (5) on actually closes the valve and isolates the accumulator (2) and the throttle side of the piston (13) from the discharge pressure on the opposite side of the piston.

As the hand throttle is opened, the cylinder (13) moves in the open direction relative to the piston. This causes a momentary increase in pressure. When the hand throttle control is fully open, the piston will be at some intermediate point between the two heads. The discharge pressure at this time will equal the pressure that had been trapped in the accumulator (see Figure 8-17).

Pump Accessories

If a discharge valve is closed, or discharge pressure rises for any reason, the pressure will rise on the tube side of the piston but will remain the same as that in the accumulator on the throttle side. The higher pressure will force the piston toward the closed position and throttle the engine down to maintain the set pressure (see Figure 8-18).

If a discharge valve is opened or the pump pressure drops for any reason, the pressure on the tube side of the piston (13) will drop below the accumulator pressure. The higher force on the opposite side of the piston will then move it toward the tube and open the engine throttle until discharge pressure again equals the preset accumulator pressure (see Figure 8-19).

If the pumps run out of water during operation, or loses prime, or if for any other reason the discharge pressure drops below 30 psig, the spring (14) will force the spool (15) down to the deactivated position. In this position, the passageways through the DFV connect the accumulator (2) with the tube side of the piston (13), and the throttle side of the piston with line (6) from discharge. Since the discharge pressure is near zero and the ac-

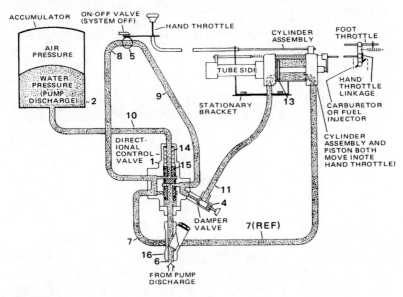

Figure 8-16 System Off (Manual Control of Pump)—System Balanced

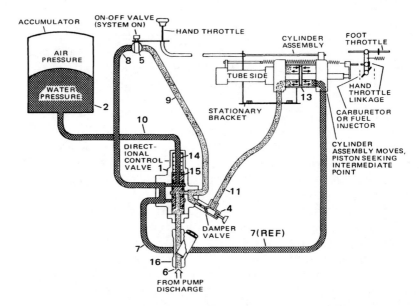

Figure 8-17 System Off (Automatic Control)—Balanced Running Condition

cumulator is still pressurized, the piston and the throttle will move toward the closed position, and the engine will return to idle (see Figure 8-20).

An Electro-Governor System

The *automatic electro-governor*, shown in Figure 8-21, is an electric control system that can be used to maintain a constant pressure over a wide range of flows. The unit shown will maintain pump pressure within 7 psi of the desired figure for pressures from 60 psi to 300 psi.

The basic system consists of two separate miniature modules that are mounted on the apparatus—a pump control module and a carburetor control module. The *pump control module* is mounted on the apparatus operator's panel. It has two controls for automatic operation; one engages the unit, while the other is a single-turn dial for selecting the output pressure. To place the system in

Pump Accessories

operation, the pump operator simply turns a dial to engage the unit, then dials in the desired discharge pressure. All internal functions take place automatically.

The *carburetor control module* is an enclosed motor that is mounted on the fire wall of the engine compartment in line with the carburetor and is connected to the carburetor. The system is able to maintain a constant output pressure by the movement of this assembly, which adjusts the carburetor setting.

The system is equipped with a semi-automatic control, in addition to the automatic mode of operation. When this type of control is selected, the operator pushes the increase or decrease switch, which causes the carburetor setting to continue to move in the desired direction as long as the switch is engaged.

An automatic safety feature is built into the system. The governor will cause the engine to idle any time the pump runs out of water. The governor will return the pump to the desired pressure once the pump is resupplied with water.

An additional feature of the system is a remote control capability. The system can be supplied with a 100-foot remote con-

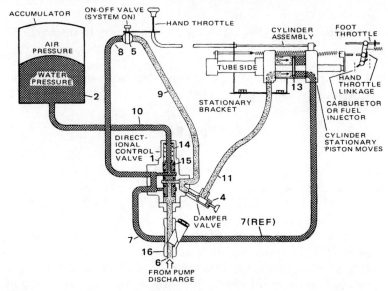

Figure 8-18 System On (Automatic Control)—Discharge Pressure Increases

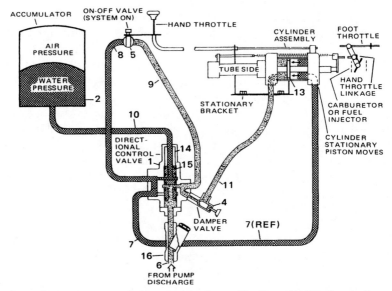

Figure 8-19 System On (Automatic Control)—Drop-In Discharge Pressure

trol unit that enables the pump operator to regulate the system from within a distance of 100 feet of the apparatus.

Combination Relief Valve

Hale Fire Pump Company markets a combination pressure control system which offers the advantages of both a recirculating relief valve or engine governor and a suction relief valve. It is designed to control the pressure on both the discharge and inlet sides of the pump.

A conventional recirculating relief valve controls discharge pressure by dumping excess water back to the inlet side of the pump. This is satisfactory provided the intake pressure is considerably less than the discharge pressure; however, this may not be the case when a pumper is connected to a strong hydrant or if the pumper is part of a relay operation.

A governor controls discharge pressure by reducing the

Pump Accessories

speed of the engine; however, this system may also be ineffective in compensating for every pumping situation.

An internal suction relief valve partially compensates for excessive incoming pressure by dumping water to the ground; however, it has its limitation as the relief pressure is usually preset and cannot be easily adjusted to compensate for every pumping situation.

Hale's total pressure system utilizes a sensing device on the inlet side of the pump which works in conjunction with a pressure master control on the pump panel to give complete control over the entire system through the regulation of the pump master control (see Figure 8-22). Small changes in pump discharge pressure are normally handled internally by the recirculating relief valve, while large changes on either the inlet or discharge side of the pump are controlled by dumping excess pressure to the atmosphere from the discharge side of the pump.

The system is designed so that the inlet and discharge sides of the pump can be set to relieve at different pressures. For instance, during an initial relay hookup the system can monitor inlet pressure and not allow it to exceed the pressure setting of the

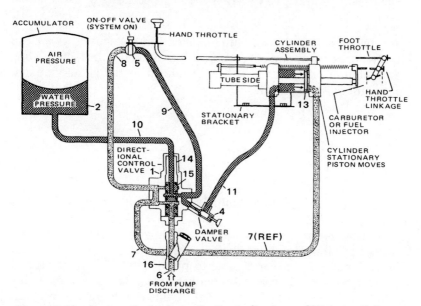

Figure 8-20 System On (Automatic Control)—Loss of Water or Prime

344   Introduction to Fire Apparatus and Equipment

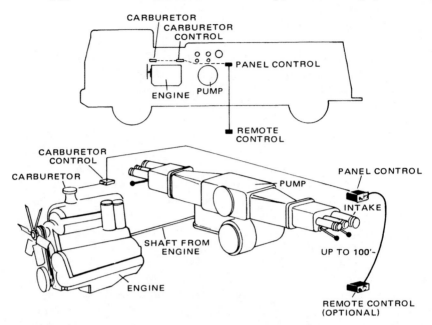

Figure 8-21   An Electro-Governor System. *Courtesy of Fire Research Corporation*

sensor. Once the relay hookup has been completed, the system can be switched over to total system control, and both the recirculating and suction relief valves will be regulated by the panel-mounted pressure master control.

PUMPING DEVICES

Main Pump Transmissions

The *pump transmission* serves a twofold purpose. It connects the pump impeller shaft to the engine power train, and it establishes the proper engine-to-pump gear ratio. Centrifugal pumps use a single gear ratio between the pump and the engine, while some positive displacement pumps incorporate several different ratios. Centrifugal pumps are high-speed pumps and, there-

Pump Accessories

fore, turn faster than the engine. Rotary and piston pumps are low-speed pumps and invariably operate at speeds lower than that of the engine.

The transmission of gear-driven centrifugal pumps generally contains at least a drive gear, an idler gear, and a driven gear. For many years these were spur gears. Spur gears are gears which are cut parallel to the shaft. Spur gears perform the job effectively but are extremely noisy; consequently, many pump manufacturers have switched to helical gearing. A helical gear is one in which the gear teeth are cut at an angle to the shaft. Helical gearing provides a greater tooth-to-tooth contact and also achieves a more uniform load transfer, because several teeth are always in contact.

At least one manufacturer provides a chain drive transmission. The chain drive transmission was developed to satisfy the demand for smoother, quieter pumps with greater power and larger capacity.

Pump transmissions are mounted either in front of or behind the road transmission. The pump transmission operates independently of the road transmission when it is mounted in front. The power of the engine is taken from the clutch and transmitted directly to the pump transmission. The pump transmission then delivers power to the pump drive shaft when the pump transmis-

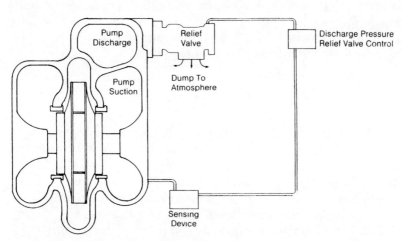

Figure 8-22 A Combination Relief Valve System. *Courtesy of Hale Fire Pump Company*

sion selector lever is in the pump position. The operator must perform two steps to put the pump in operation. First, he must place the road transmission in the neutral position, thereby disengaging the road transmission from the rear wheels and taking it completely out of the power train. Then the operator must properly engage the pump transmission.

A slightly different procedure is used when the pump transmission is located behind the road transmission. In this case, the power of the engine is transmitted through the road transmission to the pump transmission by the selection of gears. The road transmission then becomes part of the power train to operate the pump. The pump operator must perform three operations with this type of arrangement (not necessarily in the following order):

1. Disengage the rear wheels. This is usually done by changing a control level or electrical switch from the road position to the pump position.
2. Place the road transmission in the proper gear (usually direct drive).
3. Properly engage the pump transmission.

The pump transmission is engaged either electrically or with a pump gear shift handle. Some engaging controls are located in the cab; others are located at the pump operator's panel. The control handle usually has either two or three positions, depending on the location of the pump transmission, the type of priming pump engagement, and the general pumping configuration. The positions generally incorporated include:

1. Road position
2. Pump position
3. Neutral

What happens inside the pump transmission depends on a number of factors. For purposes of illustration, let us consider a transmission that is mounted in front of the road transmission and that utilizes only two positions on the gear shift handle. The two positions on the handle are road position and pump position.

The only thing that moves inside the gear box is a sliding shift collar. Figure 8-23 shows the collar in the road position. All the teeth on the shift collar are engaged with all the teeth on the coupling shaft, transmitting full engine power to the rear wheels.

Figure 8-24 shows the gear box in the pump position. The

collar has moved forward, engaging with all the teeth on the hub of the drive gear and transmitting the engine power up through the idler and driven gear on the impeller shaft.

Auxiliary Cooling Systems

An internal combustion engine operates most efficiently at a uniform temperature. Although the cooling systems of various engines are designed to maintain the temperature within different limitations, and some high-compression engines operate at higher temperatures, it can generally be said that the best operating temperature of an engine is somewhere between 160°F. and 185°F. The engine's normal cooling system may be inadequate to keep the engine within its proper operating range when pumping under a heavy load. An auxiliary cooling system is often installed on the apparatus to provide additional cooling. The control for the heat exchanger is most often located on the pump operator's panel.

The *auxiliary cooler* is usually a *heat exchanger*. It may be built into the radiator assembly or designed as a separate unit. When designed as a separate unit, it is normally installed in the line from the engine outlet to the radiator top tank inlet.

Most auxiliary coolers consist of a series of tubes or cores built inside a water jacket. Water in the jacket is circulated from the discharge side of the pump and returned to the intake side of the pump. Hot water from the engine is circulated to the tubes or cores, transmitting part of the heat from the pump to the water. With some systems the water from the pump is circulated through the tubes and the water from the engine through the jacket. The result is the same.

Since the water from the pump is confined to the auxiliary cooler system, the engine coolant is not diluted or contaminated. This is extremely important, particularly when the pump is taking draft from a salt water source.

Radiator Fill System

Some pumpers are equipped with a radiator fill system which permits the operator to add water to the radiator from the pump during pump operations. The need to fill the radiator dur-

Figure 8-23 The Transmission in the Road Position. *Courtesy of Waterous Company*

Pump Accessories 349

Figure 8-24 The Transmission in the Pump Position. *Courtesy of Waterous Company*

ing this period is normally caused by the operator allowing the water level to become low prior to the commencement of pumping operations. Pump operators should ensure that the radiator is full of water at all times; however, if the need to add water during pumping operations arises, then the system should be utilized.

The system normally consists of small tubing which extends from the discharge side of the main pump to the radiator. It contains a shut-off valve at the pump operator's panel. This valve should be kept closed at all times when the system is not being used.

Care should be taken whenever this system is utilized. The engine should be running whenever water is added to the radiator if the water temperature is excessive. Water should be added slowly. Failure to keep the engine running or adding water too quickly could result in a cracked cylinder block or head. Although the radiator is normally equipped with an overflow valve designed to provide for excess water and pressure, it is best to remove the cap when water is added through the radiator fill system. Extreme care should be taken when removing the cap, as the water in the radiator is under pressure and has most likely been converted to steam, which could result in a serious burn to the person removing the cap. Use gloves or a rag to cover the cap, and twist it only the amount required to commence the release of steam pressure. The cap can normally be safely removed when the pressure has been released. The purpose of removing the cap is to prevent additional buildup of pressure in the radiator caused by the discharge pressure of water from the pump. It is best to fill the radiator when pumping at a lower discharge pressure; however, regardless of the discharge pressure, only crack the radiator fill valve rather than open it fully. This procedure will generally reduce the pressure to a controllable level.

With some systems, the manufacturer recommends that the cap not be removed. When so recommended, the manufacturer's advice should be followed.

MONITORING DEVICES

Pump Intake and Discharge Gages

Direct pressure-reading gages on the pump operator's panel are necessary in order to keep the pump operator constantly informed as to what is happening inside the pump. An intake gage measures the incoming pump pressure or inches of mercury (vacuum), while a discharge gage measures the discharge pressure. These gages are normally at least 3½ inches in diameter so that the pump operator can tell at a glance how much pressure is coming into and leaving the pump.

Pump intake and discharge gages are calibrated in pressure relative to atmospheric pressure. The intake gage is a compound gage that provides pressure readings both above and below atmospheric pressure. The discharge gage on newer apparatus is generally a straight pressure gage, providing only pressure readings above atmospheric pressure. Pressure levels below that of the atmosphere are referred to as negative pressure; those above are referred to as positive pressure.

Negative pressure on the compound gage is calibrated in inches of mercury (Hg). One inch of mercury is equivalent to .49 psi, or about a half pound of pressure. A reading of 10 inches Hg on the compound gage, therefore, refers to a pressure reduction within the pump of about 5 psi. Pressure levels below that of the atmosphere (negative pressure) are generally referred to as vacuum. The compound gage is designed to read from 0 to 30 inches of vacuum.

Positive pressure on both the compound gage and the discharge gage is calibrated in pounds per square inch. Both gages are designed to read at least 300 psi—that is, at least 300 psi above atmospheric pressure. Most gages on newer apparatus are calibrated to read at least 600 psi.

The reduction of atmospheric pressure is approximately one-half psi per 1000 feet of elevation. The overall effect on fire department operations is that the pump gages will no longer be indicating the correct pressure when apparatus are taken to elevations above or below that for which the gages have been calibrated. For example, if a pumper is taken from sea level to 5000 feet, the

352   Introduction to Fire Apparatus and Equipment

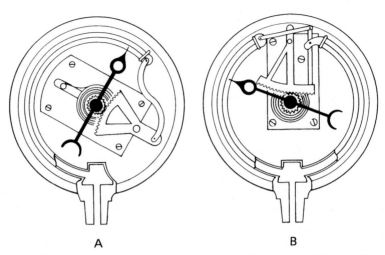

A   B

Figure 8-25 Bourdon Type Pressure Gages. *Courtesy of International Association of Fire Chiefs*

reading on the intake gage upon arrival will be approximately 5 inches Hg. This must be taken into consideration if emergency pumping operations are to be undertaken immediately upon arrival. Of course, the gages should be recalibrated if the pumper is to remain at that elevation for an extended period of time.

Compound gages and discharge gages used on fire apparatus are *Bourdon tube gages,* so named after their inventor. Bourdon tube gages are not as accurate as some of the more complicated mercury gages used in experimental work, but they are sufficiently accurate for pumping operations. However, Bourdon gages are of much more rugged construction than mercury gages and so are more capable of withstanding the type of operations to which they are subjected in the fire service.

Bourdon gages are of single- and double-tube types. While the principle of operation of both types is the same, the double tube is more often used on fire apparatus, due to its more rugged construction and greater stability. Figure 8-25A is a sample of a single-tube gage. Figure 8-25B is a sample of a double-tube gage.

The principle of operation is simple. Water under pressure enters the gage through the threaded gage fittings and passes into the tube or tubes. Each tube is curved and hollow and is closed at

Pump Accessories

the end. The water entering the tube tends to straighten it out, resulting in movement at the upper end of the tube. This movement is transmitted to the pressure-indicating hand through a series of connecting levers and a rack and pinion. A hair spring attached to the pinion shaft holds the assembly tightly together, and serves to dampen the movement of the indicating needle. Positive pressure in the tube causes the needle to move in a clockwise direction.

Negative pressure in the tube acts in the opposite direction. Rather than straightening out the tube, the negative pressure tends to increase the amount of curvature, therefore moving the indicating needle in a counterclockwise direction.

As a centrifugal pump presents a continuous pathway from the intake inlet to the discharge outlets, any positive pressure coming into the pump will be registered on both the intake and the discharge gages. This pressure will remain until the pump has been placed into operation.

The negative pressure registered on the intake gage when one is drafting will also be transmitted to the discharge gage. Some discharge gages have a manually controlled shut-off on the intake side of the discharge gage that permits the pump operator to close off the line to the gage whenever negative pressures are encountered. Other gages have a built-in check valve to prevent damage to the discharge gage as a result of the negative pressure. Still others are so designed that they can be subjected to some degree of pressure reduction without suffering any damage.

Many pumpers are equipped with separate discharge gages for each discharge outlet, in addition to the main discharge gage on the pump. These gages are of the straight pressure type and work in the same way as the pump discharge gage. Some of them are mounted directly on the discharge outlets; most are mounted on the pump operator's panel.

Most gages used on fire apparatus have a tendency to flutter during pumping operations. Some are equipped with needle valves in the line between the pump and the gage. The needle valves can be closed to a point where the vibrations are kept to a minimum; however, no attempt should ever be made to eliminate the vibration completely, as the pointer on the gage would no longer be indicating the correct pressure. The center of the needle swing should be taken as the average pump pressure when the operator is reading the gage.

Flow Meters

One of the tasks of the pump operator on the fireground is to supply the proper amount of water at the proper pressure to the personnel at the end of the hose line. The nozzle may be difficult or even hazardous to hold if too much water flows at too high a pressure. Not enough water or insufficient pressure will reduce the efficiency of firefighting operations and may even place firefighters in a precarious position.

With conventional pressure gages, the pump operator must know the size of the tip used, the size and length of the hose line, and the back pressure in order to provide the right amount of water at the correct pressure. Then he must mentally calculate the variables to determine the required discharge pressure, but even supplying the proper pressure does not ensure that everything will go well. Pressure alone is not an indication that water is actually flowing in a hose line. The pressure gage may show the proper pressure, but a kink in the line could be restricting the flow.

A *flow meter* is an instrument that measures the flow of liquids, gases, or vapors. A flow meter installed on an outlet of a fire apparatus will show the amount of water flowing from that outlet. The amount of water discharged from the nozzle tip is the same as that flowing from the discharge outlet. Consequently, all the pump operator needs to know to provide the proper nozzle pressure to the people at the end of the line is the size of the tip used. For example, a 1¼ tip at 50 psi nozzle pressure will discharge approximately 325 gpm. If a 2½-inch line is taken off a pumper and advanced up the stairwell to an upper floor of a multistory building, all the pump operator has to do is flow 325 gpm to the line if it is equipped with a 1¼-inch tip. The need to consider the length of the line and the back pressure is eliminated. The pump operator is assured that as long as the flow meter reads 325 gpm, water is flowing and the firefighters at the end of the line are operating with a 50 psi nozzle pressure.

Another advantage of a flow meter is that it allows a pump operator to become aware of a burst hose line immediately. This is

Pump Accessories

indicated by a sudden increase in flow to a hose line without a corresponding increase in the pump pressure.

Measuring Water Flow

Water develops a variable *flow pattern* across the diameter of the pipe as it passes through the pipe. This pattern fluctuates in a *velocity profile*, the form of which will depend on the flow rate and any turbulence caused by upstream or downstream ball valves, interior wall surfaces (friction loss), elbows, or bends.

Water will be flowing at a faster rate at some point in the flow pattern (usually near the center of the pipe) and at a substantially slower rate at another point (near the interior surface of the pipe). Accurate flow measurement takes into account the different flow rates across the entire diameter of the contained area. Ideally, in order to determine the velocity profile of the flow, one should take readings at approximately 40 different locations.

A device capable of such a feat would be extremely complicated to make. Bernoulli studied this problem and is credited with classifying all the quadrants of flow and their different flow rates and characteristics. Simply stated, he found that one only needs to know the size of the contained area, and the velocity of flow through this area, in each quadrant of the flow profile in order to measure accurately the volume of water passing a certain point.

Using Chebyschef calculus, it was determined that accurate flow measurement could be obtained by sensing flows in only four equal *annular quadrants* and then averaging them.

Many devices for measuring flow have been developed, but few have taken into consideration the total aspects of flow characteristics.

The pitot gage, for example, provides a flow measurement only at a single point in the flow stream. It requires accurate placement and calibration. It does not take into consideration the influence of atmospheric pressure. Also, a reading taken at one time may not be consistent with a continual fluctuation in flow rate.

An improved method of flow measurement is achieved with a refined version of the pitot gage known as the *Annubar*. The Annubar is designed with four sensing ports facing upstream. These ports monitor the flow velocity in each of the equal annular

Figure 8-26 An Auto-Scan Digital Flowmeter. *Courtesy of Fire Research Corporation*

cross-sectional areas of flow. Velocity pressures are transmitted through the hollow shell of the Annubar to a middle point, where an accumulative average is made by an interpolating tube. The tube carries the resulting average flow pressure through standard tubing to a metering point.

The average pressure is known as the *total head*. Total head is a combination of upstream flow pressure (dynamic pressure) and residual pressure (static pressure). To arrive at an accurate flow measurement, however, it is necessary to isolate the upstream flow component. This is done by using a static port or probe located next to the Annubar. This single sensing port monitors the residual pressure only. This pressure is transmitted through a separate set of tubing to the meter.

Auto Scan Flowmeters

The *auto scan digital flowmeter* (see Figure 8-26) adds a new dimension to the operation of the standard flow meter. Some of the features of this instrument include:

1. The system automatically scans up to eight discharge outlets. Only those outlets discharging water are displayed.
2. Every discharge outlet (up to eight) on the apparatus is equipped with a flowmeter. Wires run from the individual flowmeters to the control panel, where they are connected to light-emitting diodes. These diodes show which outlets are flowing water.
3. The panel has a digital display that indicates the flow of

Pump Accessories

the outlet monitored by the scanner. The diode connected to the monitored flowmeter will flash during the monitoring phase. The flow reading is displayed for approximately four seconds.

4. The next-to-last display in the scan cycle is the total flow. This position displays the total amount of water being discharged by the pump.

5. A manual override permits the pump operator to monitor constantly any one of the discharge outlets.

6. The last position on the scanner is a totalizer. The totalizer adds up all the water used from the beginning to the end of the fire.

An Apparatus Warning System

The pump operator's panel has, over the years, become more and more cluttered. This has come about through the addition of new systems and the increased need to monitor older systems. The cluttering of the panel has made it more likely that the pump operator will fail to notice a monitoring device that is giving warning of potential trouble.

The apparatus warning system includes both visual and audible warning devices (see Figure 8-27); thus it can give a warning

Figure 8-27 The Components of an Apparatus Warning System. *Courtesy of Fire Research Corporation*

of trouble to an operator who is at some distance from the apparatus. The standard system monitors the oil pressure, generator, and engine water temperature. Additional monitors can be added, such as a monitor for a dry pump. An apparatus back-up alarm can also be incorporated into the system.

A light will flash and a bell will ring (approximately one second on, one second off) for as long as the failure lasts. The bell can also be used as a back-up alarm, ringing steadily whenever the vehicle is placed in reverse. If the system includes a dry pump-monitoring capability, the bell will emit a pulsating sound any time the pump is turned on for longer than one minute without water in it.

AUTOMATION

High-rise structures, increased hazards, and the general reduction in fire company budgets have placed more and more of a burden on fire personnel. Fire companies are now required to do more with less. Consequently, many departments have turned to industrial technology for assistance in coping with the problem. While research has provided a number of ideas for the improvement of fire department operations, many of the products developed as the result of research have been abandoned. Two in particular that were featured in the first edition of this text are Grumman's radio-controlled nozzle and the radio-controlled hydrant valve. While the concept of each of these had merit, the products as developed were not fully accepted by the fire service. Despite this fact, much was learned which should prove beneficial in the development of future products. There is little doubt that the concepts, when perfected, will enhance the effectiveness of fire department operations.

Much of the success of automation research has been incorporated in the radio-controlled automatic pumper designed by Fire Research Corporation; yet it in itself has undergone many changes since its inception. The following section is a description of the pumper. The results of research such as the radio-controlled

Pump Accessories

nozzle, the electro-governor, the constant monitoring of apparatus and engine systems, and the warning system should be noted. While these features indicate that much has been achieved as a result of research and experimentation, much still remains to be done. However, only through the constant evaluation of needs and the sharing of ideas with others will the needed changes be realized.

Radio-Controlled Automatic Pumper

The radio-controlled automatic pumper (Figure 8-28) is a completely automatic system developed by the Fire Research Corporation. It incorporates a radio system and a computer to achieve desired results.

Placing the System in Operation

A supply line, preferably large-diameter hose, is laid from the hydrant to the fire and connected to the suction inlet on the pumper. When the connection is completed, the hydrant is turned on. The pump operator engages the pump in the cab which activates the full computer, brings the pump pressure up to the pre-set pressure, and commences the monitoring of all critical param-

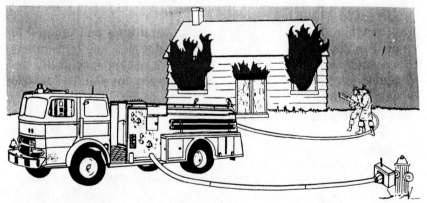

Figure 8-28 The Radio-Controlled Automatic Pumper. *Courtesy of Fire Research Corporation*

eters. While the pump operator is performing the functions described, the firefighters remove hose from the apparatus and move toward the fire. Each firefighter advances to the point of attack with his line and manpack. When ready for water, he transmits a digital radio command by pressing the OPEN button on the top of the manpack. This broadcasts a signal which activates an electric motor which in turn opens that particlar hose line's valve, thereby providing water at the predetermined pressure. If it is desired to close the discharge valve on the pumper so the line can be moved, the firefighter presses the CLOSE button on the side of the manpack.

System Operation

The system is designed to make it easier for a fire company to attack a fire with as much water as possible, as quickly as possible. Additionally, it gives each nozzleman radio control over pressure and flow.

The total operation requires the pump operator merely to engage the pump. The system then starts up, checks all the important apparatus parameters, and is ready to receive signals from the nozzlemen.

The system is activated from the cab of the pumper by the action of the pump being engaged. At this time, the engine speed will increase until a pre-set pump pressure is obtained. The pressure setting is adjustable from 70 to 300 psi. The pre-set pressure is reached within ten seconds.

When the nozzlemen take their hand lines and proceed to the fire, they also take their radio transmitters. The transmitters are color-coded to the nozzles. The nozzles are specifically designed to operate with the system; however, any standard nozzle will function.

Once the system is operating itself, the pump operator is free to perform other functions as needed or, if necessary, take a line from the pumper and advance to the fire.

The system computer also monitors the incoming hydrant supply. Any time the hydrant supply falls below a pre-set level, a warning bell sounds and water is automatically taken from the tank. If the water supply is restored, the system will again start taking water from the hydrant.

Control Panel

The control system is a panel-mounted, heavy-duty box that contains the computer and the radio receiver (Figure 8-29). The box is 9 inches wide, 15 inches high, and 8 inches deep. The controls are clearly labeled, with accurate descriptions of their functions.

The switches are of the heavy-duty, positive-feel type and are equipped with waterproof covers. The lights have daylight lenses. Only green lights will be illuminated when the system is on. If there is a failure, such as a generator going out or a low tank level, a red light will go on, and the alarm system will be activated. The nozzle warning system is indicated by individual red lights on the panel.

The system as shown contains positions for monitoring rapid water and wet water; however, other functions can be monitored in lieu of these, depending on the needs and desires of an individual fire department.

Radio Transmitter and Nozzle Radio

The radio system operates in a 450 to 470 Mhz band. It has a range of over two miles. This particular frequency band was selected because it can be transmitted out of large buildings. Since it is a high-power radio system, a license is required from the Federal Communications Commission.

Each nozzleman's transmitter is equipped with three separate buttons: one on top next to the antenna, one on the side, and one on the front. The top button is the "valve open" control, the button on the side is the "valve closed" control, and the one on the front is the "warning" control. Pushing the warning control button turns on a light on the control panel and activates the alarm system. The buttons are installed so that they will not be activated accidentally.

Each miniature radio transmitter is designed to be carried in a firefighter's pocket. The unit weighs two pounds and is constructed of a very heavy-gauge aluminum alloy, with all openings watertight. The firefighter can press the buttons right through the cloth of a turnout coat.

Figure 8-29 The Radio-Controlled Automatic Pumper Control Panel. *Courtesy of Fire Research Corporation*

Electro-Governor

The system uses the same electro-governor as the one previously explained in the section on pressure control devices, but, in addition, it ties into the apparatus warning system.

If the pump runs out of water, the alarm system will sound and the engine rpm will slowly decrease to an idle. If a second water source is obtained, or the water pressure increases, the pump will resume its normal operation and the warning will be cancelled.

If, because of lack of water, the governor cannot reach and maintain the desired pump pressure, within fifteen seconds the governor warning light will come on and the alarm system will be activated.

The alarm system is also activated if the governor is inadvertently left in the OFF or the MANUAL position. The audio alarm portion can be bypassed by depressing the GOVERNOR BYPASS switch.

Valve Control

The *valve control* is a heavy-duty servomechanism that positions the quarter-turn ball valve on the discharge outlet to obtain the flow desired by the nozzleman. Each nozzleman's radio is digitally keyed to its own servo valve. The valve opens fully or partially whenever the appropriate button on the transmitter is pressed, the degree of opening depending on the previous adjustments made in the computer.

The unit has enough power to open a standard, properly adjusted, 2½-inch ball valve against pressures of 300 psi. It is designed to be mounted right on the ball valve itself and can be mounted on most pumpers without any plumbing changes.

The sprocket is attached by a chain drive to the control handle, moving the handle as the chain moves. The amount of movement initiated by the nozzleman's transmitter is pre-set into the computer. If at any time the pump operator desires to resume manual control or to override the nozzle operator, he simply turns off the switch at the panel and manual control is resumed immediately.

Alternating Audible and Visual Alarm System

The alarm system incorporates a large six-inch bell, which is mounted externally on the apparatus. The bell rings for approximately three seconds any time there is a system failure, and then a very distinctive colored light illuminates on top of the pumper for the next three seconds. The bell and light alternate continually until the "audio bypass" switch is activated. The audio bypass switch can also be used to ignore a system problem, if the pump operator so desires. The bell will not ring for that failure, but the warning light will remain lighted.

The system constantly monitors engine oil pressure, generator, water temperature, governor pressure, water tank level, hydrant pressure, and the warning alarm of each nozzleman.

Pump Pressure Computer

Fire Research has developed a pump pressure computer which will determine within ten seconds the pressure requirement for a ground-level hose layout. The panel display consists of two pressure gages, three dials which can be manipulated by the pump operator, and operating directions.

One of the dials is used for setting the hose layout. Settings are available for single 1½-inch lines, double 2½ inch layouts, single 2½-inch lines, and single 3-inch lines.

A second dial permits the pump operator to crank in the length of the layout. Settings are available for layouts from 50 to 1200 feet.

The third dial is used for setting the size of the nozzle. Arrangements have been made for setting eleven different nozzle sizes. Settings are available for ¾-inch, 1-inch, 1⅛-inch, 1½-inch, 1¾-inch and 2-inch solid streams and 100, 120, 240, 500, and 750 gpm fog nozzles.

The right-hand pressure gage is used to set the desired nozzle pressure.

Once the hose size, length of layout, nozzle size, and nozzle pressure have been set by the pump opearator, the pump pressure required will be displayed on the left-hand pressure gage. While the pump pressure computer does not provide for all the variables found on the fire line, it should eliminate many of the errors made by pump operators in calculating the required pump pressure under the stress and strain encountered in emergency situations.

REVIEW QUESTIONS

1. What operations should a pump operator be able to perform from the pump operator's operating position?
2. Where is the pump panel generally located?
3. What is the most common pressure source of water used in the fire service?
4. What are some of the common drafting sources of water?
5. What are drafting operations?
6. What is meant by priming a pump?
7. What types of positive displacement priming pumps are in current use in the fire service?
8. What is the unit of measurement for pressure reduction within the main pump?
9. How much vacuum are most well-maintained rotary-type priming pumps capable of producing?
10. What is meant when it is said that a vacuum exists within the main pump?
11. Explain the operation of a rotary gear priming pump.
12. What are the three methods used to drive priming pumps?
13. What is the purpose of the priming valve and where is it generally installed?
14. What would probably be the first indication of an air leak in the priming valve?

15. What three general methods for activating a priming valve are in current use?

16. What would happen if the vent hole in the outlet elbow of the oil reservoir (priming tank) became plugged?

17. What things must be done to prime the main pump properly?

18. On those pumps that use a solenoid-operated valve, what generally happens when the priming switch button on the control panel is depressed?

19. How long should the priming valve be kept open during priming operations?

20. How long should the priming pump take to begin discharging a steady stream of water?

21. What two types of pressure control devices are used in the fire service?

22. How do relief valves control pressure?

23. How do pressure governors control pressure?

24. What is the pressure range of operation of relief valves and pressure governors?

25. Which pressure control device keeps the pump and engine under a constant load—the relief valve or the pressure governor?

26. Why is a relief valve an absolute necessity on a positive displacement pump?

27. What are the two components of a fire department relief valve?

28. How does a spring-loaded relief valve work?

29. What is the purpose of a suction relief valve?

30. What is the operating range of the electro-governor pressure control system?

31. How does the electro-governor system work?

32. What is a combination relief valve?

33. What two purposes does the pump transmission serve?

34. What three gears are generally included in the transmission of a gear-driven centrifugal pump?

35. What is a spur gear?

Pump Accessories

36. What is a helical gear?

37. What is the advantage of a helical gear over a spur gear?

38. What was the reason for the development of the chain drive transmission?

39. When the pump transmission is installed ahead of the road transmission, what two functions must be performed by the pump operator when putting the pump in operation?

40. When the pump transmission is installed behind the road transmission, what three functions must be performed by the pump operator when putting the pump in operation?

41. What are the three positions that might be incorporated into the arrangement for putting the pump in operation?

42. What is generally considered the best operating temperature range of an engine?

43. How does an auxiliary cooler work?

44. What does a vacuum gage do?

45. In what units is the negative pressure on the compound gage calibrated?

46. What readings is a compound gage designed to show?

47. How much is atmospheric pressure reduced per 1000 feet of elevation?

48. What is the principle of operation of the compound gages and discharge gages used on fire apparatus?

49. How does a Bourdon tube work?

50. What provisions are made to keep straight pressure gages on the discharge side of centrifugal pumps from being damaged during drafting operations?

51. What is a flow meter?

52. What does a pump operator need to know in order to provide the proper nozzle pressure to the firefighters at the end of the line when flow meters are being used?

53. What is an auto scan digital flowmeter?

54. Who was responsible for the development of the radio-controlled pumper?

55. How is the radio-controlled pumper placed into operation?

56. What are some of the warning features built into the radio-controlled pumper?

57. What does the radio-controlled automatic pumper system monitor?

58. What is the purpose of a pump pressure computer?

9

Pumper Operations

An apparatus operator is more likely to be assigned to an engine company than to a truck company, boat company, or crash company. The function of an engine company is to extinguish the fire; consequently, the responsibility of the apparatus operator is to deliver the crew and apparatus safely to the fire and then to provide the firefighters with an adequate amount of water at the proper pressure.

The *triple-combination pumper* is the most prevalent of the types to which the operator could be assigned. A triple-combination apparatus is one that carries hose and water and has a pump rated at 500 gpm or more. Very few triple-combination apparatus are also equipped with a booster pump. Others use the main pump for supplying small lines from the tank.

A pump operator may be required to work from the booster tank, from a hydrant, from draft, or as part of a relay operation in order to deliver water to the firefighters.

TANK OPERATIONS

At least 90 percent of the fires extinguished by engine companies are extinguished using water from the booster tank. Most triple-combination apparatus have a water tank with a capacity of at least 400 gallons, the 500-gallon tank being the most popular. The water from the tank will last from 5 to 14 minutes or more, depending upon the size of the hose and tip placed in operation. Normally, preconnected lines will be used whenever water is taken from the tank, thus permitting a rapid attack to be made on the fire.

Operating from the tank is the simplest of all pumping situations to which a pump operator could be exposed. The general procedure runs as follows:

1. Place the apparatus as close to the fire or emergency as is reasonably safe and prudent.
2. Place the road transmission in neutral, unless it is required that the apparatus be mobile during the pumping operation.
3. Open the valve from the tank to the pump.
4. Engage the pump.
5. Open the discharge gate to the line that has been led into the fire.
6. Increase the throttle until the proper pressure is reached.

If there is any doubt that the water in the tank might not be sufficient to extinguish the fire, the pump operator should make an effort to secure a supplementary supply of water to the tank as soon as water is delivered to the members at the fire. This might be merely a matter of opening and closing the tank-to-pump valve if a supply line had been laid going in and was charged. If, however, the first-responding pumper uses its booster tank without previously laying a supply line from a hydrant, then the department's standard operating procedure should be for the second-responding apparatus to provide a supply line for the first. A water supply might also be provided by a tanker. If the nearest hydrant is not too far away, and the second apparatus is late in arriving, then the first-responding pumper crew will have to hand-stretch a supply line to the hydrant.

Pumper Operations

Another source of water supply should be found if a supply line was not laid going in. One possibility is to pump water from the tank of another apparatus into the working tank if additional apparatus were dispatched to the emergency. If necessary, an incoming company can be asked to back up the operation by laying a supply line from the nearest hydrant.

When the water is first dropping from the tank to the pump, one problem that can arise is the formation of an air lock. The problem is caused by air concentrating in the eye of the impeller. An air lock is usually indicated by exceptionally low pressure or, in some cases, by a total lack of flow. An air lock is eliminated by priming the pump—keeping the priming pump operating longer than the operator thinks necessary. Do not be fooled by a modest pressure reading. Prime until there is full operating pressure. Some pump manufacturers recommend that the main pump be primed every time it is used for booster operations.

HYDRANT OPERATIONS

Whenever a fire appears to be too large to risk a booster tank operation, the most common method of obtaining water is to connect to a hydrant. The operator must be capable of properly spotting, or positioning, the apparatus so the connection can be made with the suction hose.

Suction Hose

Three types of *suction hoses* are used to move water from the supply source to the pump. One is the *soft suction* (or soft sleeve), the second a *hard suction*, and the third *flexible hard suction*.

A soft suction is woven-jacket, rubber-lined hose with a normal length of between 10 and 14 feet. While there is no standard throughout the United States, most departments carry one with at least a 4-inch diameter. Some pumpers carry the soft intake preconnected to a gated inlet; others carry it uncoupled in a rack. The pump operator removes the suction hose from the rack and makes the connections to both the pumper and the hydrant.

While either a soft suction or flexible hard suction is carried on most pumpers, a hard suction is not. Hard suctions are used primarily for drafting operations; however, they can be used for hydrant operations. Waterfront companies and those in other areas where drafting is frequent normally carry at least two 10-foot lengths of hard suction. Hard suction hose is made of reinforced, rubber-jacketed, rubber-lined material which is designed to withstand both internal and external pressure. The hose should be able to withstand at least 23 inches of vacuum and a pressure of at least 200 psi.

The flexible hard suction is similar in construction to the hard suction, but weighs about half as much. This hose is flexible, making it relatively easy for one firefighter to connect it to a hydrant. The flexibility allows a complete loop to be made in the hose.

The adoption by many departments of the use of large-diameter hose has resulted in a change in operations. Newer tactics permit company officers to lay long supply lines. The large-diameter hose is connected to the hydrant and the line is laid to the pumper inlet. Where water systems are strong, there appears to be little difference in the amount of water available to the pump with large-diameter long supply lines than when the pumper is connected directly to the hydrant using standard suction hose. Tests conducted by the National Fire Hose Corporation indicate that the friction loss in 5-inch triple-duty hose when flowing 600 gpm is only 0.5 psi per 100 feet (refer to Table 9-8). There is, however, a distinct advantage in overall operations as the friction losses in the lines taken off the pumper are held to a minimum because of the reduction in the length of the lines from the pumper to the fire. This permits the pump operator to achieve the same results without overworking either the engine or the pump. Another advantage is that the pumper can be spotted near the front of the fire, which permits quicker access to equipment carried on the apparatus in the event it is needed.

Hydrant Installations

It would be helpful if there were a universal method of installing hydrants; however, there is not. Some hydrants are installed so that the primary outlet is perpendicular to the curb, but outlets may be found pointed in any direction. While the majority of hydrants are located reasonably close to the curb, some are set back a considerable distance from the curb and may even be located behind fences or walls. The location of a hydrant and the manner in which it is installed have a definite effect on spotting and hookup procedures.

Another factor affecting hookup procedures is the speed with which the location of a hydrant can be seen as an approach is made to the emergency. A number of cities have installed aids that assist the pump operator. Some cities place wide bands on telephone poles or other similar objects. The bands are high enough to be seen over the tops of parked cars. Some cities paint the bands yellow; others prefer to match the band color with the color of the hydrant.

Another method used to identify hydrant locations is that of placing traffic reflector markers in the center of the street, immediately adjacent to each hydrant location. The reflectors have two possible colors, one on each side (for example, red and yellow). One color indicates that the hydrant is located directly to the right; the other that it is located to the left. Other communities use a single color reflector and locate the reflector near the center of the street but one to three feet closer to the side where the hydrant is located. These indicators are not as effective during daylight hours as they are at night, when the reflections are readily picked up by the headlights of responding apparatus. These reflectors permit hydrant locations to be easily identified, even from a block away, except during heavy rain or snow. They are of particular value to pump operators who are responding in a district other than their own first-in district.

Selecting a Hydrant

On most occasions pump operators have little choice as to which hydrant they will use at a fire; however, in some instances a choice is available. This is particularly true in those areas where the required fire flow is large, therefore requiring a close spacing of hydrants. In some high-value areas hydrants are located at every street intersection and in the center of the block. Occasionally, more than one hydrant will be installed at an intersection.

While hydrants are generally selected so that the length of

the hose layout can be kept to a minimum, this is not always the best procedure. If it is possible that extinguishing the fire will require maximum fire flow and the hydrant closest to the fire is attached to a weak main, it would be better to lay out more hose in order to connect to a stronger water supply.

Another factor that has an effect on the choice of hydrants is the number and size of the hydrant outlets. It is normally best for a pump operator to consider that the pump may have to pump at capacity, even if only a small amount of smoke or fire is visible as the approach is made. A number of fires that at first appeared to be booster-line fires have developed into greater alarms while lines were being laid. Consequently, if a choice is available between a hydrant with a single 2½-inch outlet and one with an outlet for larger hose, then by all means the hydrant with the larger outlet should be used.

An additional consideration in the selection of a hydrant is traffic. If two equally good hydrants are located on opposite sides of the street, then the hydrant on the fire side should be used. This will permit lines to be laid while the street is kept open to traffic. Taking the hydrant on the opposite side of the street would require that the line be laid across the street, in which case the firefighters must either close the street or provide some method of allowing vehicles to cross the hose without damaging it. Of course, if the fire is of any size the entire block is generally closed to traffic. In the early stages of the fire, however, traffic continues to flow, and even if the street is closed to the public, there is generally a certain amount of emergency vehicle movement.

It is also important to try to avoid selecting a hydrant that will require lines to be laid over an active railroad track. Of course, if lines must be laid over tracks, they must; in this case, arrangements should be made as soon as possible to stop all rail traffic during the progress of the fire.

Hydrant Hookups

The pumper may be equipped with intake inlets on both sides and in the front, on both sides and in the rear, or on both sides and in the front and in the rear. The hydrant may also have two or more outlets. Additionally, these outlets may be of different

Pumper Operations

sizes. Thus there are a number of possible combinations available to the pump operator when all the pumper and hydrant variables are considered. The choice of the variables is an important one. One combination may provide the pump operator with all the water needed, while another may make it difficult to use the capacity of the pump fully. While no specific rules can be established as to how to connect to a hydrant, there are some principles that should be considered when making the decisions.

1. When possible, an inlet should be used that will restrict the friction loss in the apparatus piping to a minimum. The side steamer inlets are the best for limiting friction loss, as they provide a direct path from the inlet to the eye of the impeller.

2. The suction hose should be sufficiently large to permit the pump to operate at full capacity, providing, of course, that the hydrant and main system are capable of supplying the flow. A larger suction hose also helps maintain a positive incoming pressure and, therefore, restricts the possibility of the pump's cavitating.

3. The hookup should, if possible, be made in such a manner as to keep from blocking the street.

The actual spotting of the apparatus at the hydrant for a hookup is important. The spot selected should be one where the suction hose can be stretched between the hydrant and the pump inlet in such a manner that it will form a curve when loaded. This type of hookup will generally prevent a kink. For side connections, it is generally best to stop the apparatus so that the intake inlet is a few feet short of being in line with the hydrant outlet. Front connections should be made in the same manner. The pumper should be spotted with the front intake a few feet short of the in-line position. For rear connections, the spot should be a few feet past the in-line position. Figure 9-1 illustrates some typical hydrant hookups.

It should be noted that the spot is made in relation to the hydrant, not the curb. It is not practical, for example, for pump operators to develop a procedure in which they stop three feet out from the curb and three feet before the hydrant every time they want to make a side or front connection. While this procedure might work if all hydrants were installed at the same distance

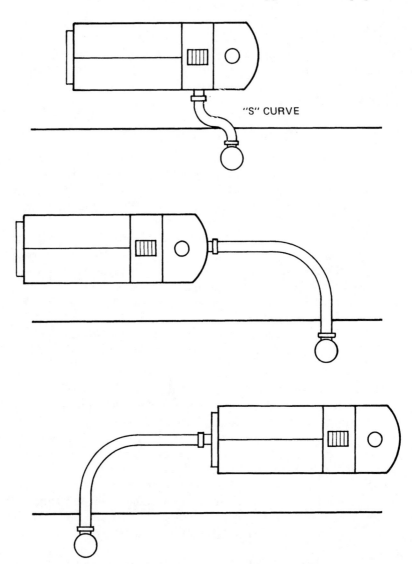

Figure 9-1 Some Typical Hydrant Hookups

from the curb and all outlets were pointed in the same direction, it will not work if any variables at all are present. In some cases the intake would be too close to the hydrant, which would probably cause a kink, while at other times it would be too far from the hydrant.

Pumper Operations

Blocked Hydrants

The term "blocked hydrant" is used to indicate any situation in which a connection cannot routinely be made between the hydrant and the pump inlet, using the standard suction hose carried on the apparatus. There are a number of conditions that may result in a blocked hydrant. Some of the most common include:

1. Hydrants too far from the curb
2. Excavations
3. Vehicles blocking hydrant
4. Freeway situations

Blocked hydrants require a pump operator to improvise rather than use a standard suction hose. There are several possibilities that have proven to be satisfactory. If the apparatus carries 3½-inch or larger hose, it might be a simple matter to use a 50-foot section of the larger hose as an improvised intake (see Figure 9-2). It should be remembered that less water will be available from the hydrant outlet when using this type of arrangement, due to the increased friction loss from the hydrant to the pumper. If a 50-foot section of larger hose is used as an improvised intake, then it will be necessary to use adapter fittings, unless the larger hose happens to be the same size as the hydrant outlet and the pumper inlet.

If larger hose is not available, then it might be necessary to use two siamesed 50-foot sections of 2½-inch hose (see Figure 9-2). This arrangement will have approximately the same friction loss and carrying capacity as a 50-foot section of 3½-inch hose.

Figure 9-2 shows a third possibility, that of coupling together the hard and soft suction hoses, if the apparatus is carrying both. This arrangement will provide a large suction line with just a slight increase in friction loss.

Few freeways are designed with water supply systems for the use of the fire department, yet a number of major fires requiring large supplies of water have occurred on these traffic arteries. While it cannot be said that a hydrant is blocked where none exists, the solution to blocked hydrants might be adaptable to some freeway incidents. The solution to a blocked hydrant situation is, in reality, a long suction layout. Occasionally water may be made available for use on a freeway by using a long suction line

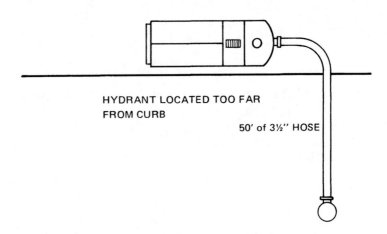

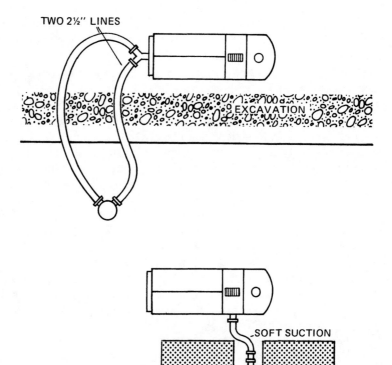

Figure 9-2 Some Blocked Hydrant Situations with Possible Solutions

Pumper Operations

from a hydrant on a freeway overpass or underpass (see Figure 9-3). While it may take more than a 50-foot section of hose to reach from the hydrant to the pumper, a fair amount of water can still be obtained if the hydrant is on a good main system with strong pressure. When the hydrant is above the pumper, as it would be on an overpass, then the forward pressure created by the head from the hydrant to the pump might be more than enough to compensate for the extra friction loss in the hose due to the long intake line.

Blocked hydrant situations do not arise frequently, but they can prove both confusing and frustrating when they are encountered. It is wise to give thought to these situations before they occur, planning on the methods to use and the fittings required. Drilling on simulated blocked hydrant situations will take the guesswork out of the problems at emergencies and reduce the hookup time to a minimum.

Burst Suction Lines

A burst suction line can occur suddenly and without warning, immediately robbing the pump of its entire water supply. This can be extremely dangerous when the firefighters are in a precarious position. The first thought of the pump operator should be to get water to the firefighters as quickly as possible, even if at a reduced pressure. Some pumpers carry a replacement suction line; others do not. In the latter case the pump operator must improvise a suction line just as if he were working with a blocked hydrant situation.

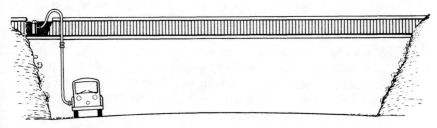

Figure 9-3 Using a Long Suction for a Freeway Incident

Estimating the Hydrant Supply

The number and sizes of lines that can be operated from a pump are restricted by both the pump capacity and the amount of water available from the hydrant. Pump capacity of an individual pump is a constant, but water availability varies from hydrant to hydrant.

Warren Kimball has done considerable research on the problem of estimating the water flow from a hydrant, and he published his results in the book *Operating Fire Department Pumpers*. Kimball proposed a method of estimating flow from a hydrant outlet which has proven to be quite helpful to pump operators. Here is how the system works (excerpted, with adaptations, by permission from *Operating Fire Department Pumpers*, Copyright National Fire Protection Association, Boston, MA).

After opening the hydrant outlet and letting water into the pump, *but before opening any discharge gate,* observe the incoming pressure on the intake gage. The pressure reading on the intake gage is *hydrant static pressure*. After the static pressure has been mentally recorded, water should be provided for a working line. The pump pressure should then be increased, to provide the desired nozzle pressure on the working line. The incoming pressure should again be observed on the intake gage when the desired engine pressure is reached. This pressure is the *residual* (remaining) *pressure* at the hydrant outlet.

If the drop in pressure between the static reading and the residual reading on the intake gage is 10 percent or less of the static pressure, three additional lines with the *same size tip at the same nozzle pressure* as the first line can be supplied, provided the total discharge does not exceed pump capacity. If the drop in pressure is more than 10 percent but 15 percent or less, two additional lines can be supplied. If the drop is more than 15 percent but 25 percent or less, one additional line can be supplied. If the drop is more than 25 percent, an additional line with the same size tip at the same nozzle pressure cannot be supplied.

It should be noted that this system is apparently based on using the same size tip at the same nozzle pressure. In reality, it is based on the delivery of a given amount of water. For example, if 250 gpm are being discharged from the first working line, additional water is available as follows:

Drop in Pressure	Additional Water Available
10% or less	3 lines at 250 gpm, or 750 gpm
more than 10%, but 15% or less	2 lines at 250 gpm, or 500 gpm
more than 15%, but 25% or less	1 line at 250 gpm, or 250 gpm
more than 25%	less than 250 gpm

It should be further noted that this system is based on the amount of additional water available from a *hydrant outlet*. If the hydrant in use has more than one outlet, and the maximum amount of water from one outlet is being used, additional water can be obtained by connecting a suction line to one of the other outlets; however, the additional supply will be limited. It is much better to seek an additional source if the limit of an outlet is reached, but, of course, if no additional source is available, then securing additional water from one of the other outlets should not be ignored.

Supplementing the Original Supply

Occasionally, a pump operator will find that the hydrant outlet to which the apparatus is attached will not supply sufficient water to meet the demands of the fire. The best indicator that the outlet capacity has been reached comes when the soft suction hose begins to collapse. At this time, the pump operator should reduce the throttle and maintain a slight positive pressure coming into the pump. If the operator does not, the pump will start cavitating, and this will ultimately result in pump damage.

The collapsing of the soft suction hose is the result of reaching the capacity of the outlet, or perhaps the pumper is connected to a weak hydrant and the capacity of the main system has been reached. As previously discussed, if a pumper is connected to a hydrant that has two or more outlets, some additional water may be available from one of the other outlets. To do this easily, the second outlet must be gated when the first line is connected to the hydrant unless the hydrant is a wet hydrant, which has valves for each outlet. Of course, if the demand is for several hundred additional gallons, then the small amount available from the additional outlets is insignificant, and the pump operator should seek an additional source. This may require laying several hundred feet of hose from another hydrant to the pump.

Connecting a second suction hose to a working pumper can present a problem. Of course, the completion of the hookup is relatively simple if the pumper is equipped with an unused gated intake; however, if a gated intake inlet is not available, then connecting a second suction hose to the pumper can be tricky.

Remember that the intake piping on a pumper is plumbed in such fashion that water is under pressure throughout the system any time the pumper is connected to a positive pressure water source. When the pumper is filled with water, water will gush out of an intake inlet once the cap is removed, if that water is being supplied to the pumper at a positive pressure. One method of reducing this flow to a minimum, or perhaps preventing it completely, is to increase the throttle to the point where the incoming pressure is at or near zero. It is not wise to operate in this condition for too long a period of time, so it is important that the new suction hose be equipped with the necessary fittings prior to removal of the inlet cap. Once the hookup has been completed, the throttle should be reduced to secure a positive incoming pressure. Connecting the supplementary line to the pumper should increase the incoming pressure to the point where the pump operator can supply the needed water to the company at the fire.

Operating the Pump

The operation of most pumps is similar, the primary differences being in the location of the controls, whether the pump transmission is ahead of or behind the road transmission, and the type of pressure control device in use. A pump operator who understands pump theory, is acquainted with the different pump accessories, and has established a good pumping procedure should be capable of operating any pumper after a short indoctrination period. The following outline is not intended for use with any particular pumper; however, it should serve as a guide for the establishment of an efficient pumping procedure, regardless of the type of pumper to which a pump operator may be assigned:

1. Spot the pumper at the hydrant.
2. Set the parking brakes.

3. Make arrangements for pumping before leaving the cab. If the pump transmission is located behind the road transmission, disengage the clutch and shift the ROAD to PUMP lever to PUMP; then shift the road transmission into the correct pumping gear (usually direct drive). Lock the road transmission in the proper gear; then engage the clutch. If the pump transmission is located ahead of the road transmission, disengage the clutch, place the road transmission in neutral, and engage the pump (if the control lever is in the cab). Then engage the clutch.

4. The first thing that should be done upon leaving the cab is to ensure that the pumper is secure. If it is parked on an incline, or if there is any possibility that the parking brakes will not hold the apparatus, then set chock blocks in place.

5. Shift the pump control lever to the OUT position, if the pumper is so equipped.

6. Make the connection from the hydrant to the pumper with the suction hose. There are two schools of thought on this connection. One says that the suction hose should be connected to the hydrant first (providing, of course, that the suction hose is not preconnected to the pump suction inlet), thus permitting the pumper to be moved in the event the spot is incorrect. The other idea is to connect the suction hose to the pumper first, so that the number of trips between the pumper and the hydrant can be kept to a minimum.

7. Open the hydrant and allow the pump to fill with water.

8. Take the kinks out of the intake hose.

9. Return to the pumper and connect the hose line or lines to the discharge gates, if this was not previously done. Do not open the discharge gates at this time if the pump is not engaged, as this will cause the pump to turn, making engagement difficult.

10. Observe the static pressure on the intake gage so an estimate can later be made of the hydrant outlet capacity.

11. If the pump control lever has been shifted to the OUT position, now shift it to the IN position.

12. Partially open the discharge gates and allow the lines

to fill with water slowly. Open the discharge gates completely when the lines are full.

13. Set the transfer valve to SERIES or PARALLEL, depending on the pumping configuration. Use the position that will give the desired result at the lowest rpm.

14. Place the governor in operation if the pump is equipped with the type of governor that must be set before the pressure is built up.

15. Slowly advance the throttle, watching both the intake and the discharge gages. Continue to advance the throttle until the desired pressure is reached, provided that a positive pressure is maintained on the intake gage.

16. Set the governor or relief valve, if this was not previously done.

17. Check all gages to make sure that all systems are functioning properly.

18. Take any kinks out of the hose lines.

19. Set the chock blocks in place, if this was not previously done.

20. Maintain a proper watch on all gages. Place the auxiliary cooler and the pump transmission cooler in operation, if the apparatus is so equipped and they are needed.

Temporary Shutdowns

Temporary pump shutdowns are made when all nozzles on the working lines become temporarily closed. The purpose of temporarily shutting down a pump under these conditions is to take the load off the pump and to keep the water in the pump from overheating. The following steps should be taken for temporary shutdowns:

1. Slowly retard the throttle.
2. Slowly close the discharge gates.
3. If the shutdown will be for longer than a few minutes, either take the pump out of gear, or open a discharge gate bleeder valve. If one of these two steps is not taken, the water in the pump will overheat from continuous churning.

Pumper Operations

Permanent Shutdowns

Permanent shutdowns are made after the fire has been extinguished and overhaul operations are completed or whenever the services of a particular pumper are no longer needed at the emergency. The shutdown is usually initiated by the company officer. Some departments use standard hand signals to initiate shutdown procedures; others depend on either a runner or the radio. Hand signals are particularly useful when the working end of the hose line is located some distance from the pumper and it is important to keep the air clear for emergency radio messages. While the sequence of operation may vary from one department to another, the following steps are necessary to complete the shutdown:

1. Slowly close the throttle.
2. Slowly close the discharge gates.
3. If the pumper is equipped with the pump transmission ahead of the road transmission, disengage the clutch and take the pump out of gear. If the pumper is equipped with the pump transmission behind the road transmission, disengage the clutch, move the lever in the cab from PUMP to ROAD, unlock the road transmission and shift it to neutral; then engage the clutch.
4. Open the discharge gates bleeder valves in order to bleed the hose lines.
5. Close the hydrant.
6. Open the pump drains.
7. Put the transfer valve in the position most likely to be used on the next response.
8. Set the pressure regulating device in the position in which it is normally carried.
9. Shut down the auxiliary cooler, if engaged.
10. Allow five minutes from the time the pump was taken out of gear for removing excess heat; then shut off the engine.
11. Disconnect all hose lines from the discharge gates.
12. Disconnect the suction hose and return it to the apparatus. Replace caps on the hydrant.
13. Replace the cap on the intake inlet, and close the control gate, if the apparatus is so equipped.

14. Make a check to ensure that all intake caps are in place, all discharge gates are closed, all drains are closed, and the apparatus is ready for the next response.

15. Return the chock blocks to the apparatus when you are ready to depart.

DRAFTING OPERATIONS

Drafting operations involve taking water from a source other than a hydrant or from a source that does not deliver water to the intake of the pump under pressure. Ponds, swimming pools, rivers, the ocean, and other such bodies of water are drafting sources. A thorough knowledge of drafting principles and procedures is required to obtain water successfully from these sources.

Absolute and Relative Pressure

Pumpers are equipped with a minimum of two gages, one connected to the discharge side and one to the inlet side of the pump. The gage connected to the discharge side is generally referred to as a *pressure gage*, while the one connected to the inlet side is generally referred to as a *compound or intake gage*. The pressure gage registers pressure above atmospheric pressure, while the compound gage registers pressure both above and below atmospheric pressure.

Absolute zero pressure refers to a complete absence of pressure, or a perfect vacuum. Absolute pressure is pressure above absolute zero and is identified as *psia*.

Atmospheric pressure is measured in absolute pressure. Normal atmospheric pressure at sea level is 14.7 psi.

Gage pressure is pressure above atmospheric pressure and is identified as *psig*. At sea level, zero gage pressure is an absolute pressure of 14.7 psi.

The zero reading on both the compound gage and the pressure gage on a pumper refers to gage pressure. The pressure readings on both pumper gages are in psi, or in pressure relative to

Table 9-1 Comparison of Absolute and Relative Pressures

Absolute Pressure	Relative Pressure
− 39.7	25 −
− 29.7	15 −
− 19.7	5 −
− 14.7 ————————————————————————	0 −
− 10	10" −
− 5	20" −
− 0	30" −

atmospheric pressure. At sea level, for example, a gage pressure of 50 psi would equal an absolute pressure of 64.7 psi (50 + 14.7).

The readings below zero on the compound gage are measured in inches of mercury. One inch of mercury is equal to an absolute pressure of .49 psi. A pressure of .49 psi at the bottom of a 1-square-inch container will support a column of water approximately 1.13 feet high.

An atmospheric pressure of 14.7 psi is equal to 29.92 inches of mercury (commonly referred to as 30 inches). A pressure of 14.7 psi at the base of a 1-square-inch container will support a column of water approximately 33.9 feet high.

Pressure above atmospheric pressure is generally referred to as "positive pressure," while pressure below atmospheric pressure is generally referred to as "negative pressure." Another way of expressing these pressures is to refer to them as relative pressures, pressures relative to atmospheric pressure. A comparison between the absolute and relative pressures at sea level is shown in Table 9-1.

The compound gage on a pumper registers pressures similar to the relative pressures in Table 9-1. It can be seen from Table 9-1 that a reading of 10 inches of mercury on the compound gage would approximate a pressure reduction within the pump of 4.7 psi (14.7 to 10), and a reading of 20 inches of mercury would approximate a reduction of 9.7 psi (14.7 to 5).

Lift

Lift is the vertical distance from the surface of the water to the center of the pump when the pump is drafting. Maximum lift is the maximum height a pumper can draft water. Maximum lift is affected by the design and condition of the pump, the adequacy and condition of the pumping engine, the size and condition of the suction hose and strainer, the atmospheric pressure, and the temperature of the water.

Theoretically, if a pump at sea level could produce a perfect vacuum, it could lift water to a height of 33.9 feet; however, no pump on a fire apparatus is capable of producing a perfect vacuum. Even if a pump could accomplish this, the friction loss in the suction hose would restrict the height of the lift. Under practical conditions, it is best to assume that the maximum lift is restricted to approximately 23 feet.

The Principle of Lifting Water

Water will move from one location to another whenever a difference of pressure exists between the locations and there is a clear path of travel between them.

In the normal procedure for drafting operations, the pump operator engages a positive displacement priming pump and then opens a line between the main centrifugal pump and the priming pump. What happens is best illustrated in Figure 9-4, where the following steps can be observed:

1. The pump operator engages positive displacement pump A. Positive displacement pumps are capable of pumping air as well as water; thus they are capable of creating a partial vacuum or a reduction of pressure.

2. The pump operator opens valve C between the priming pump A and the main pump B. This valve will open automatically on some pumps when the priming pump is engaged.

3. The priming pump turns and removes air from main pump B and discharges it through hose D.

4. The result of the removal of air from pump B is a reduction of pressure within pump B, which is registered in inches of mercury on compound gage G. For example, there is a pressure reduction of approximately 9 psi when the compound gage reads 18 inches of mercury. The remaining pressure within the pump is then 14.7 − 9, or approximately 5.7 psi.

Pumper Operations

5. The pressure differential between the atmospheric pressure of 14.7 psi at the surface of the water and the pressure of 5.7 psi within main pump B causes the water to move up the hard suction hose and into pump B.

6. The water continues through valve C into priming pump A and is discharged through hose D.

7. When a solid stream of water is discharged from hose D, priming pump A is disengaged and valve C is closed.

8. Main pump B is engaged, which makes water available at the discharge gates.

How Far Will Water Rise?

Theoretically, water will rise 2.3 feet for each pound of pressure reduction within the pump. The amount of pressure reduc-

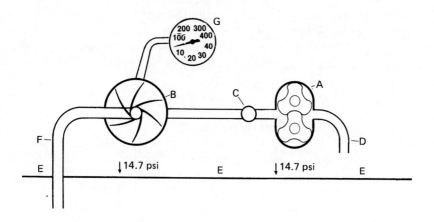

A Positive displacement priming pump.
B Main centrifugal pump.
C Valve between priming pump and main pump.
D Discharge outlet from priming pump.
E Atmospheric pressure.
F Suction hose from main pump to body of water (End of suction hose should be at least two feet below the surface of the water).
G Compound gage.

Figure 9-4 The Principle of Lifting Water

tion can be determined by multiplying the vacuum reading in inches of mercury by 0.49. Water will rise 1.13 feet for each inch of mercury reduction (0.49 × 2.3), so the height to which water will theoretically rise can be determined by the formula $h = 1.13$ Hg.

In actual practice, such heights are not obtained because of minute air leaks, friction loss in the hose, and so forth. A good rule of thumb is to estimate a rise of approximately 1 foot per inch of mercury, instead of 1.13 feet.

Effects of Altitude on Drafting Operations

There is a reduction of approximately one-half pound of atmospheric pressure for each 1000 feet of elevation. The reduction in atmospheric pressure directly affects the height to which water may be lifted, the maximum lift being reduced as a pumper moves from sea level to a higher elevation.

The ability to lift water at higher elevations is further reduced by the loss of power in the engine. The power generated by an internal combustion engine decreases approximately 3½ percent for each 1000 feet of elevation. Therefore, an engine that was just adequate at sea level would be about 14 percent deficient at an altitude of 4000 feet.

Effects of Weather Changes

Increases or decreases in the barometric pressure due to the movement of air masses will have the same effect on the drafting ability of a pump as will a change in elevation. Barometric pressure increases during good weather and decreases during poor weather. A pumper operating on a clear day may be capable of lifting water as much as a foot higher than it can on a rainy day.

Quantity of Lift

The amount of water that a pumper can be expected to lift depends upon its rated capacity, the size of the suction hose, and the lift height. The minimum discharge that should be expected of

Pumper Operations

a pumper in good condition, operating at draft at various lifts, is shown in Table 9-2. Table 9-2 shows that the full capacity of pumps rated at 1250 gpm and more cannot be achieved when any single suction hose less than 6 inches in size is used; however, in practical operations, maximum capacity can be approached or even achieved at low lifts through the use of dual 4-inch intake lines. Dual 4-inch intakes are approximately equivalent to a 5.6-inch suction hose.

Drafting Procedures

The location for drafting should be carefully selected. It should be one that will permit the pumper to be put on level ground and one that will be able to support the weight of the apparatus under conditions of considerable vibration.

The pumper should be spotted as close as possible to the water's edge, preferably with the pump operator's panel located away from the water's edge. The entire crew should assist in setting up for drafting, as the manipulation of the hard suction hose is a difficult and cumbersome task.

In most cases it takes at least two 10-foot lengths of hard suction hose to reach the water. An allowance should be made for tide movements or other changes in the water level when selecting the number of lengths to use. It is important to remember that if the source of water is limited (such as a swimming pool), the height of the lift will continue to increase, while the depth of the strainer beneath the water will continuously be reduced.

If possible, plans should be made to keep the intake strainer at least two feet below the surface of the water and at least one foot off the bottom at all times. A distance of less than two feet will allow whirlpools to form, which will cause the pump to lose its prime.

If early evaluation indicates that it will be difficult to maintain a distance of two feet below the surface of the water, then either use the booster hose stream to break up the whirlpool or decrease pump demand.

Check the suction hose for gaskets prior to coupling the sections together. A small leak in a coupling can make it impossible to prime the pump properly. Before the suction hose is con-

Table 9-2 Minimum Discharge That Should Be Expected of a Pumper in Good Condition at Draft at Various Levels

Conditions: Operating at net pump pressure of 150 psi; Altitude of 1000 feet; Water temperature of 60°F; Barometric pressure of 28.94" Hg (poor weather conditions).

Rated Capacity, Pump	500 gpm		750 gpm		1000 gpm		1250 gpm		1500 gpm	
Suction Hose Size in inches	4	4½	4½	5	5	6	6	6	Dual 5	Dual 6
Lift in Feet 4 (20" Suction Hose, two sections)	590	660	870	945	1160	1345	1435	1735	1990	2250
6	560	630	830	905	1110	1290	1375	1660	1990	2150
8	530	595	790	860	1055	1230	1310	1575	1810	2040
10	500	560	750	820	1000	1170	1250	1500	1720	1935
12	465	520	700	770	935	1105	1175	1410	1615	1820
14	430	480	650	720	870	1045	1100	1325	1520	1710
16	390	430	585	655	790	960	1020	1225	1405	1585
18 (30" Suction Hose, three sections)	325	370	495	560	670	835	900	1085	1240	1420
20	270	310	425	480	590	725	790	955	1110	1270
22	195	225	340	375	485	590	660	800	950	1085
24	65	70	205	235	340	400	495	590	730	835

Notes: 1. Net pump pressure is 150 psi. Operation at a lower pressure will result in an increased discharge; operation at a higher pressure, a decreased discharge.
2. Data based on a pumper with ability to discharge rated capacity when drafting at not more than a 10-foot lift. Many pumpers will exceed this performance and therefore will discharge greater quantities than shown at all lifts. *Copyright Insurance Services Office, 1975.*

Pumper Operations

nected to the pump inlet and lowered into the water, the suction hose sections should be coupled together, the intake strainer placed on the end, and the tie rope attached to the suction hose. Suction hose couplings must be airtight. A rubber mallet can be used to tighten the couplings.

If possible, a pump operator should avoid running the suction hose over a bridge rail or other object before placing it in the water. Air will become trapped in the suction hose if any point in the hose is higher than the pump. The trapped air will cause the pump to lose its prime once a discharge gate is opened.

Once the suction hose has been lowered into the water, the pump operator should check all drains, discharge gates, valves, bleeders, and any other openings into the pump to ensure that they are closed. Any leakage on either the intake or the discharge side of the pump, no matter how small, might result in failure to prime the pump.

After the pumper has been checked to ensure that all openings are closed and all couplings and caps are tightened, the priming pump should be engaged and the throttle advanced to the recommended rpm. The priming valve should be opened if the pumper is so equipped. This will open a line between the priming pump and the main pump. All pumpers do not have a separate priming valve control; with some pumpers, the function takes place automatically when the priming pump is engaged. Some pumpers require the main pump to be engaged during priming operations; others do not.

The pump operator should remain alert for air leaks while waiting for water. A high whistling sound is often an indication of such leaks.

Water should be received within 30 seconds on pumpers with a capacity of less than 1500 gpm. It may take as long as 45 seconds with pumpers of 1500 gpm and larger. The priming pump should be disengaged and the source of trouble sought if water is not received within the allocated time. The problem is usually one or more of the following: the priming pump is dry and needs lubrication; there is an air leak; the strainer is obstructed; or the lift is too high.

If the priming pump is running dry, a quick check of the oil reservoir for the pump will disclose the trouble.

The air leak could be at a number of locations. However, if

the pump operator has made a complete inspection of all discharge gates, valves, and so on, the leak is probably at a coupling on the intake hose. The most likely place for a leak is the first coupling below the intake inlet. This is generally caused by a bend in the intake hose near the coupling.

If the lift is too high or the strainer is obstructed, there may be an extra high reading on the vacuum gage.

The first indication that water has entered the main pump is a discharge of water from the priming pump. The first discharge will be a mixture of air and water. Allow the water to flow steadily for several seconds before closing the priming valve. The main pump should then be engaged and the priming pump disengaged, the process depending on the configuration of the pumping arrangement.

A discharge gate should then be opened slowly. At the same time, the throttle should be advanced to take up the load. Open the discharge gate or gates fully, and advance the throttle until a satisfactory stream is flowing from the working lines. Watch the pressure increase on the discharge gage until the desired pressure is reached. The intake gage should continue to show a vacuum.

The discharge pressure should continue to increase as the throttle is advanced. If the throttle is advanced without a corresponding increase in pressure, one of two things has happened:

1. The most efficient point of the pump has been passed. Back the throttle down to the point where the discharge pressure begins to drop. Operate at this pressure. It is useless to increase the throttle beyond this point.

2. The pump has not been completely primed. Back the discharge pressure down to a low pressure and shift the transfer valve from volume to pressure and back to volume. This may clear the air lock. If it does not, reprime the pump.

Once the pump is operating satisfactorily, the pressure control device should be set.

It may be necessary, once operations are under way, to shut down all discharge lines for a short period of time. The following steps should be taken during this period to avoid losing the prime:

1. Reduce the engine speed slowly until the discharge gage reads approximately 50 psi.

Pumper Operations

2. Open the tank fill valve or "crack" a discharge gate in order to keep some water flowing from the pump.

3. Once the nozzle or nozzles have been placed back in operation, close the valve or discharge gate and increase the pressure to that necessary.

Drafting from a Broken Connection

In a broken connection, water is taken from a hydrant or other positive pressure source, discharged into an open container, and then taken from the container by a drafting operation.

This type of operation may become necessary in the case of a break in a flammable liquid line, where it is necessary to pump water into the break to force the liquid back into the piping. Such an operation generally continues until the flammable liquid line has been shut down and the liquid drained off.

Some flammable liquids, such as certain liquified petroleum gases (LPGs), have extremely high vapor pressures. It is possible that the vapor pressure will force the liquid or vapor into the hose lines and subsequently back into the water mains. This can be prevented by the use of broken connections.

Several possible makeshift reservoirs can be set up for this type of operation. A hole can be dug in the ground and a salvage cover can be used as a lining; several ladders can be arranged to form a basin, with salvage covers used for linings; or barrels, if available, can be used. The makeshift reservoir can be kept full by the discharge of open-butt lines from hydrants.

Drafting procedures with this type of operation are the same as routine drafting operations. However, care must be taken to ensure that the makeshift reservoir is kept full so that the pump does not run away from the water. Care must also be taken to see

Figure 9-5 Drafting from a Broken Connection

that the line used to fill the makeshift reservoir is not directed at the strainer of the suction hose. This causes air to be introduced into the pump, and the pump will then lose its prime.

RELAY OPERATIONS

Relay operations involve moving water with two or more pumpers in series; the water discharged from one pumper passes into the intake side of another. Such operations are necessary whenever the conditions encountered require more than a single pumper to produce an effective stream. These conditions generally arise because of excessive friction loss in long hose layouts, or because of back pressure in high-elevation situations.

Relay operations are not part of the normal day-to-day operations of most fire departments, since an adequate hydrant distribution system is generally provided in built-up areas. However, any fire department may, at any moment, find it necessary to set up such an operation. While the need for relays is normally restricted to brush fires or rural fires, the need may arise in the heart of any metropolitan area, should there be damage to the water system because of an earthquake or other disaster. Relay operations may also become necessary in high-density metropolitan areas because of incidents on freeways, where the emergency is far from the nearest water supply. It is therefore imperative that all pump operators be thoroughly familiar with the theory of relay principles, and be able to operate properly in a relay situation should the need arise.

Factors to Be Considered in Relay Operations

It is doubtful that any two relay operations will be exactly the same, so it is not practical to establish rigid rules governing all such operations. It is best to consider some of the basic principles, together with some general requirements and limitations that can be tailored to fit the situation.

Time is a critical factor in most fire situations. Setting up a

Pumper Operations

relay operation is generally a slow process. The time required to set up a relay depends largely on the amount of water required, the distance between the source of water and the intended point of use, the pumping capacity of the available apparatus, the size of hose available, and the topography.

The Amount of Water to Be Relayed

The amount of water required at the fire is one of the primary factors in the determination of pumper spacing and the size of hose to be used in the relay. The officer in charge of relay operations should consider carefully the amount of water required at the fire. Once hose has been laid and pumpers have been placed, the maximum amount of flow cannot be increased without laying additional lines between pumpers. It is probably better to overestimate the amount of water required than to underestimate.

Very small quantity flows do not normally present a serious relay problem. In fact, relays may not be required even when there is an exceptionally long distance between the source of water and the point of use, except perhaps where elevation is involved, a situation that often results in back pressure problems. For example, the discharge of a ⅝-inch tip at 50 psi nozzle pressure is approximately 80 gpm. This flow will result in a friction loss of about 1.5 psi per 100 feet of single 2½-inch hose. Water could be pumped 10,000 feet with this flow and still provide a nozzle pressure of 50 psi without exceeding a pump pressure of 200 psi. However, the problem changes completely when it becomes necessary to transfer large quantitities of water over a long distance. Friction loss becomes excessive in small lines or in single lines stretched between pumpers. Large-quantity relays require larger hose, more pumpers, and more critical decisions.

The Distance Between the Source of Water and the Point of Use

The critical distance between the source of water and the point of use increases with the difficulty of the topography, the amount of water required, and the size of hose used for relay operations. For example, no difficulty would be encountered in supplying 80 gpm through a single 2½-inch line over a distance of

1800 feet, if the line were laid at ground level. A relay operation would be required, however, if the point of use were located 400 feet above the level of the pumper supplying the line. Given these conditions, an intermediate pumper would have to be put into use if it became necessary to use a 1-inch tip at the fire. Of course, if the layout were 1700 feet of single 3½-inch hose, reduced to 100 feet of single 2½-inch hose to which the 1-inch tip was attached, then an adequate stream could be provided, because the friction loss in the 3½-inch line would only be about 1¾ psi per 100 feet.

The problem would become much more critical if a master stream were used at the fire. If a 1¾-inch tip were to be used, the flow required would be 800 gpm. With this flow, the friction loss in a single 3½-inch line is about 23 psi per 100 feet. Even if two siamesed 3½-inch lines were used, the total pressure requirement would be over 200 psi.

The Capacity of Available Pumpers

The pumping capacity of the available apparatus is a primary factor influencing the amount of water that can be relayed. The amount of water is limited to the pumping capacity of the smallest pumper in the relay. When considering relay operations, it should be remembered that pumpers are rated at full capacity at 150 psi net pump pressure and at 70 percent of rated capacity at 200 psi net pump pressure. Because pumpers in a relay will be pumping at or near a net pump pressure of 200 psi, only 70 percent of the rated capacity can be expected. Thus, when the smallest pumper is rated at 500 gpm, the maximum flow that can be depended upon is 350 gpm; when the smallest pumper is rated at 1000 gpm, maximum flow is 700 gpm. The maximum dependable flows and the largest tips that can be supplied by a single pumper are shown in Table 9-3.

The maximum dependable flows shown in Table 9-3 can be increased somewhat for the pumper at the hydrant, if water is supplied from a hydrant with a strong residual pressure with the required discharge flowing. The pumper at the hydrant will be pumping at less than 200 psi net pump pressure even when the discharge pressure is 200 psi, as the net pump pressure is the difference between the discharge pressure and the incoming pressure. For this reason, it is best to place the smallest capacity

Pumper Operations

Table 9-3 Maximum Dependable Flows from Pumpers in a Relay

Rated Capacity of Pumper (gpm)	Maximum Dependable Flow (gpm)	Largest Tip to Be Supplied (inches)
500	350	1¼
750	525	500-gpm fog nozzle
1000	700	1½
1250	875	1¾
1500	1050	2, or 1000-gpm fog nozzle

Table 9-4 Maximum Efficient Carrying Capacity of Hose Lines

Hose Size (inches)	Maximum Efficient Carrying Capacity (gpm)
2½	300
3	500
3½	750
4	1000

pumper at the hydrant. This will permit maximum use of the hydrant's contribution.

The Size of Hose Available

Selection of the proper size of hose for use in a relay greatly influences the effectiveness of the operation. If possible, hose that will result in excessive friction loss should not be selected. Because of the importance of reducing friction loss to a minimum, it is recommended that relay operations use hose of 2½ inches or larger. Table 9-4 lists the maximum amount of water that various hose lines can carry without introducing excessive friction losses.

An evaluation of the carrying capacity in hose indicates that single 2½-inch hose should not be used in a relay if the total consumption of water at the fire exceeds 300 gpm. Therefore, the use of single 2½-inch hose between pumpers in a relay limits tip size; either a 1¼-inch tip, which produces an effective stream while discharging 300 gpm, or a 300-gpm fog nozzle can be used. Of course, a number of smaller tips can be used if the total flow does not exceed 300 gpm.

The adoption by many communities of the use of large-diameter hose has not only changed hydrant connection tactics but has also simplified relay operations. The reduced friction loss in this hose (refer to Table 9-8) permits the transfer of small flows over longer distances without the necessity of resorting to relay operations and also provides for the movement of large flows in relay operations utilizing a limited number of pumpers.

Topography

If the flow is small and lines are laid at ground level, it may be possible to lay several thousand feet of hose without setting up a relay; however, requirements change rapidly for fires in hills or mountains. A fire at a high elevation may require a pumper every 300 or 400 feet, whereas the same layout on level ground could be completed without resorting to a relay.

Back pressure or forward pressure will almost always need to be considered when long lines are used at brush or other fires in mountainous areas. In fact, where the ultimate consumption of water is less than 200 gpm, consideration of back pressure is more important than that of friction loss in establishing the positioning of pumpers in a relay.

It should be remembered that relay operations will generally be made along dedicated streets or highways, and that a definite effort has been made to hold the grade on these travelways to 12½ percent. In the absence of indications to the contrary, the officer responsible for establishing the relay is usually safe in assuming a grade of 10 percent, which is a rise of 10 feet in 100 feet, and allowing a back pressure of 4½ psi for each 100 feet of hose laid out. This will amount to an allowance of 45 psi for back pressure for each 1000 feet of hose laid.

The spacing of pumpers in a relay is affected by the friction loss in the hose, the topography, and the general restriction of 200 psi discharge pressure. With the objective of providing 10 psi to the suction inlet of the next pumper in the relay, only 190 psi is available for overcoming back pressure and friction loss in the hose. Of course, if any lines are laid downhill, then this will have the effect of increasing the allowable distance between pumpers. These conditions can be summarized in the formula:

Pumper Operations

$$\text{Maximum distance between pumpers} = \frac{190}{FL + BP - FP}$$

where FL = friction loss per 100 feet of hose layout
BP = back pressure per 100 feet of hose layout
FP = forward pressure per 100 feet of hose layout

As an example, if the layout is at ground level and the friction loss in each 100 feet of hose between pumpers is 25 psi, then by the formula the maximum distance between pumpers should be:

$$\frac{190}{25} = 760 \text{ feet}$$

Therefore, pumpers should be placed a maximum of 750 feet apart. However, if the layout is up a 12 percent grade, the back pressure for each 100 feet of hose in the layout would be approximately 5.2 psi. Then the maximum distance between pumpers by formula would be:

$$\frac{190}{25 + 5.2} = \frac{190}{30.2} = 629 \text{ feet}$$

and the pumpers should be placed a maximum distance of 600 feet apart.

The formula for determining the distance between pumpers in a relay can only be used if terrain conditions permit such spacing; the spacing indicated by the formula is not usually possible in moutainous areas. Occasionally lines are laid along narrow roads that must be kept open for other apparatus. Many times relay pumpers must be placed at passing points; since such points are short of the ideal spacing, more pumpers will be required than the formulas indicate. If the officer in charge has time prior to committing the crew, he should survey the area and determine the logical locations for placement of pumpers. Without such a survey, a pumper laying a line may exceed the maximum distance allowed between pumpers before the company officer locates a place where the pumper can be spotted.

Terrain is not always a negative factor. Sometimes it becomes an ally. If lines are laid downhill, the forward pressure will assist the operation. Under certain conditions, in fact, the need for a relay operation may be completely eliminated by the extra advantage gained by the forward pressure. The same method used for estimating the back pressure should be used for estimating the forward pressure.

Of course, formulas for spacing pumpers in a relay are essentially theoretical. They can be used to develop a pre-fire plan that requires a pumper relay, but, in real life, each engine will drop its hose load and pump.

In rural areas, pumper relays are common. Generally, the attack engine goes directly to the fire and uses its booster tank water for the initial attack. At the start of a long driveway or a farm lane, the attack pumper will drop hose at the main road. The other engines will begin stretching from the attack pumper's hose to the water source. The luxury of properly spacing pumpers is out of the question except when a target hazard has been pre-fire planned. The source engine may stretch only a few lengths of hose to reach its drafting spot.

Each pump operator should know the length of hose in the bed and how much water he can pump through the line he has laid.

When large-diameter hose (4- or 5-inch) is used, hose loads are limited so that no engine has to operate at more than 150 psi. When 2½ or 3-inch line is used, the starting engine pressure desired is 200 psi. Obviously, if an operator knows he dropped a shorter than normal line, he will reduce his initial pump pressure accordingly.

After the relay has provided water to the firegound, a water supply officer will then refine the engine pressures throughout the relay to provide optimum use of each pump.

Relay Standards

The following are certain standards and guidelines that should be considered in setting up relay operations.

1. Where possible, all pumpers in the relay, except the one supplying water to the fire, should pump at or near the same pressure.

2. If possible, pumpers in a relay should restrict their

discharge pressure to a maximum of 200 psi for double-jacket hose (2½-, 3-, and 3½-inch), but the recommended desirable pressure for polymer, single-jacket hose (3½-, 4-, and 5-inch) is 150 psi, even though hose is normally tested annually at a pressure of 250 psi. Otherwise, a sudden closing of a line or a discharge gate could increase the pressure above 250 psi, and a hose line could rupture before the pump operator had time to act. A ruptured hose line is possible even if relief valves are properly set.

3. Each pumper in the relay should receive water at a pressure of 10 to 20 psi at the pump inlet. Relay relief valves generally are set at 10 to 20 psi; they dump water received at pressures of approximately this range.

4. The intake inlet that will restrict friction loss to a minimum should be used. The friction loss in one intake inlet can be as much as 10 or 15 psi greater than that in other inlets, due to elbows and other restrictions in the plumbing. The large inlet on the side entering directly into the pump is generally the best.

5. All pumpers in the relay except the one at the supply source should be equipped with internal or external suction relief valves.

Operational Considerations

In addition to the items already discussed, there are a number of operational factors to consider with regard to relay operations. These include:

1. Setting up the relay
2. Pumper failure
3. Detecting ruptured lines
4. Shutting down lines
5. Looking for unaccountable pressure increases
6. Establishing communications

Setting Up the Relay

The following procedures should be followed when setting up relay operations:

1. The pumper at the water source should be spotted and water should be taken into the pump as soon as possible.

2. The operator of the first pumper should open the pumper's discharge gates slowly and start building up pressure as soon as he is informed that connections to the second pumper have been made.

3. The operators of the second and succeeding pumpers should partially open a discharge gate not only to vent air but also to provide a means of discharging an undue surge of water, if necessary. The gate can be opened to the degree necessary to divert the undesired flow of water.

4. All succeeding pump operators should load their lines as soon as they are informed that hookup has been completed to the next pumper in line.

5. Discharge pressure should be built up and maintained at 200 psi for 2½- and 3-inch hose and at 150 psi for large-diameter (4-inch) hose on all pumpers, except the one supplying the working lines. The operators should not reduce the pressure until notified to do so.

6. The operator of the last pumper should proceed just as if the pumper were connected to a hydrant. He should determine the engine pressure required to provide good working streams and set the pump accordingly. If the relay has been properly organized, the operator should be able to provide good working streams without reducing the intake pressure to less than 10 psi.

7. All operators must remain continuously alert once the relay has been initiated. They must remember that they are members of a team and that their operations must be coordinated with those of all other pumping units in the relay. The successful discharge of water at the end of the line depends on the efficiency of every operator in the relay.

Pumper Failure

There are several valves available that can be used in relay operations which will permit the removal of a pumper in the event of mechanical failure without interrupting the flow of water.

The four-way relay valve is generally connected into the line one length of hose before the pumper inlet. Figure 9-6 shows the configuration of the hookup in the normal relay position. Water from the previous pumper in the relay passes through the four-way valve and into the pump. A section of hose is attached to the discharge outlet of the pumper and to the four-way valve. The water then passes from the pumper through the four-way valve and on to the next pumper in the relay.

A four-way relay valve can be placed in a line being laid so that a later-arriving engine can take its place in the relay without disrupting the movement of water.

If it becomes necessary to remove a pumper from the relay due to mechanical failure, then the handle of the four-way valve should be shifted to the opposite position and water will flow as shown in Figure 9-7. The lines connected to the faulty pumper can then be taken off, another pumper placed in the relay, and the lines connected to the new pumper. During this period it is normally necessary to restrict the flow in the relay in order to reduce the friction loss to an acceptable amount in order to compensate

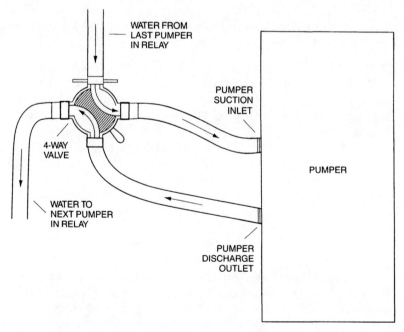

Figure 9-6   Four-Way Valve in Normal Relay Position

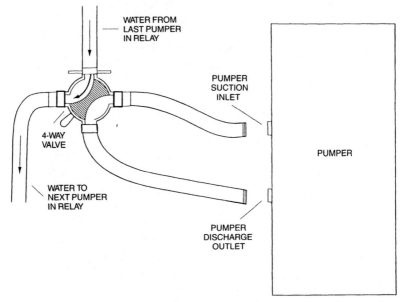

Figure 9-7   Four-Way Valve in Position to Remove a Pumper from the Relay

for the increased distance between two relay pumpers. Once the lines are connected to the replacement pumper, the handle on the four-way valve can be shifted and relay operations returned to normal.

While there is an advantage in using this type of valve in relay operations, there is also a disadvantage. The use of the valve will restrict the amount of water that can be relayed. When it is necessary to flow large amounts of water, this restriction should be given serious consideration by the officer responsible for the relay.

Detecting Ruptured Lines

Ruptured hose lines must be detected immediately and steps taken for their replacement. A ruptured hose line can first be detected by the operator who is supplying the line. The rupture will cause a rapid drop in the discharge pressure. If two or more lines are laid between pumping units, the discharge gate supply-

Pumper Operations

ing the ruptured line should be slowly closed. There will be substantial increase in the discharge pressure over the original pressure if the proper gate has been closed. The discharge pressure on the good lines should be increased to 200 psi, assuming the previous operating pressure was below this limit.

The operator of the first pumper toward the nozzle from the ruptured section of hose will probably have to operate at or near zero intake pressure until the ruptured line has been replaced. This will tend to increase the supply of water through the remaining lines, but it is doubtful that the original supply can be matched.

Shutting Down Lines

All pump operators in the relay should be notified, prior to the closing, that a nozzle is to be shut down. If time does not permit such notification, then the nozzle should be closed slowly, to allow operators time to observe the change and make necessary adjustments. Sudden closing of a nozzle or a discharge gate anywhere in the relay could result in severe water hammer, with resultant burst lines and possibly damage to the pumpers.

Unaccountable Pressure Increases

Unaccountable increases in pressure may indicate flow-restricting debris in the hose, the closing of a discharge gate on a pumper near the nozzle, or the closing of a nozzle.

Communications

The best means of communication between operators in a relay is by radio. Radio communications are fast and accurate, permitting the minor adjustments that result in overall smoothness of operation.

Despite the fact that radios are available on all apparatus and that radio communications are the most effective in relay operations, each fire department should establish standard hand signals that can be used in relay operations. It is possible that full use of radio communications cannot be made at all relays because of overloaded airways or communication blocks caused by difficult terrain or multi-story structures.

Simplified Relay Operations

At many emergencies, time does not permit the calculations necessary to determine the best way to establish a relay operation. Relays must be started immediately, in order to get water on the fire as quickly as possible. Except where extraordinary conditions exist, some general procedures will usually produce good streams on the fire. These simplified procedures are based on the following assumptions:

1. A department or mutual aid group should decide on the volume of water to be supplied. Large-diameter hose can provide an initial 500 gpm minimum. With 4-inch hose, it is possible to supply 1000 gpm over a distance of a mile once the relay has been fully established.

2. The largest size tip used for heavy-stream appliances can be greater than 1½-inch when the standard initial flow with large-diameter hose provides a sufficient rate of flow.

3. In suburban and rural areas, where relays are common, apparatus carry 1600 to 2000 feet of 2½- and 3-inch hose. It is not unusual for 2000 feet of 4-inch hose to be carried on a reel. Reel trucks may carry 4000 feet of 4-inch hose.

Based on these assumptions, relay operations can be set up relatively quickly at fires on level ground or where the terrain presents no particular problem.

There is a difference of opinion among authorities regarding what action should be taken by the first pumper to arrive. Some authorities prefer that, while relay operations are being set up, the first pumper respond directly to the fire, making an immediate attack on it using water available from the apparatus tank. With a tank capacity of 400 gallons, a single 1½-inch line can be used on the fire for approximately five minutes without depleting the water in the tank. This type of attack can be advantageous in some situations.

Other authorities prefer that the first pumper attach to the last available hydrant or water source and make connections as soon as possible. The second pumper lays out its entire bed of hose from the first pumper with either a single line or dual lines, depending on whether hand-held lines or a heavy stream will be used at the fire. The pumper then clears the road and connects the lines that have been laid to the intake side of the pump. As soon as connections are completed, the operator of the first pumper is notified to start the water flowing. Each succeeding pumper follows the same procedure as the second pumper, laying lines from the last pumper in the relay. This procedure is continued until sufficient hose has been laid to reach the fire. A pumper is then spotted near the fire and the working lines are taken from it.

Whether the first pumper should proceed directly to the fire or connect to the hydrant can be decided by the officer in charge at the time of the incident, or departmental policy regarding this type of operation can be established prior to the actual incident. In either case, it is wise to give thought to the problem before the situation occurs.

PUMPING TECHNIQUES

Smooth pumping operations can be accomplished by pump operators who acquire pumping techniques that are adaptable to either hydrant or drafting situations. Some procedures that have proven to be effective follow.

Supplying a Single Line

Supplying a single line is the simplest of the pumping situations. After connecting to the hydrant, the pump operator should let water into the pump, checking the static pressure on the compound gage prior to charging the line. After the static pressure has been mentally recorded, the pump operator should slowly open the discharge gate and allow the working line to fill with water. The discharge gate should be fully opened once the line is full. The throttle should then be slowly advanced until the desired discharge pressure is recorded on the discharge gage. At this time, the residual pressure on the hydrant should be observed so that a calculation can be made as to the amount of water still available from the hydrant. Of course, if this is a drafting situation, or the intake connection has been made to a strong water source, then the mental gymnastics required to determine the amount of water available from the hydrant are not necessary. The pressure control valve should be set once the working line has been stabilized.

Supplying Two Lines

If the pumper is connected to a strong water supply where it is not necessary for the pump operator to estimate the amount of water available from the hydrant, and if both lines are of the same length and size and have identical tips attached, then supplying two lines is not much different from supplying a single line. The discharge gates to both lines should be opened slowly and the lines allowed to fill with water. Both discharge gates can be opened fully once the lines are charged. The throttle can then be advanced to the desired pressure and the pressure control device set.

If there is doubt as to the capability of the hydrant to supply both lines, then it is best to estimate the amount of water available from the hydrant prior to placing the second line into service. Coordination will be required, under these circumstances, between the opening of the discharge gate and the throttle manipulation as the second line is placed in operation. It is necessary that the throttle be advanced as the discharge gate to the second line is slowly opened, the amount of throttle advance being sufficient to maintain the discharge pressure that was established for the first line. If the operation is completed correctly, the discharge pressure will remain constant until the discharge gate to the second line is fully open. At this time, the pressure control device can be set.

A different procedure is required if the two lines to be supplied each require a different discharge pressure, or flow, or if the apparatus is equipped with flow meters. The discharge gates to both lines should be opened slowly and the lines filled with water, then the discharge gates opened fully. Slowly advance the throttle until the discharge pressure requirement (or flow) for the lesser line is reached. At this time, the pump operator should start "feathering" the discharge gate of the line requiring the lesser pressure, while continuing to advance the throttle until the pressure required for the second line is reached. Then the pressure control device can be set.

Feathering a discharge gate refers to closing the gate slowly so as to restrict the flow. This type of operation is best performed with one hand on the handle of the gate to be feathered and the other hand on the throttle. At the same time, the operator should be closely watching the pressure gages or flow meters. With practice, this type of operation can be effectively coordinated.

Placing a Second Line in Operation

Many times it is necessary to take an additional line from a pumper after a working line has been placed in operation. The objective, in these instances, is to add the line without disturbing the flow or pressure to the working line. Disturbing either of these factors may place the firefighters at the nozzle in jeopardy.

The procedure to use when adding another line depends on whether the new line will require a higher or lower pressure than the working line. It is assumed that the pump operator has set the pressure control device for the first line, and that the device is working properly. It is not necessary to disturb the setting on the pressure control device if the pressure requirement for the second line is less than that of the first.

Assume that the original working line required a discharge pressure of 150 psi, and that the pressure requirement for the line to be added is 135 psi. The throttle should be advanced to maintain 150 psi on the working line, as the discharge gate to the second line is partially opened. During the operation, it is necessary for the pump operator to watch the discharge pressure gages to both lines carefully as the throttle is advanced and the discharge gate to the second line opened. The advancing of the throttle and the opening of the discharge gate to the second line should be continued until the pressure gage for the second line reads 135 psi and the discharge gage for the first line reads 150 psi. At this point, the discharge gate on the second line will be partially feathered while the discharge gate to the first line is fully opened. It is not necessary to readjust the pressure control device.

The operation is different if the pressure requirement for the second line is greater than the operating pressure for the working line. Assume that the pressure requirement for the first line is 150 psi and the pressure requirement for the second is 180 psi. It is first necessary to readjust the pressure control device so that it is set above 180 psi. Once this has been done, the pump operator can slowly open the discharge gate to the second line, advancing the throttle at the same time to maintain a pressure of 150 psi on the pressure gage of the first line. This coordinated effort should be continued until the discharge gate on the second line is fully open. At this point, the pressure gages for both lines should read 150 psi. It is now necessary to start feathering the discharge gate on the original line as the throttle is further advanced, maintaining 150 psi on that line. The coordination of feathering and throttle advance should be continued until the gage to the second line reads 180 psi. The pressure control device should then be set to maintain 180 psi.

SUPPLYING MASTER STREAMS

Supplying a master stream is one of the most challenging pumping tasks a pump operator must face. Not only must the pump be operated at or near capacity, but extreme care must be taken not to exceed the discharge capability of the pump. Master streams are those projecting from heavy-stream appliances. A heavy stream is one that is too large to be hand-held by firefighters and is generally considered to be one discharging 400 gpm or more. Some of the heavy-stream appliances used in the fire service are ladder pipes, wagon batteries, portable monitors, deluge sets, and turret nozzles.

To supply a master stream properly, a pumper should be placed close enough to the appliance so that the discharge pressure does not exceed a net pump pressure of 175 psi. At this discharge pressure, the pump should be capable of supplying at least 80 percent of its rated capacity.

Remember that the net pump pressure is the actual pressure produced by the pump. If the pumper is connected to a positive pressure source, such as a hydrant, the net pump pressure is equal to the discharge pressure minus the incoming pressure. Consequently, the net pump pressure would be 175 psi if the discharge pressure was 185 psi while the hydrant residual pressure was 10 psi. If the hydrant residual pressure is 25 psi, the pump discharge pressure can be advanced to 200 psi before reaching a net pump pressure of 175 psi.

The amount of water that can be depended upon and the largest tip that can be supplied by various capacity pumpers at a net pump pressure of 175 psi are shown in Table 9-5. Table 9-6 illustrates the pressure requirement at the inlets of various appliances when smooth-bore tips with a nozzle pressure of 80 psi are used. Assuming that the pumper is connected to a strong water supply where the residual hydrant pressure is 25 psi when the net pump pressure is 175 psi, then the pump could be operated at a discharge pressure of 200 psi when supplying the appliance. Under this condition, the total friction loss in the hose between the pumper and the appliance should not exceed 95 psi for portable monitors and wagon batteries, and 40 psi for ladder pipes and elevated platforms.

The friction loss limitation places a definite restriction on the maximum distance that can exist between the appliance and the pumper supplying it. The maximum allowable distance de-

Table 9-5   Pumping Capacity at 175 psi Net Pump Pressure

Rated Capacity of Pumper	gpm	Maximum Tip Size to Supply (inches)
750	600	1½
1000	800	1¾
1250	1000	2, or 1000 gpm fog nozzle

Table 9-6   Pressure Requirement at Inlet of Various Appliances at 80 psi Nozzle Pressure

Appliance	Pressure Requirement with Smooth-bore Tips (psi)
Portable monitor (Deluge set or deluge gun)	105
Wagon battery (Deck gun or deck pipe)	105
Ladder pipe	160
Elevated platform	160

Table 9-7 Friction Loss for Various Appliance Hose Layouts

Layout	Friction Loss per 100 Feet (psi)		
	1½"	1¾"	2"
two 2½"	20	36	53
three 2½"	10	17	26
four 2½"	7	11	18
one 3"	31	54	84
two 3"	9	15	23
three 3"	4	7	11
one 3½"	13	23	35
two 3½"	3	6	10

pends on the hose layout between the pumper and the appliance. Table 9-7 shows the friction loss for various layouts when supplying either a 1½-inch, a 1¾-inch, or a 2-inch tip. Table 9-8 shows the friction loss for various layouts using large-diameter hose.

Following is an example of the maximum distance that should exist between the pumper and the appliance. Suppose that a portable monitor is equipped with a 1¾-inch tip while the pumper is connected to a hydrant with a residual pressure of 25 psi. Table 9-5 shows that at least a 1000-gpm pumper would be needed to supply this tip. With a hydrant residual pressure of 25 psi, the discharge pressure could be 200 psi, which would allow 95 psi for friction loss in the hose. If the layout between the pumper and appliance consisted of 2½-inch lines, then the maximum distance would be limited to 250 feet (95 divided by 36). Of course, if three 3-inch lines were used between the pumper and the appliance, then the layout could be extended to 1350 feet (95 divided by 7).

The operation becomes more critical when supplying ladder pipes or elevated platforms. Under the same conditions, only 40 psi would be available for friction loss in the hose. If the layout consisted of two 2½-inch lines, then the maximum distance between the pumper and appliance would be limited to 100 feet (40 divided by 36). Even if three 3-inch lines were used, the distance between the pumper and appliance would be restricted to 550 feet (40 divided by 7).

Supplying a 2-inch tip with a single pumper is an extremely critical pumping situation. Consequently, it is best to limit the tip

Table 9-8 Friction Loss in Triple-Duty Hose

Flow in gpm	450	500	600	700	800	1000	1100	1200	1400	1500	1600
Friction loss, psi per 100 feet											
4-inch hose	1	1.5	3.5	4.5	6	11	14.5	17	22	—	—
4½-inch hose	—	—	1	2	3.5	6.5	8	10.5	14.5	17	20.5
5-inch hose	—	—	.5	1	2.5	4	5	7	9	11.5	13.5

Majority of information courtesy of National Fire Hose Corporation

size to 1¾ inches, if possible. Table 9-5 indicates that the pump capacity must be at least 1250 gpm in order to supply a 2-inch tip. With a pump discharge pressure of 200 psi and a residual hydrant pressure of 25 psi, 95 psi is available for friction loss in the hose if a portable monitor or wagon battery is used, and 40 psi is available in the event a ladder pipe or elevated platform is used. If two 2½-inch lines are used, the maximum allowable distance when supplying a portable monitor or wagon battery would be 150 feet, with only 50 feet allowed if a ladder pipe or elevated platform is used. This type of layout would be difficult to set up, unless a hydrant was ideally located in relation to a desired spot for the appliance. If larger hose and more lines are used, then more leeway can be given to the hydrant selection. For example, a distance of 350 feet is permissible if three 3-inch lines are used between a pumper and a ladder pipe or elevated platform.

REVIEW QUESTIONS

1. What is a triple combination pumper?
2. What is the simplest of pumping situations to which a pump operator could be exposed?
3. What is the general procedure for operating from a tank?
4. When should a pump operator secure a supplementary supply of water to the tank when pumping from it?
5. What are some of the various methods of providing a supplementary supply of water to the tank?
6. What are some of the methods of breaking an air lock when pumping from a tank?
7. What are the three common types of suction hose currently in use?
8. What is hard suction hose?
9. What change has taken place in hydrant hookup pro-

Pumper Operations

cedure as a result of the adoption by many communities of the use of large-diameter hose?

10. What are some of the factors affecting hydrant hookups?

11. What are some of the methods used to identify the location of hydrants?

12. What are some of the considerations given to selecting a hydrant?

13. What are some of the hydrant hookup variables?

14. What are some of the factors that should be considered when making a decision as to how to connect up to a hydrant?

15. If possible, which intake inlet should be used?

16. If possible, to which side of the apparatus should the suction or supply hose be connected?

17. In general, where should the apparatus be spotted for a side hookup? For a front hookup? For a rear hookup?

18. What is meant by the term "blocked hydrant"?

19. Give some of the common blocked hydrant situations.

20. What are some of the methods available for improvising an intake hose?

21. What should be the first thought of a pump operator when an intake hose bursts?

22. What method can be used to estimate the amount of water available from a hydrant outlet?

23. What is the best indicator that the outlet capacity of a hydrant has been reached?

24. Explain how a second supply hose can be connected to a pumper in the event an unused gated intake is not available.

25. Explain the procedure for putting a pump into operation.

26. What is the first thing the pump operator should do upon leaving the seat of the apparatus in preparation for pumping?

27. Explain how to make a temporary shutdown when working from a hydrant.

28. Explain the steps necessary for a permanent shutdown when pumping from a hydrant.

29. What is meant by drafting operations?

30. What is the difference between the gage connected to the intake side of a pumper and the one connected to the discharge side?

31. What is absolute pressure?

32. What is the normal atmospheric pressure at sea level?

33. How are readings below atmospheric pressure measured on the compound gage?

34. What is the equivalent in inches of mercury of an atmospheric pressure of 14.7 psi?

35. What is the definition of lift?

36. Theoretically, how high could a pump lift water at sea level?

37. What is the practical maximum lift that should be expected from a pumper at sea level?

38. What causes water to move from one location to another?

39. Explain what happens when a pump operator drafts water.

40. What is a good rule of thumb for estimating the height that water will rise due to the pressure reduction within the pump when drafting?

41. What effect does altitude have on drafting operations?

42. What effect do weather changes have on drafting operations?

43. What factors determine the amount of water a pumper can be expected to lift?

44. If possible, how far below the surface of the water and above the bottom should the suction strainer be kept when drafting?

45. How soon should water be received when drafting?

46. What are some of the possible problems if water is not received within the allocated time?

47. Where is the most likely spot for an air leak in the suction hose when drafting?

Pumper Operations

48. What are the possible problems if the throttle is advanced without a corresponding increase in pressure?

49. What is the general procedure for a temporary shutdown when drafting?

50. What is the purpose of establishing a broken connection situation at an emergency?

51. What is meant by a broken connection?

52. What are some of the possible makeshift reservoirs that can be set up for broken connection situations?

53. What are relay operations?

54. What are some of the factors that should be considered when establishing a relay?

55. Where should the smallest pumper in a relay be placed?

56. What is the maximum flow that should be attempted through a 2½-inch line in a relay operation?

57. What effect has the adoption of large-diameter hose had on relay operations?

58. Why can't the maximum benefit available from the use of large-diameter hose normally be achieved in relay operations?

59. What grade should be assumed in a mountainous area if the grade is not known?

60. What is the formula normally used for determining the maximum distance apart that pumpers should be spaced in a relay operation?

61. In general, what should be the maximum discharge pressure of pumpers used in a relay?

62. At what pressure should water be provided to the next pumper in a relay?

63. What method can be used to replace a defective pumper in a relay operation without shutting down the relay?

64. At what point in a relay is a ruptured hose line first detected?

65. What action should be taken by a pump operator in coping with a ruptured line?

66. What might be indicated by an unaccountable increase in pressure?

67. Explain the method for setting up a simplified relay operation.

68. Explain the method for supplying a single line when pumping.

69. What procedure should be used to supply two lines of the same size and length and having the same size tips? Of supplying two lines each requiring a different discharge pressure?

70. What is meant by the term "feathering"?

71. What is the procedure for placing a second line in operation, if the second line requires less pressure than the first line? If it requires more pressure?

72. What is the primary problem with supplying master streams?

73. If possible, what should be the largest tip used on a master-stream appliance?

10

Aerial Ladder Operations

The aerial ladder has been used in the fire service for over a century. The first aerial ladders were made of wood and were transported to fires on horsedrawn apparatus. These ladders were manually raised; the efforts of the entire crew were required to place them into operation. The earlier models were restricted to a height of 85 feet.

The first spring-loaded aerial ladders were placed in service shortly after the turn of the century. The compressed springs were used to carry the ladder out of the bed; however, raising the ladder to the vertical, extending it, and rotating it still required the cranking action of the entire crew.

The power from the engine was first used to assist in aerial ladder operations in the late 1930s. Continuous improvements finally resulted in the development of a fully hydraulic aerial ladder. Such an aerial ladder uses only one individual to perform the various functions, thus freeing the remainder of the crew for other assignments.

The length of an aerial ladder is measured by a plumb line which extends from the top rung of the ladder to the ground, when the ladder is fully extended at its maximum elevation. The maximum elevation with most ladders is at an angle of approximately 85° from the ground. Maximum length is achieved through the combined use of three or four ladder sections. The most common ladder lengths are 75, 85, and 100 feet; however, some specially built ladders for use in the United States extend to a height of 135 feet. There are a few aerial ladders in service in Europe that extend over 200 feet. With an increasing number of high-rise buildings springing up around the country, a closer look will have to be taken to the possibility of extending the maximum length of the aerial ladder past the 135-foot mark in the United States.

TYPES OF AERIAL LADDER APPARATUS

There are three general types of aerial ladder apparatus in common use. The particular type a department adopts will depend upon the specific needs of the department. Street layouts, terrain, traffic, district hazards, and engine house storage space all influence the decision. The three types are referred to as the *rear-mounted aerial*, the *midship-mounted aerial*, and the *tractor-trailer combination* (see Figure 10-1). While each of these basic types performs the same function, each appears to have certain advantages.

The rear-mounted aerial is built on a single, two-axle chassis. The turntable, gages, and controls are located at the rear of the apparatus.

Like the rear-mounted aerial apparatus, the midship-mounted aerial is built on a single, two-axle chassis. The turntable, gages, and controls are located directly behind the cab, making it a simple matter for the driver to move from the cab to the operating position.

The tractor-trailer apparatus is a six-wheel, three-axle piece of equipment with a fifth wheel at the rear of the tractor. The fifth

Aerial Ladder Operations

wheel is used to connect the tractor to the trailer. The tractor houses the engine, the driver's cab, and seating arrangements for the crew. The trailer carries the aerial ladder, the ground ladders, and all the miscellaneous firefighting equipment. Maneuvering this type of apparatus requires the combined efforts of the driver and tillerman.

The tractor-trailer combination (see Figure 10-1A) is more maneuverable than the other two; it can be taken into spots that

Figure 10-1A   A Four-Section Seagrave Tractor-Trailer Ladder Apparatus. *Courtesy of Seagrave Fire Apparatus, Inc.*

Figure 10-1B   A 100-Foot, Rear-Mounted Aerial Ladder Apparatus. *Courtesy of Seagrave Fire Apparatus, Inc.*

Figure 10-1C A 65-Foot Midship-Mounted Aerial Ladder Apparatus.
Courtesy of Peter Pirsch & Sons Co.

are inaccessible to either the rear- or midship-mounted apparatus. Because it can be jackknifed, the tractor-trailer combination can also be positioned in a manner that provides greater stability, when the aerial ladder must be operated in other than an in-line direction. The tractor-trailer combination is much longer than either of the other two, however, and thus requires additional space for housing or storage.

Most aerial ladder apparatus carry a sufficient number of ground ladders, firefighting tools, and equipment needed to comply with national recommended standards. If the apparatus also carries hose, a water tank, and a pump, it is called a *quint*.

THE DUTIES AND RESPONSIBILITIES OF AN AERIAL LADDER OPERATOR

The duties and responsibilities of an aerial ladder operator can be compared with those of a pump operator. The responsibility for operating an aerial ladder should only be assigned to an individual who has the skill and knowledge to operate the equipment in a smooth, safe, and efficient manner. To carry out these responsibilities effectively, an aerial ladder operator must:

Aerial Ladder Operations

1. Be capable of spotting the apparatus properly for raising the aerial to a given location
2. Be able to stabilize the apparatus properly prior to raising the aerial ladder
3. Know the location of every operating control
4. Know what each control does and how it works
5. Be able to operate all controls smoothly and safely
6. Know the location of every safety device, how each one works, and how to place each in operation
7. Be familiar with the maximum loads that can be safely placed on the aerial ladder under various operational conditions
8. Be aware of the various methods available for operating the aerial ladder in unusual circumstances

When you think of the seriousness of making aboveground rescues and the possible negative results of improperly spotting, stabilizing, or operating the aerial ladder, the importance of practice and training becomes clear. A firefighter cannot learn to operate an aerial ladder effectively by reading a book or participating in classroom training sessions. This can only be accomplished by taking the apparatus out of quarters and practicing, supplementing the practice with experience on the fireground.

AERIAL LADDER CONSTRUCTION

It is essential that aerial ladders be as light as possible, yet be constructed in such a manner and of such a material that they have sufficient strength when operated in either a supported or an unsupported position. The type of construction best suited for these operations is referred to as *truss construction*. The type of truss construction normally used in aerial ladders consists of an upper and lower beam supported by a rigid framework of bars and rods that run perpendicular to and/or at an angle to the two beams (see Figures 10-2 and 10-3). This method of construction permits tension and compression stresses to be distributed for maximum strength and provides almost equal strength in either a tension or a compression situation.

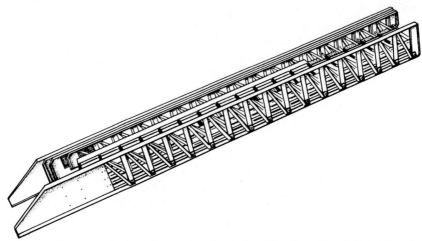

Figure 10-2 A Close-up View of a Pirsch Four-Section Aluminum Alloy Aerial Ladder. *Courtesy of Peter Pirsch & Sons Co.*

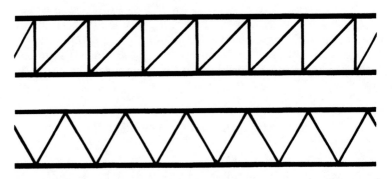

Figure 10-3 Two Types of Truss Construction Used on Aerial Ladders

Tension strength is the strength required to resist breaking when a material is stretched. A good example of tension strength is that in a rope during a game of tug-of-war. *Compression strength* is that necessary to prevent a material from being squashed when force is applied. The studs in a house are subjected to compression from the weight of the roof resting on the upper plates. When an aerial ladder is operated in an unsupported position, the upper beam is subjected to tension and the lower beam to compression (see Figure 10-4). The situation is reversed when the ladder is operated in a supported position—the upper beam is subjected to compression and the lower to tension.

While an aerial ladder has almost equal strength when operated in either a tension or a compression situation, the construction is such that the ladder can only support its designed weight when the forces are acting in a direction perpendicular to the rungs. A force applied in any other direction will tend to twist the ladder; consequently, side stresses should be avoided. A twisting force can be applied if the ladder comes to rest on one beam only, or, when the ladder is being used for ladder pipe operations, whenever the stream is projected in a direction other than one perpendicular to the rungs. The ladder should never be handled, overloaded, or used in a manner for which it was not designed.

Aerial ladders are fabricated of either rust-resistant steel and steel tubing or heat-treated aluminum and aluminum tubing. Fabrication may be by either welding or riveting. The ladder beams are of hollow reinforced I-beam or similar construction. While the beams are by no means fragile, ordinary care should be exercised in handling; an excessive blow on the bottom of the beam, such as would occur if the ladder were dropped violently over a window

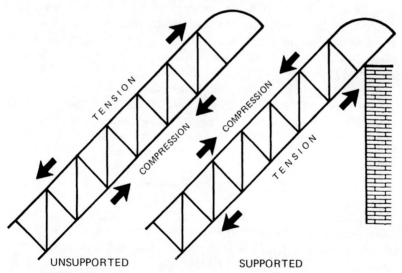

Figure 10-4 Compression and Tension Forces Working on Aerial Ladders

ledge, might dent the beam and therefore affect its smooth action over the rollers.

The rungs of the ladder are usually made of hollow tubing covered with rubber. The rubber provides insulation and makes climbing easier.

OPERATING CONTROLS AND SYSTEMS

Aerial ladder apparatus are a lot like fires. While they are not exactly alike, all are somewhat similar. Provisions are made for stabilizing the apparatus, and various controls and systems are provided for raising, rotating, and extending the ladder. Additional provisions are built in for securing the ladder in position once the operating height has been reached and for ensuring a degree of safety when things go wrong.

It is not the intent of this text to provide operating instructions for every type of aerial ladder in general use; however, the intent is to familiarize students with the various types of controls and safety features that are used to achieve desired results. It should not be difficult for students to transfer the general knowledge presented here to the operation of a particular apparatus, once they fully understand the principles involved. Of course, it is essential that aerial ladder operators be thoroughly familiar with the operating instructions, safety procedures, and maintenance procedures as established by the manufacturer for the particular apparatus to which they are assigned, rather than depend on their general knowledge of how an aerial ladder works.

The Power Take-Off

The hydraulic pump for fully and partially hydraulically powered aerial ladders is turned by means of a *power take-off* from the road transmission. The interconnect between the transmission and the pump is a universal joint shaft.

The power take-off is engaged by the operator prior to leaving

Aerial Ladder Operations

the cab. The power take-off control may be either a lever or a push-pull type control. The control is generally marked AERIAL.

Some apparatus are equipped with an auxiliary hydraulic pump for use in the event of loss of power from the main pump. The auxiliary pump is usually powered by a 12-volt battery system with an actuating switch at the control panel. Other apparatus have a hand crank for powering the hydraulic pump in an emergency.

The Turntable

The *turntable* is the base on which are mounted the aerial ladder and all the mechanisms for operating and controlling its various movements. It is normally of sufficient size to provide standing and walking room for the ladder operator and those who will use the ladder. While the construction of various manufacturers' turntables differs to meet individual needs, the general principle of design and function is basically the same.

The turntable used by Seagrave, for example, is constructed of several pieces of steel plate, electrically welded into one section of plate having great strength and rigidity. On the bottom are machined ball races that register with similar races in a large-diameter heavy steel ring bolted to the chassis frame. A large number of hardened steel balls in these races carry the weight of the ladder and the mechanism and permit free rotation of the turntable. Suitable gear teeth are cut on the outer periphery of the ring and are engaged by gears driven by the rotation motor, which furnishes power for rotating the turntable. Another steel ring, bolted to the underside of the turntable, has a ball race that engages inversely with another in the gear ring. Steel balls in this race take the reaction of the tipping load of the aerial ladder. A grease fitting for lubricating the ball races is located on top of the rear of the turntable.

The primary function of the turntable is to provide for rotation of the aerial ladder. Turntables are designed so that they can be turned on a horizontal plane for 360° in either a clockwise or a counterclockwise direction. They can be operated continuously in one direction.

Figure 10-5 The Location of the Operator's Control Pedestal on the Turntable. *Courtesy of Peter Pirsch & Sons Co.*

The rotation of the turntable is controlled by a lever located on the control pedestal. Most ladders are locked in position whenever the operating lever is in neutral; however, some aerial ladders are designed with an additional rotation lock. Care must be taken to ensure that this lock is disengaged before attempting to rotate the ladder.

Most aerial ladders are equipped with some type of mechanism that permits hand rotation of the turntable under emergency or selected conditions. Some manufacturers recommend that hand rotation be used whenever a ladder pipe is in operation. This provides a slow, smooth movement that eliminates almost entirely the possibility of overcontrolling.

The Operator's Control Pedestal

The operator's *control pedestal* (see Figures 10-5 and 10-6) is located on the turntable, either to the right or to the left of the

Aerial Ladder Operations

aerial ladder. All controls and instruments necessary for normal operation of the aerial hoisting mechanism and ladder are usually located on or near the control pedestal. An explanation of the more common controls and instruments is included in this section.

When standing in front of the control pedestal in the operating position, the aerial ladder operator faces the end of the ladder. This arrangement provides the operator with complete visual control of the ladder at all times. From this position, he can sight

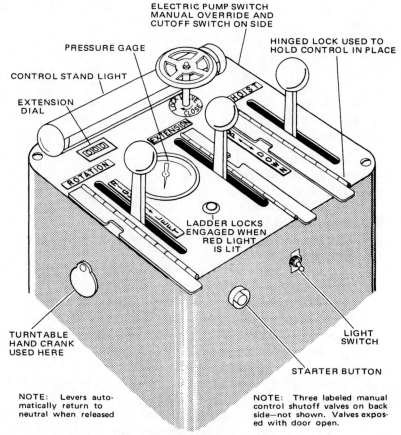

Figure 10-6   The Location of the Controls on a Pirsch Operator's Control Pedestal. *Courtesy of Peter Pirsch & Sons Co.*

down the ladder in preparation for extending and lowering into a building. The pedestal itself is about waist high and is generally designed with a slope toward the operator. The three main controls on the pedestal are the levers for raising, extending, and rotating the ladder. Some of the other controls or instruments that may be located on or near the pedestal are a starter switch, a light switch, a hydraulic pressure gage, a cylinder pressure gage, an extension indicator, a ladder lock control, a ladder hold-down lock control, and provisions for manually controlling certain ladder movements.

There is no standard for the location of the controls on the pedestal, nor is there any standard method of labeling the various controls. For purposes of explanation, the three primary controls will be referred to as the hoisting, extension, and rotating controls.

Operating Controls

While each of the operating controls performs a different function, there are certain characteristics or procedures that apply to all. These include:

1. The movement of the control requires a positive action on the part of the operator, beyond merely moving the handle back and forth. Some controls require that the lever be pushed down before it can be moved from the center or neutral position. This is a safety feature that prevents the lever from being accidentally engaged in the event someone leans against it.

2. Levers are usually self-centering, returning to neutral in the event the operator's hand slips off the control.

3. The speed of movement depends on the distance the lever is moved from the center position. With hydraulic systems, the movement of the lever causes an increase in engine speed to provide the proper hydraulic pump speed and flow. There is one exception—speed does not increase when the ladder is being lowered, as the lowering speed is controlled by metering valves in the hoisting cylinders.

4. All controls should be moved slowly and continuously. They should never be jerked or slammed.

5. Normal procedure is to use one control at a time; how-

ever, with practice, an operator can learn to use two controls simultaneously.

Some apparatus are designed with manual controls for each of the ladder control functions; others are not. The manual controls are used in the event of power failure and sometimes in critical routine operations, such as rotating the turntable when a ladder pipe is in operation.

The Hoisting Control

The hoisting control is used for raising and lowering the ladder. Some apparatus are equipped with a hold-down locking device which must be disengaged before the ladder can be lifted out of the bed. Control of the movement of the ladder is accomplished with hoisting cylinders. The hoisting cylinders are constructed of heavy seamless steel tubing electrically welded into suitable steel flanges, with the hole on the inside accurately bored and honed to a smooth finish. The lower flange is equipped with trunnions over which are fitted suitable bronze trunnion boxes. These boxes are bolted into a machined recess in the turntable.

The hoisting cylinders for Seagrave aerials have a 6-inch inside diameter and a 24$\frac{5}{16}$-inch stroke. The cylinder barrels are hard chrome plated inside to resist wear. The piston rod is 2½ inches in diameter and is also hard chrome plated. The piston is fitted with two synthetic sealing cups, and the center of the piston is fitted with a non-stick-surface ring to prevent the piston from rubbing the cylinder walls, thus providing for long cylinder life.

The cylinder head and the seal carrier on the rod end of the piston are sealed to the barrel and to each other with O-rings. The piston rod is provided with an outer dirt seal and a multiple-ring inner hydraulic oil seal. The piston rod is hard chrome plated to aid in resisting wear. Operating pressurres range from 850 to 1000 pounds, although the system can operate at higher pressures where required.

There is a safety check valve installed in the inlet in the bottom of each cylinder. This valve has a calibrated orifice which prevents a rapid falling of the ladder if the connecting tubing bursts or is completely severed. The bottom inlet is connected to the control valve through a Chiksan swivel and steel tubing. The top or low-pressure side has a hydraulic hose connection to the control valve.

Hydraulic pressure to the cylinder is controlled by the hoisting lever. When the hoisting lever is moved slightly from neutral toward the hoisting or up position, a simple cam and lever mechanism is actuated, resulting in a buildup of the pressure of the oil to that at which the relief valve has been set. Further movement of the hoisting lever moves the hoist control valve to a position that permits the oil under pressure to flow to the bottom of the hoist cylinder, where it acts upon the pistons, forcing the piston rods out and thereby raising the ladder. The oil that was in the cylinders above the pistons flows out of passages to the reserve tank and back to the pump. If the pump supplies oil in excess of that required for hoisting the ladder, the surplus passes through the relief valve and directly back to the reserve tank and pump. Whenever the hoisting lever is moved to neutral, the oil flow is stopped and the ladder is locked at whatever angle of inclination it may have attained. The procedure is reversed when the hoisting lever is moved to the down position for the purpose of lowering the ladder.

Some hydraulically operated systems are equipped with a hand-operated lock valve that will lock the oil pressure in the cylinder when it is closed. This locks the ladder in position, preventing it from lowering even if lock pawls are not positioned and the hydraulic pump is not pumping. The lock valve should be used as a precautionary measure whenever the ladder is to remain in a set position for any period of time. The lock valve must be disengaged before the hoisting control can be used.

The method recommended for raising the ladder is first to move the hoisting lever slowly from the neutral to the hoisting or up position. The raising speed can be increased by further movement of the lever, once the ladder has cleared the bed. The raising movement should be slowed as the ladder reaches its desired height.

The Extension Control

The ladder is extended and retracted by the use of cables. The cables are wound on a cable drum that is powered mechanically, hydraulically, or electrically, depending on the manufacturer's design. On a fully hydraulically operated aerial, the movement of the extension control to the extension position causes hydraulic pressure to rotate the drum and therefore extend the ladder. When the lever is returned to neutral, the oil circulation is stopped and the ladder will remain at the extended height. Moving the hoisting control to the retraction position reverses the direction of the extension drum, which retracts the ladder.

Some ladders are equipped with extension gages that show the operator the length that the ladder is extended. Automatic stops are also provided on some systems, to stop the ladder at full extension and full retraction positions. At least one manufacturer uses a warning horn to inform the operator as the ladder approaches full extension. An automatic switch cuts the throttle to a predetermined idling speed when the warning horn is activated.

Movement of the extension control lever should be gradual and smooth. The extension of the ladder can be speeded by increased lever movement. The movement should be slowed as the ladder approaches the desired extension height by returning the lever toward the neutral position.

With some systems, a manual override switch is used to speed up the engine in the event of an emergency.

The Rotation Control

Rotation of the turntable and thus the aerial ladder is accomplished through the use of the rotation control lever. Movement of this lever in one direction causes the turntable to rotate in a clockwise direction. Movement in the opposite direction results in a counterclockwise rotation. Continuous travel of the turntable in either direction is possible.

The degree of lever movement from the neutral position determines the speed with which the ladder and turntable rotate. The lever should be moved slowly so as to maintain a positive but smooth operation. Rotation should be slowed as the ladder approaches its intended mark in order to prevent any whipping of the ladder.

A Load Indicator Gage

Some manufacturers install a *load indicator gage* on or near the operator's control pedestal. This gage shows the stress on the beams of the ladder when operated in either the supported or unsupported position. A green mark is used for safe, a yellow for caution, and a red for danger. This gage eliminates the guesswork as to how many people can be supported on the ladder or the effect of ladder pipe operations.

Hoisting Cylinder Pressure Gage

Information similar to that provided by the load indicator gage is available to the operator of some aerial ladders from a gage that indicates the pressure at the bottom of the hoisting cylinder. This gage provides an indication of the load on the ladder when the ladder is unsupported at the outer end. The pressure on the gage increases as the load on the ladder is increased and decreases as the load is decreased.

It is important that the operator not allow the pressure on this gage to exceed the working pressure in the hydraulic system. Such pressure could occur if too many people were allowed on the ladder at one time. This condition could cause the ladder to fall, as the working pressure in the system would no longer be able to hold it up.

Insufficient pressure on the gage is also a sign of a dangerous situation. This condition could cause the ladder to extend beyond the vertical position and tip in the opposite direction. A good operating practice is to ensure that the pressure does not fall below that recommended by the manufacturer.

The Hydraulic System Pressure Gage

A hydraulic system pressure gage is generally provided on the pedestal. It indicates the operating pressure in the hydraulic system. The system pressure should be maintained in accordance with the manufacturer's recommendation.

The Extension Gage

The extension gage, when provided, shows the height reached by the ladder when at approximately an 80° angle from

Aerial Ladder Operations

the horizontal, or the equivalent extension at any angle. It is usually driven by gearing with the extension motor.

The Rung Alignment Indicator

Some apparatus have an indicator light on the pedestal to inform the operator that the rungs are aligned for climbing.

The Lock Valve Control

The lock valve is normally carried in the open position. When the valve is closed, oil is locked in the bottom of the hoist cylinder; this prevents the ladder from being lowered. This valve should be closed if the ladder is to be left in an elevated position for an extended period of time.

The Rotation Lock Control

The turntable can be locked against undesired rotation by the closing of the rotation lock valve. The valve is normally carried in the open position.

The Hand Rotation Control Valve

The hand rotation valve should be kept closed, except when it is desirable to rotate the turntable by hand. Opening this valve on a Seagrave apparatus causes hydraulic oil to bypass the rotation motor so the apparatus can be operated manually.

The Starter Button

A starter button is provided to enable the operator to start the engine from the control pedestal. It is used primarily for restarting the engine in the event of a stall.

The Throttle Button

The engine speed can be raised at any time by the use of the throttle button. This might be necessary when testing the hydraulic system by means of the test valve.

The Light Switch

The light switch is used to light up the control stand light and additional accessories.

The Master Power Switch

The master power switch is used to control electrical power to various subcontrol switches.

Stabilizing Devices

An extended aerial ladder is a lever with the fulcrum centered on or near the turntable, somewhere near the base of the ladder. The weight of the ladder itself places certain stresses on the apparatus; however, the degree of stress varies with the positioning of the ladder. The least amount of stress is placed on the apparatus when the ladder is raised to its maximum extension angle, with the tip supported on a building. If the tip of the ladder is not supported when the ladder is raised to its maximum elevation angle, then the stress on the apparatus is due to the weight of the ladder plus any ladder loading. The greatest amount of stress is placed on the apparatus when the ladder is fully extended in a cantilever position horizontal to the plane of the earth. Various stresses take place between this maximum stress position and the fully elevated position and will shift as the ladder is rotated.

The stress at the fulcrum is transmitted from the turntable to the apparatus. If some type of stabilizing device is not used, the forces may be sufficient to damage the springs and possibly turn the apparatus over. Damage would also occur if the apparatus were allowed to move when the ladder was extended in a supported position. Three types of devices are used to assist in stabilizing the rig: wheel chocks, axle jacks, and stabilizing jacks.

Wheel Chocks

Wheel chock blocks are used to prevent forward or reverse movement of the apparatus in the event of failure of the emergency or hand brake. They should be properly positioned every time the aerial ladder is to be placed in operation. Failure to position the chock blocks could result in severe damage to the ladder, should the apparatus move after the ladder is set in position in a window, on a roof ledge, or against a wall. The wheel chocks should be placed both in front of and behind the wheels if the apparatus is spotted on level ground, and on the downhill side when it is spotted on a slope. It is good operational practice to assign the duty of placing the wheel chocks to a particular company member in order to ensure that this task is performed.

Axle Locks

Axle locks are used on tractor-trailer apparatus to lock the tractor frame rigidly to the tractor rear axle. The axle locks make the chassis springs inoperative, which prevents excessive sag to the side when the ladder is in use. These locks are used in conjunction with stabilizing jacks to provide complete apparatus stability.

The axle locks are placed in operation by engaging the lock control. It is normally better to engage the controls prior to jack-knifing the apparatus, as the locks are more easily engaged at this time. Prior to lifting the ladder out of the bed, a visual check should be made to ensure the locks are engaged, as sometimes only one side will engage when the apparatus is sitting on an uneven surface. Raising the ladder prior to engaging both locks may seriously affect the stability of the apparatus.

As an example of the general construction of axle locks, let us look at those used on Seagrave apparatus (see Figure 10-7). The locks are applied by moving the lever on the left-hand side of the tractor to engage the notches on the quadrant. The clamps consist of a pair of discs on each side of the tractor frame, one of the discs attached by a lever and link to the rear axle and the other attached

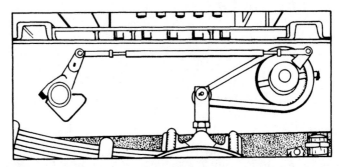

Figure 10-7 Axle Locks Lock the Tractor Frame Rigidly to the Tractor Rear Axle, Making the Chassis Springs Inoperative. *Courtesy of Seagrave Fire Apparatus, Inc.*

to the frame. A cam is so arranged as to force these discs together. The discs have a large number of serrations or teeth on their outer edges which engage and securely lock each to the other when the hand lever is forced to the lock position. If engagement does not occur readily, it may be necessary to stand on the running board in order to cause a slight shift in the discs, permitting the teeth to engage.

Stabilizing (Ground) Jacks

Stabilizing jacks are used to transmit the load placed on the ladder through the ladder fulcrum at the turntable to the ground. In addition, they increase the carrying capacity of the apparatus by providing a steady and widened operating platform. Stabilizing jacks may be either manual or hydraulic. Both types are used on some apparatus.

Screw-type, hinged ground jacks are attached to heavy horizontal beams or other supporting devices which are normally stored so that they do not extend beyond the width of the apparatus (see Figure 10-10). Extension beams are held in place by a steel pin that locks the extension channel beam to the permanently mounted beam. The pin is pulled and the beam extended to its maximum to place the jacks in position for lowering. The wider the beam, the more stability provided. The locking pin is replaced to hold the extended beam in position once the maximum exten-

Aerial Ladder Operations

sion is reached. An additional pin holds the jack in place under the beam. The pin is removed to lower the jack.

Metal foot plates are generally provided with the ground jacks. These foot plates increase the area of ground contact and therefore provide for a better transmission of the load from the frame to the ground. The plates should always be used when operating on soft ground. While their use may not be absolutely necessary when operating on pavement, it is good standard procedure to use them at all times.

Screw jacks should always be set in a straight up-and-down position. If the jacks are set at an angle and the ladder is extended to one side of the truck, the jack on the other side will crawl toward the perpendicular; then, when the aerial is swung in line with the truck, the trailer will be forced into a tilt. When set vertically, as is correct, the off-side jack may leave the ground an inch or more, but it will reset itself correctly when the ladder is brought in line with the truck.

Hydraulically operated ground jack outriggers are used on

Figure 10-8 A 135-Foot Aerial Ladder Apparatus Using Underslung, Crisscross Outriggers. *Courtesy of Emergency One, Inc.*

Figure 10-9 A Wider Base for the Aerial Ladder is Achieved Through the Use of Ground Jacks. *Courtesy of Pierce Manufacturing Inc.*

some apparatus (see Figures 10-9 and 10-10). These outriggers are stored in such a manner that they do not extend beyond the width of the apparatus when they are not in use. The outriggers are lowered hydraulically until they come into contact with the ground, removing the load from the frame and providing a wider base to absorb the leverage and back thrust of the ladder when it is

Aerial Ladder Operations

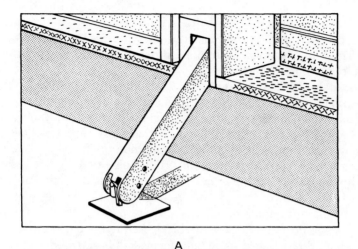

A

A Pirsch "A"-Frame Type
Hydraulic Outrigger Jack.

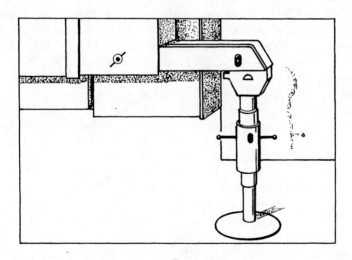

B

This screw extendible-type
ground jack is used on
Seagrave tractor-trailer
apparatus.

Figure 10-10   Ground Jacks May Be Either Manual or Hydraulic

raised at various angles out of line with the chassis. At least one manufacturer uses underslung, crisscross outriggers (see Figure 10-8). These outriggers are mounted under the frame, which minimizes loss of compartment space.

Operators must use extreme caution when retracting the hydraulic ground jack outriggers. The aerial ladder operator must be certain that all persons are a safe distance away from the truck before retracting the outriggers. The truck body may have been elevated a few inches from ground level when the outriggers were set in place. If so, the rear of the truck body may settle down quickly when they are retracted, severely injuring any knee or leg that is in the path of the settling vehicle body or rear step. As a safety precaution, some apparatus are equipped with a warning light on each outrigger jack assembly. These lights come on automatically and begin flashing when the system is actuated.

Ladder Cables

Ladder cables are used on some apparatus to extend and retract the ladder sections. The cables are usually made of galvanized plow steel that is many times stronger than necessary to carry the load imposed. The cable is attached to an extension cable drum which may be mechanically, hydraulically, or electrically powered. The following example shows the installation of a cable system used on Seagrave apparatus (see Figures 10-11 and 10-12).

The main hoisting cable comes off the bottom of the extension drum. It passes under the main ladder section, through a pulley mounted on the upper main head block, and down between the rungs of the main and second sections, where it attaches to the lower head block of the second section. This applies power directly to hoist the second section.

The *fly ladder*, or third section, is hoisted by means of a cable loop, one end of which is attached to the upper head block of the main section. It passes over a pulley in the upper head block of the second section and is brought down and attached to the lower head block of the fly. Any outward motion of the second section will carry the fly out at twice the speed of the second section.

The second section is retracted by means of a cable attached

Aerial Ladder Operations

to its lower head block and to the extension drum. The fly is retracted by means of a small-size cable loop, the reverse of the extending loop (see Figure 10-11).

On four-section ladders, the cables are arranged in the same manner but with an additional cable loop for extending the fourth section and an additional cable loop for retraction (see Figure 10-12).

It is not necessary that the cables be tight throughout the ladder travel, nor should they be so loose as to drag excessively on the pulley sheaves or wind improperly on the cable drum.

The cables that lead off the drum—that is, the main hoist and

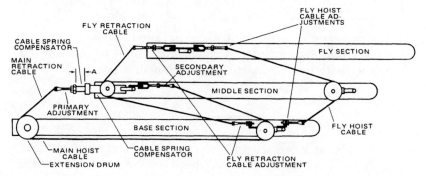

Figure 10-11 A Three-Section Aerial Ladder Cable and Adjustment Diagram. *Courtesy of Seagrave Fire Apparatus, Inc.*

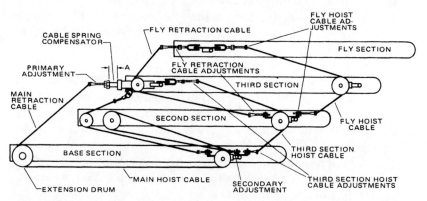

Figure 10-12 A Four-Section Aerial Ladder Cable Arrangement and Adjustment Diagram. *Courtesy of Seagrave Fire Apparatus, Inc.*

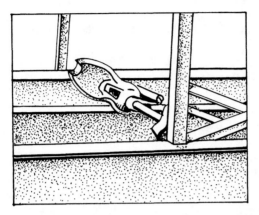

Figure 10-13 When the Ladder Locks Are Engaged, All Sections Are Accurately in Line for Climbing. *Courtesy of Seagrave Fire Apparatus, Inc.*

main retraction cables—can both be adjusted at the spring compensator (see Figure 10-11). Adjustments should be made in accordance with the manufacturer's recommendations.

Ladder Locks

Ladder locks are used to keep the ladder sections from drifting down, once the ladder has been extended to the desired height, and to take the load off the extending cables. The locks also align the rungs of all the sections. The ladder locks are normally controlled by a small lever located at the base of the ladder, near the control pedestal (see Figure 10-14).

On a three-section ladder, the lever is connected by cable to *dogs* or *pawls* near the top of the main ladder section and near the top of the second section. The indicator near the lever will generally read IN or OUT or OFF or ON. When the lever is placed in the OUT or OFF position, the pawls (Figure 10-13) are held clear of the rungs of the ladder. This position is generally used when the ladder is being extended or lowered. Placing the lever in the IN or ON position will result in the pawls coming to rest on the ladder rungs.

The normal use of the ladder locks is as follows:

The lever is placed in the OFF or OUT position when the ladder is to be extended. The lever is moved to the IN or ON position when the desired extension is reached. The ladder is then retracted slowly until the pawls come to rest on the ladder rungs,

Aerial Ladder Operations

and the retracting pressure is continued until the tension is taken off the cables. The lever should be kept in the IN or ON position as long as the ladder is extended.

At least one manufacturer uses a red indicator light on the control stand to show that the ladder locks are engaged. The light remains lighted until the locks are disengaged, giving warning in the event the locks fail to release when the operator is preparing for lowering.

When the ladder is to be retracted it should first be extended until the pawls are clear of the supporting rungs. The lever should then be placed in the OUT or OFF position. The pawls will remain clear of the ladder as it is retracted.

The ladder can be extended when the level is in the IN or ON position; however, this should normally only be done when a short extension is to be made.

The Inclinometer

Limitations on the use of aerial ladders depend on the angle of extension and whether the ladder is operated in a supported or an unsupported position. A device provided on most aerial ladders with a length of 75 feet or more informs the apparatus operator of operating limitations. It is referred to as an *inclinometer*. The most common type of inclinometer is mounted on the aerial ladder and is free to swing with the elevation of the ladder (see Figure 10-14). The apparatus operator should inspect it occasionally to ensure that it is free to swing. The inclinometer indicates the maximum safe extension of the ladder at various angles when operated supported, unsupported, or as a water tower. The inclinometer readings are conservative and provide for perfectly safe operation under all conditions. Indicated operating limitations should only be exceeded with care and good judgment. Some inclinometers are lighted for night use.

Figure 10-15 shows an inclinometer for a Pirsch 100-foot tractor-trailer aerial ladder. This inclinometer is equipped with a light for night operations. It shows the safe extension limits when the apparatus is properly jackknifed (60°). The readings shown indicate that the ladder is raised to a 40° angle. At this angle the ladder should not be extended beyond 60 feet when operated

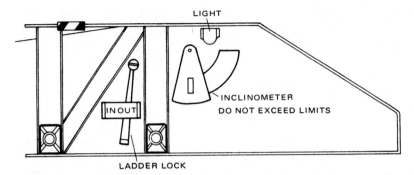

Figure 10-14 The Ladder Lock Control Lever Is Located at the Base of the Ladder. *Courtesy of Peter Pirsch & Sons Co.*

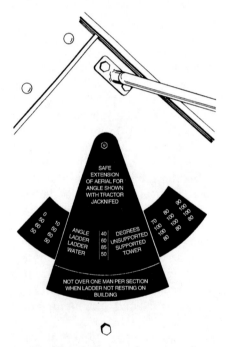

Figure 10-15 An Inclinometer for a Pirsch 100-Foot Tractor-Trailor Apparatus. *Courtesy of Peter Pirsch & Sons Co.*

unsupported. It may, however, be extended to 85 feet if supported. It should not be extended at this angle beyond 50 feet if operated as a water tower.

Similar inclinometers are used on four-wheel aerial ladder apparatus; the basic difference is that they do not carry the reference to jackknifing.

Aerial Ladder Operations

The information provided by the inclinometer is normally presented by some other method on those apparatus not equipped with an inclinometer.

OPERATIONAL PROCEDURES

The true test of aerial ladder operators is in their ability to spot and stabilize the apparatus properly and to operate the aerial ladder under the variety of emergency conditions they might encounter. This requires a keen ability to size up the situation in regard to overhead obstructions, electrical wires, traffic, road width, terrain, parked cars, and other, similar factors.

As the approach is made to the emergency, these factors must all be considered by the company officer who is selecting the aboveground location where the aerial ladder should be raised. A quick estimate should be made as to the proper distance the turntable should be placed from the building, in order to provide a climbing angle of between 70 and 80°. One method of estimating this distance is to allow 2½ feet for every floor above the first that the aerial will be raised. For example, if it is desired to bring the ladder to rest in a window of the seventh floor, then the distance the turntable should be placed from the building would be approximately 15 feet (the seventh floor is 6 floors above ground level, and $6 \times 2½ = 15$). Of course, because of the distance of the curb from the building, and the presence of parked cars, it is not always possible to place the turntable exactly in the desired location. However, this guideline gives the operator some idea as to where to spot the apparatus. If the operator is driving a tractor-trailer apparatus and has a choice, this spotting guide may help him to decide whether to use an inside or outside jackknife. The outside jackknife will place the turntable closer to the building than the inside jackknife.

Jackknifing is the term used to denote the turning of the tractor out of line with the trailer for the purpose of providing better apparatus stability. The ideal jackknife is an outside one which provides a 60° angle between the tractor and the trailer (see Figure 10-16). In an outside jackknife the ladder is raised away

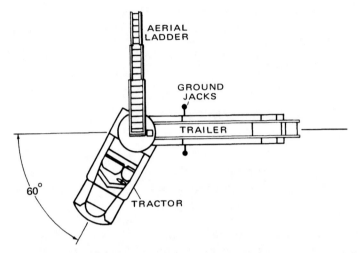

Figure 10-16 Correct Truck Setting for Maximum Stability. *Courtesy of Seagrave Fire Apparatus, Inc.*

from the jackknife angle, while an inside jackknife has the ladder raised toward the jackknife angle.

If streets are narrow, it may not be possible to jackknife the apparatus to the full 60° position. It should be remembered, however, that any jackknife whatsoever is better than none, if the ladder is to be operated out of line with the apparatus. For example, a 30° jackknife provides approximately twice the stability of the in-line position, when the ladder is operated at 90° from the in-line position.

With four-wheel apparatus, the apparatus should be turned in such a manner that the aerial will not be operated at right angles to the truck. This is particularly important when the ladder is operated at maximum reach at low angles of elevation.

When selecting the spot for raising the aerial, the operator should choose the most level spot available, considering the need for the climbing angle and other emergency requirements. Figure 10-17 indicates the stability provided by various positionings of the truck.

If the aerial ladder must be operated on a steep incline, it is best to place the truck in such a position that the ladder may be extended at an angle of approximately 45° from the center line of the truck chassis and that the ladder will be pointed uphill when

Aerial Ladder Operations

extended to the position required. This will, of course, cause the ladder to be at an angle with a window ledge or roof coping, but it will reduce the sideways tilt of the ladder and will be safer under heavy loads.

Once the apparatus has been properly spotted, the apparatus operator should, before leaving the cab, set the hand brake and engage the power take-off for driving the hydraulic oil pump. The

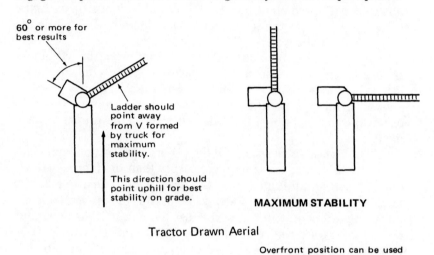

Tractor Drawn Aerial

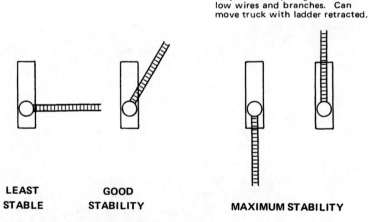

Straight Chassis Aerial

Figure 10-17 Aerial Ladder Truck Using Outrigger Jacks and Jackknifing. *Courtesy of Peter Pirsch & Sons Co.*

clutch pedal should be depressed when doing this, as in shifting gears. Most apparatus have a manual control for the automatic lock-up spring-set brakes, used as a parking brake. Other apparatus may have a drive shaft brake in addition to, or in place of, the manually controlled spring brakes.

Upon leaving the cab, the operator should make sure the chock blocks have been properly placed by the person assigned this task. Failure to place the chocks could result in damage to the ladder in the event the truck should accidentally drift while the ladder is against a wall.

If the apparatus is of the tractor-trailer type, the axle locks should be engaged to lock the trailer frame rigidly to the tractor rear axle. Care should be taken to ensure that both locks properly engage. It might be necessary to stand on the running board in order to cause a slight shift in the discs to secure proper engagement.

The stabilizer jacks should then be securely set on both sides of the truck. The truck is no more secure than the footing; consequently, soft ground under the stabilizers should be avoided. The large support plates should always be used if the apparatus is not on solid pavement, and it is best to use them even if the apparatus is spotted on a paved roadway. The aerial ladder is ready to be placed in operation once the stabilizing jacks have been properly set.

Hoisting Operations

Some tractor-trailer apparatus require that the tiller operator's seat be swung clear before the aerial can be raised. This function is normally performed by the tillerman, after the apparatus has been jackknifed. The aerial ladder operator should be sure that the tiller seat and wheel are clear before attempting to raise the ladder out of the bed.

If the ladder is equipped with hold-down locks, these should be disengaged before the hoisting control is operated. The ladder is ready to be lifted out of the bed once the hold-down locks have been disengaged.

The ladder should be raised in a slow, positive manner through the use of the hoisting control. Movement should be continued until the ladder is elevated to a position slightly higher

Aerial Ladder Operations

than is intended. Raising the ladder to this position will enable it to be lowered into position once it has been rotated to the operating location and the ladder has been extended.

It is permissible to operate the hoisting, rotating, and extending controls one at a time. While some operators are sufficiently proficient to operate two or more at the same time, little is gained by their simultaneous use, and confusion at a critical moment could result in considerable damage.

Rotation Operations

The hoisting control should be returned to the neutral position when the ladder has been raised to the desired height. The operator is then ready to rotate the ladder.

Rotation should be steady and smooth, with the movement slowed as the ladder approaches the point of intended use. Operating the rotation control in this manner will prevent the ladder from whipping when it is stopped.

The operator should sight along the ladder beams to see that the ladder is properly positioned and that it has been elevated above the point of intended use. If the ladder has not been elevated sufficiently, proper adjustment should be made at this time.

Extension Operations

The ladder is extended by the use of the extension control. The ladder lock should be moved to the OUT position prior to operating this control. The fly should then be extended until the tip is somewhat over the roof coping or wall. Extending the ladder to this position will enable the firefighters who will climb the ladder to step safely from the ladder to the roof. It also makes it easier to find the ladder in smoky or nighttime conditions. The tip of the ladder should be painted white, both for the benefit of the ladder operator in judging position, and for the benefit of firefighters searching for the ladder in obscure situations.

If the ladder is to be raised to a fire escape balcony or window ledge, then it should be extended to a height that will permit it to come to rest one foot above the railing or windowsill.

The ladder lock should be moved to the IN position once the

ladder has been extended to the desired height. The extension lever should then be moved to the retraction position, and the ladder allowed to drift back against the locking pawls. The ladder sections should not be jammed into the locks. Physical damage to the rungs has been caused by such jamming. Retraction should be continued until the tension has been removed from the extension cables. The extension lever should be returned to, and left in, the NEUTRAL position as long as the ladder is extended.

If the ladder is to be used in an unsupported position for any considerable period of time, it is advisable to close the lock valve if the system is so equipped. This will prevent any possible leakage back through the control valve, with consequent slow drifting of the ladder. The lock valve must, of course, be opened before the hoisting control can be operated.

Lowering to the Building

The hoisting control should be carefully depressed to allow the ladder to drift gently to a position against or near the point to be reached. Most manufacturers recommend that the ladder be positioned approximately 6 inches from the touching surface. The weight of a person climbing the ladder will cause the ladder to drift against the building, putting the trussing in tension and providing the maximum load-carrying configuration. The ladder should not be lowered against the building with power, as this could overload the ladder. Care should be taken to ensure that the ladder does not strike the building violently, as damage to the ladder could result.

The aerial ladder hydraulic pump should be disengaged if the ladder is to be left in the same position for an extended period; however, in cold weather, or if it is anticipated that the ladder will be moved within a short period of time, the pump should be left in gear and allowed to idle.

Load Limitations

The suggested limits for use of the ladder are shown on the inclinometer or a similar device. These suggested limits are rea-

Aerial Ladder Operations

sonable figures for general use and are not maximum use conditions. It is impossible to set absolute maximum use figures, since they will vary with ground stability, amount of wind, wind direction, and many other variables. However, extra caution should be exercised when using the ladder beyond the limits suggested.

The operator should be familiar with the meaning of the inclinometer readings on the apparatus to which he is assigned. For example, on Pirsch apparatus, when the ladder is supported against the building, the inclinometer is based on six people on the aerial ladder up to 45°, eight at 60°, and ten at 70°. When the ladder is not supported against the building, the human load should be limited to one person on the top section and one on the bottom section.

The construction and mounting of the aerial ladder also have a bearing on the load limits that can be allowed. Consequently, limits will vary according to apparatus and manufacturer. The recommendations of the manufacturer should be carefully followed in regard to ladder operations. The limitations imposed by Seagrave for use of their ladders are as follows:

The best angle for the ladder is from 70 to 80° from the horizontal. When fully extended at 70°, and *with the outer end of the ladder supported*, the ladder will safely carry a person on every sixth rung, if the load is approximately evenly distributed. Allowing for a normal amount of load concentration, and for a live hose line or other equipment, the following table shows safe loadings on Seagrave steel aerial ladders at various hoisted angles from the horizontal.

Ladder Size	70°	45°	30°	Horizontal
65-foot	7 people	4 people	3 people	3 people
75-foot	9 people	5 people	4 people	3 people
85-foot	10 people	5 people	4 people	4 people
100-foot	12 people	6 people	5 people	4 people

With the ladder unsupported (cantilever position) there is very little need for many people on the ladder. The inclinometer extension readings are safe for one person at the top of the ladder and one halfway up.

If the ladder is to be rotated with people on the ladder, it should be rotated by means of the hand crank. The sudden starts or stops of power rotation may cause whipping that may be dangerous to the ladder or the people.

The operator should not elevate or lower the ladder while people are climbing it. Not only does this place a serious live load on the apparatus, but it also subjects the people to possible injuries from contact with moving parts.

Lowering the Ladder

The aerial ladder may be lowered to change its operating location on the building, or it may be lowered for bedding when the need for its use is over. When the location on the building is being changed, it may be advantageous to lower the ladder in an extended position. Remember, however, that an extended ladder should never be lowered unless the fly sections have been retracted to the allowable extension for the new angle wanted. If this precaution is not taken, the ladder may place the truck in an overbalanced situation, which could cause it to tip over.

When the ladder is to be lowered, first raise it away from the building a short distance, then extend the ladder a few inches to permit releasing of the ladder locking pawls. Then move the ladder lock lever to the OUT position. Before applying power to lower the fly, make certain that the ladder locks are off and the sections have been extended adequately to free the pawls. Many ladders have been seriously damaged because of failure to heed this precaution.

If the ladder is to be bedded, the extension lever should be moved to the retract position, allowing the ladder to retract until it comes to rest against the limit stops. The turntable may be rotated to the bedding position while the ladder is retracting.

The hoisting lever is used to lower the ladder. With most ladders, the speed of descent of the ladder is limited by metering valves built into the hoisting cylinders. This is a safety feature which prevents a dropping of the ladder in the event the piping is ruptured. The ladder would, in this case, continue to lower at a speed no greater than that available with the control lever.

When ready to lower the ladder into the bed of the truck, the

Aerial Ladder Operations

operator should check the arrow pointers on the turntable to make sure that they are properly aligned. This indicates that the ladder will lower properly into the guides. Always slow the movement of the ladder as it approaches the truck bed guides, to prevent damage if it is slightly off line. The hold-down lock should be engaged and the hoisting lever returned to neutral when the ladder is properly bedded.

Unusual Operations

Occasionally, an aerial ladder is called upon to perform unusual functions in emergencies. For example, it may act as a tower for rigging a breeches buoy for rescue work from swift streams. In such cases always bear in mind that the ladder is much stronger in a direction at right angles to the rungs than it is in a sideways direction. Also, the truck is much less apt to tip over in the direction opposite to that in which the ladder is leaning, and it is practically impossible to tip it over if the load is applied in a direction lengthwise to the truck. Therefore, if possible, position the truck facing either directly toward or directly away from the load.

Do not extend the ladder higher than necessary. The same principles apply to any unusual conditions or ladder loading. When properly handled, the ladder will take astonishing loads, in emergencies, without damage.

Never use the ladder for pulling down walls or structural members. The ladder is neither designed nor constructed for this type of operation. The ladder can be used to secure a line to the object to be pulled down, and then a winch or other source of power can be used for demolition. The truck should be moved a safe distance away prior to the commencement of demolition operations, in order to prevent it from being struck by falling debris.

Operating in High Winds

Extreme caution should be used when operating the aerial ladder in high winds. This is particularly true when the winds are blowing sideways against the ladder or when the ladder is extended beyond 75 feet.

It is advisable to attach a guy line to the top of the fly before the ladder is raised out of the bed whenever winds are blowing at 35 mph or more. If possible, the ladder should be raised into or against the wind. Tension on the rope should be maintained in the direction from which the wind is blowing as the ladder is raised, and particular care should be taken as the ladder is swung out of line with the wind. The danger is lessened once the ladder is in a secure position against the building; however, for safety's sake it is best to tie the rope in such a manner as to maintain tension on the rope as long as the windy conditions continue. The same type of care should be taken in lowering the ladder, with positive tension maintained on the rope in the direction from which the wind is blowing.

Operating Close to Electrical Wires

The best procedure when operating near electric wires is to de-energize the circuit. Of course, this is usually not possible during the early stages of the emergency, especially when the wires are used to service an extended area. Consequently, it is best to assume that all electrical wires are "hot" and to avoid contact with the wires by either the apparatus or the aerial ladder.

Both the aerial ladder and the apparatus are good conductors of electricity and will become immediately charged when placed in contact with energized equipment. Unfortunately, grounding devices are not generally provided for aerial ladder apparatus. One common error is to think that an apparatus is grounded once the ground jacks have been placed in position. This may be true if the jacks are resting on a rail or similar device; however, under normal circumstances it makes little difference whether the ground jacks are in place or not.

The best procedure for the apparatus operator to follow in the event the ladder is accidentally placed in contact with energized wires is to move away from the controls and touch nothing until the circuit has been de-energized. The fact that the ladder comes to rest on wires may cause the circuit to be de-energized because of the circuit breaker; however, the circuit may have an

Aerial Ladder Operations

automatic re-set feature, which would result in the circuit's being re-energized within a short period of time. Do not assume that the circuit has been cut off until informed of such by a reliable person from the electric company.

Any person on the ladder at the time contact is made with the wires should keep hands and feet on the rubber-covered rungs until the circuit is de-energized. The important point is not to complete the circuit. This would happen if an individual on the ground came in contact with the apparatus or if a person on the apparatus came in contact with the ground. Firefighters on the ground must remain clear of the apparatus, and those on the apparatus should remain on the apparatus. If it becomes necessary for firefighters on the apparatus to leave, then they should jump clear, making sure that they do not fall back onto the apparatus as they hit the ground.

Operating at Extreme Angles

The load an aerial ladder can support varies according to the angle of elevation, ladder extension, and whether the ladder is operated in a supported or an unsupported position. The safe use of the ladder is indicated on the inclinometer. The readings on the inclinometer reflect the principle that aerial ladders can carry maximum loading when operated near the vertical and minimum loading when operated in an unsupported horizontal position. The most critical operating position would occur when the apparatus is in line, with the ladder fully extended perpendicular to the in-line position. The maximum strain in this position is centered at the fulcrum near the base of the ladder. If the operation is changed from an unsupported to a supported position, the maximum stress would move from the fulcrum to the center of the ladder.

If possible, ladder operations should not be conducted at low angles, but if such operations must be performed, then certain precautions must be followed:

    1. Take extreme care to ensure that stabilizing jacks have been properly set and that chock blocks are in place.

    2. Operate with the ladder in line with the apparatus or as close to the in-line position as possible.

3. With tractor-trailer apparatus, be sure the apparatus is jackknifed, if it is practical to do so.

4. Position the ladder in the direction of intended use and then extend the fly section, rather than extending the fly and then lowering the ladder into position.

5. Restrict ladder loading to the absolute minimum.

6. If the ladder cannot be supported, then use ground ladders near the tip to assist in supporting it.

7. If the angle of the elevation is to be increased, retract the fly section and then elevate the ladder, rather than attempting to elevate it in the extended position.

Ladder Pipe Operations

A *ladder pipe* is a heavy-stream appliance which is attached to an aerial ladder in order to supply an elevated hose stream. The ladder pipe is supplied by hose laid up the ladder, 3-inch being commonly used. The hose should be laid on the ladder rungs near the middle, between the ladder beams; it should never be hung over the side of the ladder.

Limitations on use of ladder pipes are imposed by aerial ladder manufacturers in accordance with ladder and apparatus construction and operating conditions. Some fire departments equip their aerial ladders for use with dual ladder pipes, one operating from the fly section and one from the bed section. Ladder pipes operated from the bed section are normally permanently mounted below the ladder, while those used on the top fly are carried separately and attached when needed. When dual ladder pipes are placed in service, it is generally considered better to use a smaller nozzle on the fly ladder pipe than on the bed section appliance.

Ladder pipes are designed to be operated by a person on the ladder or to be controlled by guy lines or other devices from the ground. Some ladders are equipped with folding steps on which a firefighter may comfortably stand while directing the stream. For direct hand operation, the ladder pipe is swiveled to provide slight lateral movement as well as full vertical movement. It is recommended that the person directing the ladder pipe be secured to the ladder with a safety belt. When possible, it is best to control the ladder pipe nozzle from the turntable or ground, so as to avoid

Aerial Ladder Operations

exposing a firefighter unnecessarily to heat, flame, heavy smoke, severe cold, or windy weather.

Ladder pipes in service can use straight tips up to 1¾ inches in size; however, smaller tips are normally employed. More and more departments are converting to constant flow fog nozzles that can provide straight streams, if needed. These nozzles have the combined ability to provide either good reach (if required) or a heavy spray (if a close approach can be made to the fire with the ladder pipe).

The effect of the nozzle reaction from the larger tips can impose dangerous stresses on the ladder if the ladder pipe is improperly operated. As an example of the forces produced, the nozzle reaction from a 1¾-inch tip at 80 psi nozzle pressure is nearly 370 pounds. If improperly used, such force can affect the stability of the truck or damage the ladder itself. It should be remembered that the ladder is designed to absorb stresses acting perpendicularly to the rungs of the ladder, and that it has little strength to resist torsional or twisting effects. Consequently, the direction of ladder pipe streams should be limited to a maximum of 15° in either direction from that perpendicular to the rungs. If it is necessary to direct the stream beyond this, then the turntable should be rotated to achieve the desired result. In this case, it is best to rotate the turntable by use of the manual control and therefore to reduce the possibility of creating any whipping action.

Guidelines for limitation on ladder extension at various angles when using a ladder pipe are given on the inclinometer. The inclinometer readings are conservative and provide for perfectly safe operations under all conditions. They should, however, only be exceeded with caution and good judgment.

In general, during ladder pipe operations, a fully extended ladder should be hoisted to not more than 80° from horizontal (and preferably, if space permits, to not more than 75° or less than 70°). A good rule provided by Seagrave, and one easily remembered, is "75-80-85," which represents 75° hoist angle, 80 feet extension, 85 pounds nozzle pressure (1¾-inch tip).

When preparing to place a ladder pipe in operation, secure the appliance to the ladder before the ladder is hoisted. Hose lines can be attached and hoisted with the ladder. Once the ladder pipe has been raised to the point of intended use, the hose lines should be properly secured to the ladder by the use of ropes or hose straps. Attachment should be made in a manner that will not interfere with the operation of the rungs. The lines can be charged once the hose has been secured to the ladder and the nozzle directed toward the fire.

The task of the aerial ladder operator, once the ladder pipe has been placed in operation, is to ensure that the stability of the ladder is maintained. Some apparatus are equipped with gages that indicate the safe operating limits of the ladder. Other apparatus have hydraulic back pressure gages that show the pressure on the system. Safe operation requires that this pressure be maintained at 100 psi or more. If the back pressure is so great as to reduce the pressure below 100 psi, then the nozzle pressure can be reduced or the elevation of the ladder changed to a point where the stability of the ladder is restored.

One of the critical points in the operation of a ladder pipe occurs during the shutdown. An instantaneous removal of the back pressure on the ladder, caused by a quick shutdown of the nozzle or by the bursting of a line, will make the ladder whip. Such whipping can cause the apparatus to tip over. Thus, when a shutdown is to be made, the information must be relayed from the nozzleman to the aerial ladder operator and hence to the pump operator supplying the ladder pipe. A slow shutdown can be accomplished better at the siamese or control point at the base of the ladder than at either the nozzle or the pump. If the ladder pipe is being operated by a person on the ladder, it is best that he return to the turntable prior to commencement of shutdown operations, providing, of course, that the direction of the nozzle stream can be locked in a secure position.

REVIEW QUESTIONS

1. When were the first spring-loaded aerial ladders placed in service?
2. How is the length of an aerial ladder measured?
3. What are the most common lengths of aerial ladders in general use?

Aerial Ladder Operations

4. What are the three general types of aerial ladder apparatus?

5. Which of the types is the most maneuverable?

6. What are some of the things aerial ladder operators must know or be able to perform in order to carry out their responsibilities effectively?

7. What is the general type of construction used on aerial ladders?

8. What is tension strength? Compression strength?

9. Does an aerial ladder have more strength when operating in a tension or compression situation?

10. What type of construction is used on the beams of an aerial ladder?

11. What is the function of the power take-off on an aerial ladder truck?

12. What is the primary function of the turntable?

13. Where is the operator's control pedestal located?

14. What are the three main controls on the pedestal?

15. What are some of the other controls or instruments that are located on or near the pedestal?

16. What are some of the characteristics that apply to all three of the main controls?

17. How is control of the movement of the ladder accomplished during the hoisting operation?

18. What is the recommended method for raising the ladder?

19. How is the ladder extended and retracted?

20. What is the purpose of a load indicator gage?

21. What is the purpose of the hoisting cylinder pressure gage?

22. What does the extension gage show?

23. What is the purpose of the lock valve control?

24. When does extension of the ladder place the greatest amount of stress on the apparatus?

25. What are the three types of devices used to assist in stabilizing an apparatus?

26. What is the proper position for setting wheel chocks?

27. What is the purpose of axle locks?

28. What is the purpose of stabilizing jacks?

29. What types of stabilizing jacks are used on aerial ladder apparatus?

30. When should metal foot plates be used?

31. How should screw jacks be set?

32. What is an outrigger?

33. What is the purpose of ladder cables?

34. What is the purpose of the ladder locks?

35. What should be the position of the ladder locks when the ladder is being extended?

36. Describe the normal use of the ladder locks.

37. What is the purpose of the inclinometer?

38. What method can be used to determine the approximate distance the turntable should be spotted from the building to which the ladder will be raised?

39. What is a jackknife?

40. What is the difference between an inside jackknife and an outside jackknife?

41. What is the ideal jackknife?

42. How should four-wheel apparatus be spotted in preparation for raising the aerial ladder?

43. How should the apparatus be spotted if the aerial ladder must be operated on a steep incline?

44. What should the driver do before leaving the seat of the apparatus, when the aerial ladder is to be raised?

45. Describe the hoisting operations.

46. Describe the rotation operations.

47. Describe the extension operations.

48. How far above a windowsill should the aerial ladder be extended?

49. Describe how to lower the ladder into a building.

50. What is the best angle for climbing the aerial?

51. Describe the operations necessary to lower the ladder.

52. What would happen if a hydraulic line were to break when the ladder was being lowered?

Aerial Ladder Operations

53. What are some of the principles that should be considered when using an aerial ladder for unusual operations?

54. What precautions should be taken when operating an aerial ladder in high winds?

55. What is the best procedure for the apparatus operator to follow in the event the ladder is accidentally placed in contact with energized wires?

56. What precautions should be taken in the event the apparatus becomes "charged"?

57. What are some of the precautions that should be followed when operating an aerial ladder at extreme angles?

58. What is a ladder pipe?

59. How are ladder pipes mounted to aerial ladders?

60. What size tips can be used on ladder pipes?

61. What is the approximate nozzle reaction from a 1¾-inch tip?

62. What is the maximum angle that a ladder pipe should be directed from a line perpendicular to the rungs? Why?

63. What is a good rule to remember concerning proper operation of a ladder pipe with regard to hoist angle, extension, and nozzle pressure (1¾-inch tip)?

64. What is the task of the aerial ladder operator once a ladder pipe has been placed in operation?

65. What is one of the most critical points in the operation of a ladder pipe?

66. How should a ladder pipe be shut down?

11

Elevated Platforms

Elevated platforms have been in service for a number of years, primarily in industry, where they have been employed for a variety of uses. Utility companies have been among the primary users of elevated platforms, employing them for installing and repairing high-voltage lines and for washing the dust from high-voltage insulators.

Commissioner Robert J. Quinn of the Chicago Fire Department is credited with first adopting the elevated platform for use in the fire service. He placed the first apparatus in operation in 1958. The first unit was a 50-foot Pitman, mounted on a GMC chassis. The apparatus was equipped and modified for firefighting operations in the Chicago Fire Department shops. This articulating type of unit quickly proved its worth in providing elevated master streams. Chicago continued purchasing this type of unit until there was an entire fleet working in various areas of the city.

For several years, the use of the elevated platform in the fire service was confined primarily to the Chicago Fire Department. The apparatus received national recognition when it was pictured in newspapers throughout the country in operation at Our Lady of The Angels school fire, which took the lives of ninety-two children and three teachers.

Early models of the platform were restricted to a working height of 90 feet. In 1966, Calavar Corporation of Santa Fe Springs, California, developed an aerial platform called the Condor, for use in the aerospace industry and by utility companies. This platform reached the then unheard-of height of 125 feet. The principle used in the Condor was later refined and adopted for use in the fire service. In 1969, the Philadelphia Fire Department placed in service a Calavar 125-foot aerial platform called the Firebird. Three years later, the Tucson Fire Department purchased and placed in service a 150-foot Firebird.

The elevated platform went through a controversial period during which fire officials nationwide debated its efficiency, compared to that of the aerial ladder. Most fire officials today agree that both the aerial ladder and the aerial platform have certain advantages and disadvantages; however, each of them has its place in the fire service, and both are here to stay.

The first elevated platforms placed in service by fire departments were used primarily to provide elevated master streams. Platforms quickly took the place of water towers, as the platforms proved to be much more maneuverable than the traditional units. As time progressed, fire officials found more and more uses for the equipment. The uses have expanded to include such tasks as:

1. Aiding physical rescue
2. Transporting heavy equipment and firefighters to upper floors
3. Providing ventilation
4. Serving as an elevated base for fireground reconnaissance
5. Serving as a portable standpipe system

The objective of this chapter is to introduce apparatus operators and potential operators to the various types of elevating platforms, the operating capabilities and limitations of these units, the functional controls, and examples of operating procedures.

TYPES OF ELEVATED PLATFORMS

An *elevated platform unit* consists of three basic parts—the platform, the booms, and the turntable. Elevated platforms are classified according to their boom arrangement. There are three general types that have been adapted for use in the fire service—*articulated, telescopic,* and *combination telescopic-articulated* (see Figure 11-1). The size and type a community should use will depend on community needs and intended use of the unit. Strong consideration should be given to factors such as traffic conditions, street widths, building construction, and fire spread potential.

In addition to being identified by boom arrangement, elevated platforms are designated as to effective working height. The effective working height is measured by a plumb line from the floor of the platform to the ground, when the platform is raised to its maximum working height. Elevated platforms are available in sizes ranging from 55 to 150 feet. The longer platforms are of the combination telescopic-articulated type.

It should be noted that when an elevated platform is used to provide a master stream, it can be operated up to its designated height. This is not true of all aerial ladders.

Articulated Platform

The articulated platform consists of two or three booms hinged or connected together, with a means provided to operate each boom independently. The end of the upper boom is equipped with a platform.

When the booms are bedded, the upper boom rests on the lower boom, with the platform overhanging the front or the rear of the apparatus, depending on the manufacturer's design.

When the apparatus is in operation, the lower boom rises almost to a vertical position, and the upper boom can be raised to its working height. The platform can be moved in an almost completely spherical path from maximum elevation to ground level. The turntable is capable of operation through 360°, with rotation continuous in either a clockwise or a counterclockwise direction.

470 Introduction to Fire Apparatus and Equipment

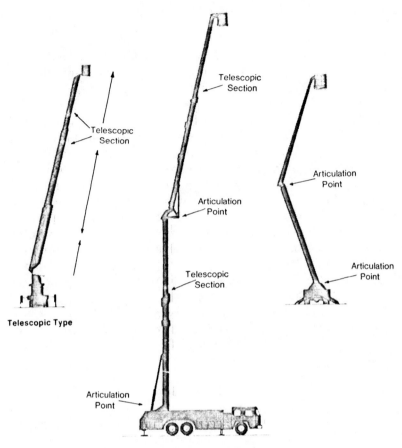

Figure 11-1 Types of Elevated Platforms *Courtesy of Calavar Corporation*

There are some limitations to the path of travel of the booms. For example, the lower boom is restricted to a certain vertical angle. The maximum horizontal reach is obtained when the platform is brought down to a point approximately level with the top of the lower boom. Above and below this point of reach the platform path travel comes closer to the vehicle; thus at ground level the platform is adjacent to the vehicle.

Elevated Platforms

The most common sizes of articulated platforms in service are 55, 65, 70, 75, and 85 feet; however, at least one manufacturer is providing units extending beyond 100 feet.

Telescopic Platform

The telescopic platform operates in much the same way as an aerial ladder, with the booms telescoping out from the main boom in a manner similar to the way in which the fly sections of an aerial ladder telescope out from the main section. As with the articulated type, the platform can be rotated continuously in either direction.

Depending on the manufacturer, the boom sections may be fabricated in a box-type manner of either open or closed construction, or in ladder-type sections complete with rungs and handrails. It matters very little, from an operational standpoint, whether the units are constructed with boom or ladder sections. The major difference among units is in the capacity of the platform. The most popular sizes currently in service are the 65-, 75-, and 85-foot units; however, 100-foot and longer units are available to those cities requiring increased reach.

Combination Telescopic-Articulated Platform

The telescopic-articulated platform combines the features of the other two types. It not only articulates like a two-boom platform, but it contains within each boom an additional section that is designed to telescope. This combination offers increased versatility and maneuverability.

Height and Range Comparisons of Types

A comparison of the working height and horizontal reach of various sizes of the three types of elevated platforms and aerial ladders is shown in Figures 11-2, 11-3, 11-4, and 11-5. The horizontal reach is just as important, if not more important, than the

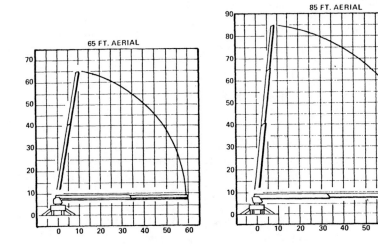

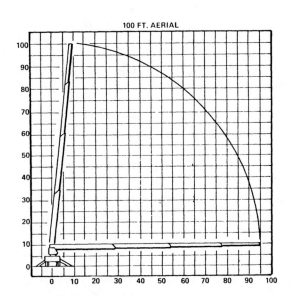

Figure 11-2 Aerial Ladders—Working Heights and Horizontal Reach. *Courtesy of Calavar Corporation*

Elevated Platforms

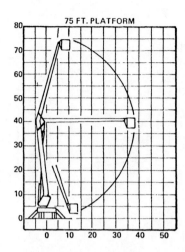
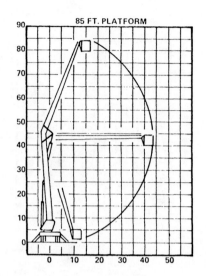
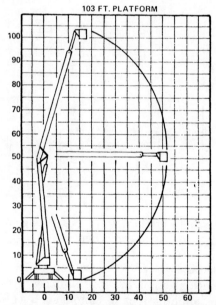

Figure 11-3 Articulated Platforms—Working Heights and Horizontal Reach. *Courtesy of Calavar Corporation*

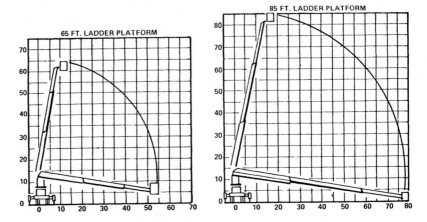

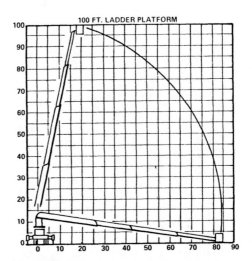

Figure 11-4 Telescopic Platforms—Working Heights and Horizontal Reach. *Courtesy of Calavar Corporation*

working height. Both of these factors must be considered when evaluating the various configurations under which the unit might work, including combinations of factors such as building height, building setback, sidewalk widths, street widths, and overhead obstructions.

Elevated Platforms

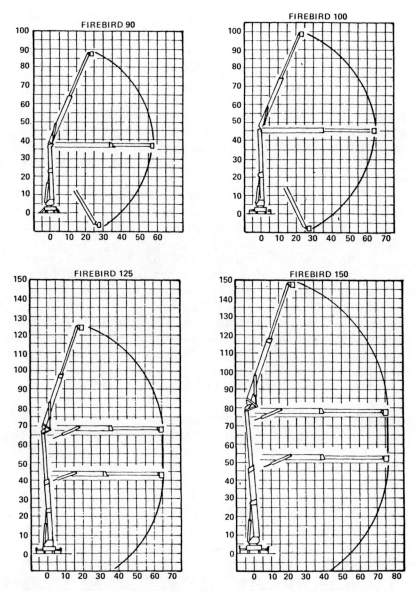

Figure 11-5 Telescopic-Articulated Platforms—Working Heights and Horizontal Reach. *Courtesy of Calavar Corporation*

ELEVATED PLATFORM FEATURES

The Platform

The purpose of an elevated platform apparatus is to place a working platform at various heights and positions on the fireground. The platform is, therefore, the business end of the apparatus. The platform must be of sufficient size, have an adequate carrying capability, and be properly equipped to achieve the desired results.

The physical size of the platform varies from one apparatus to another. Accepted standards call for an area of at least 14 square feet. The larger platforms have a working area in excess of 20 square feet.

Platform carrying capacity varies from 700 to 1000 pounds, depending on the manufacturer and the rated working height. This rating prevails at both the maximum height and the maximum reach; however, the rating decreases on some types when heavy streams are placed in operation. The rating of one manufacturer's platform is reduced approximately 50 percent during the operation of a master stream, the decreased rating caused by the negative effect of the nozzle reaction. At least one manufacturer uses a Stang "Intelligiant" monitor to restrict the nozzle reaction to a minimum.

Every platform is equipped with a heavy-stream appliance, most often one with a discharge capacity of 1000 gpm. Control of the water flow for these appliances is in the platform. Some platforms are equipped with dual monitors. The monitors on some apparatus have the capability of operating through a 180° radius in both the vertical and horizontal directions. Because of its ability to rotate the platform 45° right or left of center, independent of the turntable rotation, the Calavar Firebird is capable of providing a 270° sweep of the monitor without rotating the turntable.

In addition to providing a heavy stream, most platforms are equipped to operate as portable standpipes. Most platforms have 2½-inch gated outlets, from which lines can be extended into the building or onto the roof. Some platforms are equipped with preconnected 1½-inch hose.

Several features are designed into platforms to provide safety

Elevated Platforms

for the firefighters assigned to operate them. Air masks are stored in the platforms for use of the firefighters when irritating or hazardous smoky conditions are encountered. Air is piped from one or more 220-cubic-foot cylinders located on the apparatus or on booms somewhere near the turntable. The size of the bottles permits the firefighters to work in smoky conditions for an extended period. The air supply from the bottles can also be used effectively to provide air to aboveground fire victims.

The platforms are equipped with one or two spray nozzles installed beneath the platform. These can be activated to provide a water curtain to protect the people in the platform from heat. Some departments use the spray nozzles for active firefighting, as well as to provide protection for the people on the platform. The spray nozzles are activated by either a hand-operated valve or a foot pedal. In addition to the spray nozzles, reflector curtains are used on some platforms to protect the firefighters from radiant heat.

Other features designed into platforms to increase firefighting or rescue capability include:

1. Weatherproof 110-volt receptacles which can be used for supplying power to smoke ejectors, power rescue saws, lights, and other electrical equipment
2. A two-way speaker system for platform-to-ground communications. Some systems are of the no-hands type.
3. Spotlights and floodlights for illumination of both aboveground areas and the fireground
4. TV cameras to provide command officers on the ground with a view of aboveground problems
5. A winch for lifting operations

There is no standard as to the equipment that is stored in the platform. One may find axes, ropes, life belts, short ladders, hand lanterns, hose packs, and the like.

Stabilizing Devices

The stabilizing devices on elevated platform apparatus are much like those on aerial ladder trucks. Elevating platform apparatus are equipped with flopover, underslung crisscross, and telescoping outriggers, swing-down jacks, and hydraulic jacks. An apparatus may be equipped with all outriggers, all jacks, or a combination of the two. The stabilizing devices should be set in accordance with the manufacturer's instructions. However, the apparatus operator should remain alert to ensure that all personnel are clear before engaging the outrigger controls.

One major difference between those devices on aerial ladder trucks and those on elevated platform vehicles is that some of the latter vehicles have jacks located at each corner of the truck to lift the entire vehicle off the ground. The hydraulic jack cylinders are normally held in place by holding valves. Mechanical locks are provided for additional security.

Most elevating platform vehicles are equipped with some type of interlock device which prohibits the booms from being raised out of the bed until the stabilizing devices are in place and secured. Despite this safety feature, the apparatus operator should physically check to ensure that all devices are in place before the boom system is activated.

Basket Leveling System

Every elevated platform is equipped with some type of device that will maintain the platform in a level operating position, regardless of the working height, reach, or position of the booms. Standards require that this leveling system be fully effective when the platform is loaded to full rated capacity.

Deadman Control

A *deadman control* (see Figure 11-6) is a device to which pressure must be applied before another system can be placed in operation. A deadman control is provided on elevated platforms for the operation of the boom system. The booms cannot be activated until pressure is applied to the deadman control, and the system will be deactivated in the event the pressure is released. This means that positive pressure must be maintained on the deadman control at all times to operate the boom system. The

Elevated Platforms

objective of this control is to stop operations in the event of the collapse of the operator.

Operating Controls

Operating controls are found in three locations—the cab, the platform, and at ground level. The controls in the cab are the parking brake, the throttle adjustment, and the power take-off. These should be engaged before the apparatus operator leaves the cab.

The boom controls are located both in the platform and at ground level, permitting operation of the booms from either position. The platform operation, however, can be taken over by the ground control operator in the event that this becomes necessary. The two control units in general use are the single-handle type and the multiple-handle type.

Figure 11-7A shows the single-handle control used in the platform of the Mack Aerialscope. A multiple-lever control is used

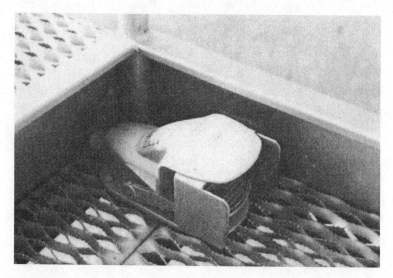

Figure 11-6 A Foot-Operated Deadman Control Switch. *Courtesy of Calavar Corporation*

480 Introduction to Fire Apparatus and Equipment

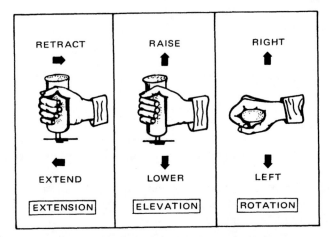

Figure 11-7A A Single-Handle Control. *Courtesy of Mack Trucks, Inc.*

Figure 11-7B The Firebird Platform Control Console. *Courtesy of Calavar Corporation*

at ground level on this apparatus. This single handle is moved forward or backward to extend or retract the booms. It is raised up or down to increase or decrease boom elevation. Moving it to the right or left rotates the booms.

The single handle used on an articulated platform is similar in appearance. The lower boom is controlled by the forward and backward movement of the control. Moving it forward raises the boom; pulling it back lowers the boom. The upper boom is controlled by raising and lowering the control. Pulling it up raises the boom; pushing it down lowers the boom. The turntable is rotated in a clockwise direction by rotating the handle to the right. Rotating the handle to the left causes a counterclockwise rotation of the turntable.

Figure 11-7B shows a multiple-handle platform control. The various boom movements are controlled by the individual handles shown on the console.

Most of the control systems themselves are of the open-center hydraulic type. Essentially, this means that hydraulic fluid flows through each valve whether or not the valve is actuated. The hydraulic pump is of the fixed or constant volume type, which means the pump produces the same volume of oil regardless of the number of functions being performed. The valves are of the directional type, with a standard open or closed operation. The operator should expect some loss of speed with this type of system whenever two or more controls are operated simultaneously.

The system shown in Figure 11-7B is of the electric-hydraulic type. This type of system uses electrically actuated valves, which open in proportion to the amount of voltage induced. For example, if the control valve is only actuated halfway, the valve only opens halfway. This type of system provides for a fine metering control that can be used to vary the amount of voltage induced into the controller even if the control lever is in the full ON position. When a variable volume hydraulic pump is added to the system, oil is put out on demand only. For example, if only one function is in operation, and it takes 5 gallons of oil to operate it, the pump delivers 5 gallons. However, should a second function requiring 10 gallons be operated simultaneously, the pump will deliver a total of 15 gallons. The overall effect is that the movement speed is the same whether only one or all five controls are being operated.

The controls for operating the outriggers and hydraulic jacks on the various apparatus may be located in close proximity to the ground boom controls or in a separate compartment. Normally, each outrigger and jack has a separate control lever. Switches for activating the holding locks are generally found in the same compartment.

THE MACK AERIALSCOPE

A particular apparatus is purchased by a fire department to meet the operating needs of its city. The New York Fire Department has found the Mack Aerialscope to be particularly effective in the city's narrow and traffic-congested streets. Over 130 of these apparatus have been purchased since the first was placed in service in 1964.

The information in this section is on the Aerialscope. It is presented through the courtesy of the Baker Equipment Engineering Company and the Mack Truck Company. The material here is for general information only. It is presented with the objective of introducing students to the construction, operating controls, and operational procedures of the Aerialscope. Any apparatus operator assigned to this equipment should become thoroughly familiar with the operating instructions issued by the manufacturer and not depend on the information presented here.

The Aerialscope is a four-boom telescoping type of aerial platform. A four-section aluminum extension ladder is permanently mounted along the top of the boom sections, so as to provide a ready route for emergency evacuation. The Aerialscope is a manually controlled, hydraulically operated unit which derives its power from the apparatus engine. The general layout of the unit is shown in Figure 11-8.

Construction

Body Assembly

The body is a welded and fabricated galvaneal steel assembly, designed for general fire department use. The interior area of

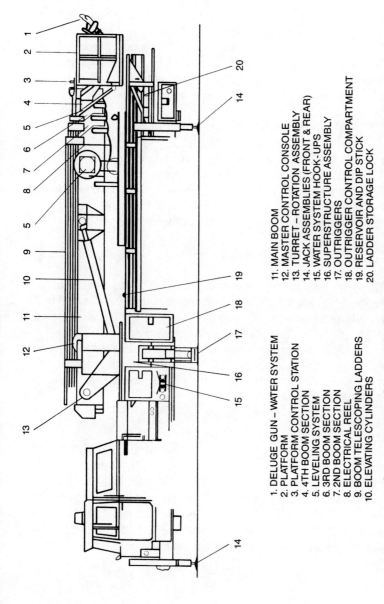

1. DELUGE GUN – WATER SYSTEM
2. PLATFORM
3. PLATFORM CONTROL STATION
4. 4TH BOOM SECTION
5. LEVELING SYSTEM
6. 3RD BOOM SECTION
7. 2ND BOOM SECTION
8. ELECTRICAL REEL
9. BOOM TELESCOPING LADDERS
10. ELEVATING CYLINDERS
11. MAIN BOOM
12. MASTER CONTROL CONSOLE
13. TURRET – ROTATION ASSEMBLY
14. JACK ASSEMBLIES (FRONT & REAR)
15. WATER SYSTEM HOOK-UPS
16. SUPERSTRUCTURE ASSEMBLY
17. OUTRIGGERS
18. OUTRIGGER CONTROL COMPARTMENT
19. RESERVOIR AND DIP STICK
20. LADDER STORAGE LOCK

Figure 11-8 Aerialscope (Designed for Standard Installation Directly Behind the Truck Cab). *Courtesy of Mack Trucks, Inc.*

the body consists of ladder-racking provisions for storing ladders of various sizes. Levers on each side near the rear of the body are for locking and unlocking the ladders. Four large compartments on each side are for equipment storage. Three compartments in front of and behind the main superstructure assembly serve for further equipment storage. Automatic compartment lights illuminate each compartment when the doors are open.

Superstructure Assembly

A welded and bolted steel structure is provided for attachment of the turret, rotation assembly, boom assembly, and platform to the truck chassis. This assembly also includes two outrigger stabilizer assemblies, the hydraulic collector block, and the electrical slip-ring assembly necessary for continuous operation.

Turret Rotation Assembly

The turret rotation assembly is a unitized weldment of high-strength steel which supports the boom assembly. A shear ball bearing with a continuous row of steel balls rolling in heat-treated races supports the turret, booms, and platform, and allows for rotation. Power for rotation is produced by a worm gear unit driven by a hydraulic motor. An integral gear on the outer race of the bearing meshes with a drive pinion on the worm gear shaft to rotate the unit.

The Platform

The platform is of all-aluminum construction, with two access gates. Its rated capacity in all positions is 1000 pounds. This capacity of 1000 pounds includes both personnel and equipment. Inside dimensions of the floor area are 65 inches by 36 inches. The platform is hinge-pinned to the outer end of the fourth boom section. A hydraulically operated, positive-leveling system automatically keeps the platform level through all boom movements.

The Booms

The unit consists of four booms, a main boom and three additional sections. The main boom is constructed of steel and is hinge-pinned to the turret ears at the lower end. Two hydraulic

Elevated Platforms

cylinders, which can be operated from either control station, raise and lower this boom through an arc of approximately 75°. The three-stage compound cylinder assembly is enclosed within the boom sections. This assembly extends or retracts the sections from either control location.

The first, second, and third boom sections are constructed from high-strength aluminum alloy tubes. The third boom section has reinforcing weldments and terminal weldments on the end for mounting the platform assembly. Brackets are attached for carrying the extension ladder and water line assembly. Phenolic-impregnated canvas wear pads are mounted at strategic locations and act as bearing surfaces for each boom to ride on the other.

The Water System

The water system consists of four telescoping sections of aluminum pipe, operating through conventional U-cup seals and packing glands, Suitable swivel elbows, fittings, and additional pipe connect this telescoping system from the platform-mounted deluge gun, through the rotating swivel in the superstructure, to the water line connections on either side of the truck frame.

Stabilizing Devices

Both outriggers and jacks are used for stabilizing the apparatus. The two steel tubular outriggers operate independently. They are hydraulically operated from the outrigger control station. The outriggers are extended and retracted by hydraulic cylinders (totally enclosed for protection). Holding valves lock the outrigger cylinders at any up or down position. A mechanical safety lock is mounted on the outriggers as an additional safeguard against any possible outrigger failure.

Four independently operated hydraulic jacks are located at each corner of the vehicle. These jacks are operated from the outrigger control station. They are used to lift the entire vehicle off the springs, so as to balance the operation of the unit with the mass of the truck. Holding valves lock the jack cylinders at any up or down position. Mechanical locks are provided for additional safety.

Hydraulic System

The hydraulic system is divided into two separate circuits, the *outrigger and jack circuit* and the *boom operating circuit*. The system is powered by a hydraulic pump which is flange-mounted to a power take-off on the truck transmission.

The outrigger and jack circuit incorporates an open-center, six-section control valve, which provides a separate section for control of each of the two outriggers and each of the four jacks. This arrangement makes it easy to bring the Aerialscope to as near a level position as possible, when positioning it for operation. Two or more valve sections can be actuated simultaneously, so that any two of the units up to both outriggers and all four jacks may be lowered or raised at the same time. It must be kept in mind, however, that because of differences in tolerances or because of other factors, one or more of these units may not operate until others have reached the ground on the down movement, or have reached the end of the stroke on the upward movement. This is normal in some instances and should not be considered a malfunction.

The boom operating circuit incorporates a closed center, servo-type electric-hydraulic proportional control. This system uses a variable electrical signal to actuate the main control valve on the master console. The main control valve may be manually actuated at the master control console by extending the movement of the control levers beyond that necessary for normal operation. This becomes necessary in case of electrical failure. The valves are self-centering (they return to neutral automatically). Valves direct the flow of hydraulic oil for actuation of all boom movements (rotation, elevation, and extension). A deadman control is incorporated for protection in the event of the collapse of the operator. All hydraulic components, including lines and fittings, are rated well above minimum recognized standards. This circuit also incorporates an accumulator and unloading valve arrangement to maintain required hydraulic operating pressure.

Elevating Cylinders

Double-acting elevating cylinders are mounted on either side of the turret ears and the main boom to raise and lower the boom

Elevated Platforms

and platform assembly. The cylinders are equipped with pilot-operated holding valves on each end, to prevent leakdown. The holding valves also act as a safety measure to keep the booms from falling, should a hydraulic line failure occur.

Platform Gravity Leveling
(Aerialscopes Serial No. 70059 and 70062 and up)

The platform is maintained in a level position by two hydraulic cylinders operated by pressure from the main hydraulic system of the unit. Previous units incorporated a master-slave system, which derived the power and control for leveling the platform from the boom movement only.

This present system depends on the use of a rotary-type hydraulic valve, which is actuated by a pendulum-type gravity-motivated control. A change in the position of the boom assembly, either up or down, requires a change in the relationship of the platform to the boom assembly, if the platform is to maintain a level position. During these changes in boom elevation, the pendulum, since it constantly seeks a vertical position, actuates the rotary valve in the proper direction to cause the two cylinders to maintain the platform in a level position. Pilot-operated check valves are provided in both systems to prevent a sudden drop of the platform in the event of a hydraulic line failure.

Operating Devices

Three control stations are found on this unit; one to operate the jacks and outriggers, one to operate the booms from the master console on the turntable assembly, and one to operate the booms from the platform. The master console controls will override the platform controls for emergency operation. Elevation and extension movements are activated by the hydraulic motion of cylinders, while the rotation movement is activated by the hydraulic motor through reduction gears.

Master Console Station

Figure 11-9 shows the layout of the master console station. The functions of the various controls and indicators are as follows:

488 Introduction to Fire Apparatus and Equipment

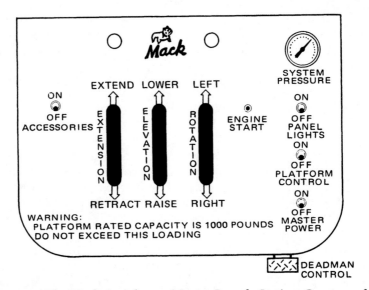

Figure 11-9 The Mack Aerialscope Master Console Station. *Courtesy of Mack Trucks, Inc.*

Master Power This ON-OFF switch supplies electrical power for the master control valve, master console panel lights, engine start, deadman control, and platform controls.

Engine Start This button allows the engine to be restarted from the master control console in the event of a stall.

Platform Control This ON-OFF switch activates the platform control station.

Accessories Switch This ON-OFF switch controls the light on the master console front and optional accessories.

Elevation Control This control handle operates the two elevating cylinders to raise or lower the booms through an arc of approximately 90°.

Extension Control This control handle operates the three extension cylinders, which are plumbed together for extension or retraction of the boom sections.

Elevated Platforms

Rotation Control   This control handle operates the rotation motor, moving the turret and booms in either a clockwise or a counterclockwise direction. Continuous rotation is possible.

Deadman Control   A slight movement of this pedal at the base of the console must be made and held to activate the system. This movement also deactivates the platform controls, shifting control to the master console station.

Pressure Gage   This gage indicates the system pressure during operation of the booms.

Platform Station

The platform control is mounted on the left side of the platform. The control consists of a single handle with a red trigger-squeeze safety switch (see Figure 11-7A). Depressing the red trigger activates the deadman system. It must be depressed during all handle movements in order to create any boom motion. Handle movements are spelled out on the decal located on the control box. The handle is moved forward or backward to extend or retract the telescoping booms. It is pulled up or pushed down to raise or lower the boom. The handle is moved in the direction of travel to rotate the boom. All three boom movements can be operated simultaneously.

Outrigger Control Station

The outrigger control station is located in the compartment on the left side of the apparatus, just behind the outrigger. The six controls for operating the two outriggers and the four jacks are in this compartment. Controls can be operated independently or simultaneously. A selector valve is located in this compartment to direct the flow of hydraulic fluid to either the outrigger and jack circuit or to the boom function circuit. This switch prevents operation of both circuits at the same time. A pressure gage for monitoring the hydraulic system accumulator pressure (boom operation circuit) is also located in the compartment. One quick disconnect fitting is located at the outrigger and jack control station for connection of an emergency auxiliary power unit.

General Operating Instructions

Following are the general operating instructions for the Aerialscope. These instructions apply only on grades of 4 percent or less. Instructions for operating on grades in excess of 4 percent are provided in the next section.

1. Spot the truck. Predetermine by survey the location that can best take advantage of the truck and platform capabilities. The truck should be spotted on as level terrain as possible.
2. Set the parking brake.
3. Engage the power take-off to start the hydraulic pump.
4. Set the manual throttle control for the engine speed specified on the dashboard in the cab.
5. Turn on all warning lights before leaving the truck cab. If additional precautionary signs or lights are required, they should be set up before operating the unit.
6. Check the apparatus to see that all mechanical locks on jacks and outriggers are disengaged. At the same time give the entire unit a last-minute inspection.
7. Check to ensure that the selector valve handle in the outrigger control compartment is in the DOWN jacks position.
8. Lower all jacks and outriggers until each one reaches the end of its stroke.

Important: If the truck is not level, adjust the jacks and outrigger on the high side of the truck until the unit is as level as possible.

9. Insert the outrigger lock pin in the last clear opening toward the outrigger foot, entered from the rear side. Insert the jack locks, being careful to place them in the highest position that is clear.
10. Once jacks and outriggers are set and locks are in position, push the outrigger interlock button and lift the selector valve to the UP position.
11. At the master control station, move the master control switch to the ON position. The boom functions, including extension, rotation, and elevation, can now be operated from the master control station on the turntable. Power is now

Elevated Platforms

available for panel lights, engine start, and the platform controls.

12. Move the platform control switch on the control panel to the ON position to energize the platform controls.

13. Now the unit can be operated from the control station in the platform; however, the lower controls will still override the platform controls should the need arise.

14. Select the platform travel movements required to reach the work location. The first motion should be to raise or elevate the booms to clear the boom rest and body structure. Use a feathering technique when operating the control levers at either the platform or master control station to ease the unit to smooth stops and starts. Always continue looking in the direction of travel.

15. Be careful to fasten the gates on either side of the work platform. Avoid carrying loose objects in the platform whenever possible. Many accidents have occurred because objects fell out of the platform and hit someone below.

16. To secure the unit, first retract the booms, then rotate the unit until the booms are over the boom rest. Lower the booms to the point of just entering the boom rest, but not to the point of being seated.

17. Move the selector valve in the outrigger control compartment to the DOWN position. Now disengage all mechanical locks on jacks and outriggers. It may be necessary to lower them slightly in order to disengage the locks if they are bound in the locked position. This operation is much easier with two people, one to operate the control levers and one to release the locks. Store all jacks and outriggers.

18. Return to the master control console and move the elevation control lever to the DOWN position. A boom limit switch is provided that will stop movement at a properly seated position. This switch is always energized when the power take-off system is engaged. Keep the deadman control activated until the system pressure gage reads zero.

19. Disengage the power take-off to stop the hydraulic pump and release the throttle control. This returns the engine to an idle speed.

20. Unhook all water hoses, store all equipment, and lock

down the ladders in the body compartment. The Aerialscope is now ready for travel.

Operating on Grades in Excess of Four Percent

The Aerialscope was designed to meet all operation specifications and requirements while in a level configuration. Data on stability and load-induced stresses are all based on a level configuration for the vehicle.

When the machine is in other than a level configuration, its performance and stability are reduced in proportion to the degree of deviation from a level setup.

For instance, the rotation system was designed for level operation. When the chassis is on a slope (from front to rear) the rotating functions now include the work of lifting the boom system uphill. If the boom is retracted and raised to its greatest elevation, this added load is at its minimum value. When the boom is extended and in a horizontal position, this added load is at its maximum. The degree of slope adds to the lifting force required in either configuration. Therefore, under certain conditions the operator may have to retract and/or raise the boom in order to rotate uphill. Because of this added load, when on a slope the operator must never rotate at high speeds and must always feather the controls when starting or stopping in order to avoid shock loads. Careful, cautious operation is essential when one is using the machine on other than level ground.

Survey charts and data concerning grades of streets and roads generally describe an average condition. Because of crowns, dips, and the like, the published or recorded value may not apply to a given spot where the Aerialscope is to be positioned. Because it is seldom possible to determine a grade angle exactly, the maximum slope for the Aerialscope operation has been set at 15 percent grade.

When working on a maximum-grade limit, the operator should exercise maximum care and remain aware, at all times, of the special precautions required for safe performance.

The following procedures are to be practiced when using the Aerialscope on slopes in excess of 4 percent. Slopes of less than 4 percent will permit the same operating practices as on level ground.

Setting Up

1. Always position the machine so that its longitudinal axis is parallel to the direction of slope and in a location that will permit working from the uphill quadrants. If possible, the machine should be headed downhill.
2. Set the parking brake.
3. Plant the jacks and outriggers firmly, but maintain ground pressure on the uphill (rear) tires so that braking action can be introduced. The use of adequate wheel chocks is recommended.

Operating

1. Perform all functions at reduced speed. Be exceptionally cautious of downhill rotation; downhill rotation should be executed at half speed or less.
2. Do not make sudden starts or stops. These functions should be performed smoothly and gradually.
3. Avoid all violent reactions so that excessive shock loads are not applied to the structure.

Note: Items 2 and 3 are good practice for *all* conditions.

4. Retract and/or raise the boom as required for uphill operation. Retracting and/or raising the boom allows the load to be rotated on a shorter radius, thereby reducing the torque requirements for uphill rotation.

The Aerialscope was designed to operate at its maximum performance specifications. However, as with all machinery, its useful life and performance characteristics will be extended through the use of intelligent, expert operating practices. This is especially true with regard to operations performed when the Aerialscope is subjected to forces and stresses beyond the normal.

Disregard for the above recommended procedures and practices may cause serious damage to the machine and could cause an accident involving human life.

Emergency Operation

Should normal hydraulic power fail, an emergency power system is provided. This system consists of an electrically driven hydraulic pump and a control valve permanently installed in the main hydraulic system. This auxiliary power system is provided primarily to put the Aerialscope in a road-ready condition or to get it out of a dangerous position rather than to serve as a source of power for normal use of the machine.

Procedure

1. Place the selector valve in the UP position for operation of boom movement. Operate boom rotation, up and down, extension and retraction, as needed, using the main system console control valve by moving the emergency system toggle switch on console panel to the ON position and holding until movement is completed. This switch turns to the OFF position when released. Keep the emergency system pump running only while movement is required in order to conserve electric power.
2. Should circumstances make it impossible or desirable not to operate through the console control valve for boom and turret movement, proceed as follows:

   a. Connect one end of each of the two hose assemblies to the two quick-disconnect fittings on the emergency power system control valve in the jack and outrigger control station.

   b. Connect the other end of the hose referred to in (a.) above to the two quick-disconnect fittings on the circuit to be operated (that is, rotation and boom lift up and down circuits on turntable, and boom extension and retraction circuit under the end of the boom over the engine compartment).

   c. Close the shut-off valves in the regular hydraulic lines that connect the regular hydraulic system to the function to be operated.

   d. After moving the emergency system toggle switch in

the jack and outrigger control compartment to the ON position and holding, operate the particular boom function by moving the emergency control valve in the direction needed to produce the desired movement.

Caution: When lowering the booms, be sure the selector valve is in the UP position to maintain pressure on the accumulator. Keep the deadman control activated during boom movement. This is necessary to keep the platform level, especially during the approach to the stored position and while putting the booms in the boom rest.

When storing the boom assembly into the boom rest with the auxiliary power system, avoid extreme pressure against the boom rest. Only slight pressure is required. The boom limit switch, which normally provides this protection, is eliminated from the circuit when the auxiliary power system is used in this manner.

e. After completing the emergency movement of each individual boom function, *open the shut-off valves for that circuit to the main hydraulic system* and remove the hoses connecting that circuit to the emergency control valve.

3. Place the selector valve in the DOWN position to operate jacks and outriggers. Operate jacks and outriggers as needed, using the main system jack and outrigger control valve in a normal manner by moving the emergency system toggle switch to the ON position and holding, until movement is completed. This switch returns to the OFF position when released. Keep the emergency system pump running only while movement is required, to conserve electric power.

Note: If the vehicle engine is operable, it is desirable to keep it running during operation of the emergency system. In this manner, the vehicle battery charge level will be reduced to a lesser degree.

REVIEW QUESTIONS

1. Which city was the first to place an elevating platform into service?
2. What is the maximum vertical reach of any elevating platform presently in use?
3. What was the first use of elevating platforms in the fire service?
4. What are some of the present uses for elevating platforms?
5. What are the three general types of elevating platforms in use in the fire service?
6. What are the three basic parts of an elevating platform?
7. How is the effective working height of an elevated platform measured?
8. What type of elevating platform provides the highest working height?
9. What is the articulated elevating platform?
10. What are the most common sizes of articulated platforms in service?
11. What is the principle of operation of the telescopic elevating platform?
12. What is the minimum size platform that meets accepted standards?
13. What is the carrying capacity of elevated platforms?
14. What is the discharge capacity of most heavy-stream appliances used on elevating platforms?
15. What are some of the features designed into platforms to provide safety for the firefighters assigned to operate the platforms?
16. What are some of the features designed into platforms to increase firefighting or rescue capability?
17. What types of stabilizing devices are used on elevating platform apparatus?
18. What is one of the major differences between the sta-

Elevated Platforms

bilizing devices used on aerial ladder trucks and those used on elevated platform vehicles?

19. What is a deadman control?

20. Where are the boom controls on elevating platform apparatus located?

21. How does the single-handle control on an articulated platform work?

22. What is an open-center hydraulic control system?

23. What are some of the unusual features of the Mack Aerialscope?

24. Describe the hydraulic system on a Mack Aerialscope.

25. How does the platform leveling system on a Mack Aerialscope work?

26. Describe the general working instructions for a Mack Aerialscope when working on grades of 4 percent or less.

27. What extra precautions must be taken when operating the Mack Aerialscope on grades exceeding 4 percent?

28. What is considered the maximum slope for operation of the Mack Aerialscope?

29. What provisions have been made for emergency operation of the Mack Aerialscope?

30. Which is the most versatile type of elevated platform?

31. Which is more important—vertical or horizontal reach of an elevating platform?

Index

accelerating pump system, carburetor, 53
accessories, pump, 317–368
aerial ladder operations, 421–465 (see also aerial ladders)
    around electrical wires, 458
    extension, 453
    at extreme angles, 459
    in high winds, 457
    hoisting, 452
    ladder pipe, 460
    lowering, 454, 456
    rotation, 453
    spotting, 449–452
    unusual, 457
aerial ladders, 195–196, 421–465 (see also aerial ladder operations)
    cables, 444–446
    construction, 425–428
    length, 422
    load limits, 454
    locks, 446
    midship-mounted, 422
    operating controls, 428–435
    rear-mounted, 422
    spring-loaded, 421
    testing, 195–196
aerial platforms, see elevating platforms
Aerialscope, 479–495
    construction, 482
    hydraulic system, 486
    master console station, 487
    operating devices, 487
    operating instructions, 490–495
    outrigger control station, 489
    platform station, 489
    stabilizing devices, 485
    water system, 485
after glow, 67
air brake lag, 229
air brake systems, 126–131, 157–159, 226
    application valves, 128
    automatic draining device, 127
    chambers, 128
    components, 126
    compressor, 126
    draining devices, 127–131
    dryer, 127
    governor, 126
    maintenance, 130
    reservoir, 127
    slack adjusters, 129
    troubleshooting, 157–159
air dryer, 127
air-fuel mixtures, 48
air leaks, 393
air locks, 371
air reservoir, 127
alternator, 83
altitude, effect of on drafting, 390
Annubar, 355
apparatus spotting, 375, 449–452
apparatus testing, 167–199
    aerial ladders, 195
    elevating platforms, 195
    pumpers, 168–195
    safety, 196–199
apparatus warning system, 364
articulated elevating platforms, 469–471, 473

Index

atmospheric pressure, 48, 286, 287
attitude, 208
automatic pressure control test, 169, 171
automatic transmissions, 106–111, 229–236
    driving tips, 231–236
        accelerator control, 231
        automatic downshifts, 110
        automatic hydraulic control, 108
        automatic upshifts, 109
        auxiliary transmission, 234
        constant mesh planetary, 108
        downshift or reverse inhibitor feature, 231
        driving on ice or snow, 234
        gearing system, 108
        hydraulic system, 109
        lockup clutch, 108
        operating in cold weather, 232
        operating procedures, 110
        parking brake, 234
        rocking out, 235
        temperatures, 235
        torque converter, 109
        towing or pushing, 234
        two-speed axle, 234
        using the engine to slow the apparatus, 232
        using the hydraulic retarder, 232
    range selector positions, 230
    troubleshooting, 160–162
automation, 358–365
auto scan digital flow meter, 356
auxiliary cooling system, 347
axle locks, 439
axles, *see* rear axles

backing up, 224
back pressure, 400, 401
back pressure gages, 461, 462
basic speed law, 247
basket leveling system, 478, 487
battery, 58, 88–89
    safety precautions, 89
    servicing, 89

bearings
    main, 25
    thrust, 26
Bernoulli, 355
blocked hydrants, 377–379
booms, 197, 469, 470, 471, 482
booster pumps, 257, 294
bottom dead center (BDC), 3, 4
Bourdon-type gages, 352
brake control, 225
brake horsepower, 34
brakes, 123–135
    air, 126–131, 226
    disc, 131
    hydraulic, 124
    linings, 123
    master cylinder, 125
    operating principles, 123
    parking, 132
    retarders, 133
    stopping distance, 227
broken connection, drafting from, 395
burst intake hose, 379

cables
    aerial ladder, 444
    ignition, 62, 72
camber, 137, 138, 139
cams, 22
camshaft, 24
carburetor, 47–54
    accelerating pump system, 53
    float system, 49
    idle system, 51
    main metering system, 49
    power enrichment system, 53
    troubleshooting, 149–151
caster, 137, 139, 140
    negative, 137
    positive, 137
    zero, 137
cavitation, 283–286
centrifugal pumps, 175, 192, 257, 266–292, 293
    cavitation, 283–286
    displacement, 269–273
    multiple-stage, 275–277
    parallel operation, 280

Index

priming, 192, 388
series operation, 278–280
series-parallel operation, 175, 277–283
series parallel variations, 289–292
single-stage, 274
theoretical capacity, 286
theory of, 267
Certification of Inspection form, 175, 176
certification tests, 169–175
cetane number, 44
chassis, 95
Chicago Fire Department, 467, 468
Chiksan swivel, 433
chock blocks, see wheel chocks
chockes, automatic, 54
Class A pumpers, 168
Class B pumpers, 168
clutch, 26, 27, 98–103, 151–152
    adjustments, 100
    brake, 102, 216
    coaxial spring damper, 100
    control, 212
    failure, 102
    misuse, 214
    operating principle, 98–99
    operation, 213
    troubleshooting, 151–152
coaxial spring damper, 100
coil, ignition, 59, 71
combination relief valve, 342
combustion, 5–7
    abnormal, 6
    detonation, 7
    normal, 6
    preignition, 7, 66
combustion chamber, 18, 30
compound gage, see intake gages, 352, 386
compression ratios, 8, 18
    diesel engines, 8
    gasoline engines, 8
compression strength, 426
compression stroke, 5, 10
condenser, 61
Condor, 468
connecting rods, 27

control pedestal, 430
coolers, oil, 75
cooling system, 75–81
    conditioner, 80
    engine coolant, 76
    fan, 80
    filter, 80
    principle of operation, 75
    radiator, 79
    thermostat, 78
    water manifold, 78
    water pump, 77
corrosion inhibitors, 81
crankshaft counterweights, 26
crankshafts, 25
cross steering levers, 136
cross steering socket assembly, 136
cross steering tube assembly, 136
curves, 240
current regulator, 85, 87
cut-out relay, 85
cylinder displacement, 12
cylinder head, 29
cylinders, 12

darting, 138
dead axles, 117
deadman control, 479, 489
defensive driver, 207
delivery tests, 177
detonation, 7
dielectric tests, 197
diesel engines, 8–11, 145–148
    compression ratios, 8
    fuel, 8
    glow plugs, 67
    operating principle, 8
    troubleshooting, 145–148
diesel fuel injection systems, 57
    distribution system, 57
    multiple-pump system, 57
    unit injection system, 58
differential, 116
disc brakes, 131
discharge formula, 195
discharge gages, 175, 184, 351, 352, 353
distributor, 60
diving, 138

double-clutching, 216, 219
downhill driving, 241
downshifting, 218
drafting, 386–396
    air leaks, 393
    from broken connection, 395
    distance water will rise, 389
    effect of altitude, 390
    effect of weather changes, 390
    lift, 387
    principle of lifting water, 388
    procedures, 391–395
    quantity of lift, 390
drag link, 136
driver characteristics, 205–210
driving conditions, 236–246
    curves, 240
    downhill driving, 241
    fog, 244
    freeway driving, 236
    intersections, 240
    night driving, 242
    one-way streets, 242
    rain, 244
    skids, 237
    uphill driving, 241
    wet weather, 244
    winter driving, 245
driving procedures, 203–252
driving techniques, 210–236
    clutch control, 212
    clutch operation, 212
    gear shifting, 215–220
        after a stop, 218
        downshifting, 218
        double-clutching, 216
        during a turn, 220
        effects of terrain, 219
        upshifting, 217
        using a clutch brake, 216
    starting procedures, 211
        diesel engines, 211
        gasoline engines, 211
    throttle control, 212
driveline, 114
drivetrain, 80, 152
driveshafts, 101, 114

eddy currents, 198
efficiency
    engine, 37
    mechanical, 37
    overall, 38
    thermal, 38
    volumetric, 36, 261
electrical systems, 81–89
    alternator, 83
    battery, 88, 89
    current regulator, 85, 87
    cut-out relay, 85
    generator, 83
    regulator (D.C. circuit), 84
    starting motor, 81
    voltage regulator, 85, 87
electric governor, 340–341
electronic control, 111
electronic ignition system, 71–72
    ballast resistor, 72
    components, 71
    control module, 71
    ignition cables, 72
    operating principle, 71
    spark plugs, 72
elevating platforms, 195–196, 467–495
    articulated type, 469–471
    basket leveling system, 478
    booms, 197, 469, 470, 471, 482
    controls, 479–482
    deadman control, 478
    features, 468, 477
    history of, 467
    Mack Aerialscope, 482–495
    platform basket, 476
    platform controls, 479
    stabilizing devices, 477
    telescopic-articulated type, 471
    telescopic type, 471
    testing, 195–196
    working height, 471
emergency response, 246–252
    aerial truck operators, 251
    apparatus not in use, 251
    apparatus speed, 247
    approaching the emergency, 250
    elevating platform operators, 254

Index

exemptions from the vehicle code, 248
idling motors, 257
positioning apparatus, 250
pumper operators, 251
response routes, 246
use of red light and siren, 248
engine company, 369
engine coolant temperature gage, 175
engines, 1–38, 43–89
compression ratio, 8, 18
construction, 11
coolant, 76
cooling fan, 80
diesel, 8–11, 145–148
displacement, 12, 31
efficiency, 37
four-stroke-cycle, 3, 9
gasoline, 2–7
oils for, 73
performance standards, 31–38
systems, 43–89
two-stroke-cycle, 2, 9, 10, 23
engine speed control, 221
escape plan, 237, 238
estimating hydrant supply, 380
exhaust manifold, 30
exhaust oxygen sensor, 69
exhaust stroke, 5, 10
extension control, 434
extension gages, 436
extension operations, 453

fan
engine-cooling, 80
thermo-modulated, 80
fifth wheel, 422
filters
coolant, 80
fuel, 47
oil, 75
Firebird, 468, 475
fire pumps, 257–311
booster, 294
capacity, 286
cavitation, 283–286
centrifugal, 266–292
displacement principles, 269–273
front-mounted, 293, 294
impellers, 266–269
main, 257, 292
monitoring devices, 351–358
piston, 262
positive displacement, 258–266
priming, 296, 320, 321
rated capacities, 293
ratings, 168, 288
rotary gear, 262–264
rotary lobe, 264
rotary, 262
rotary vane, 265
series-parallel, 277–283
slippage, 274
testing, 168, 195
theoretical displacement, 286
transmissions, 344–347
troubleshooting, 301–311
volumetric efficiency, 261
flammable limits, gasoline, 44
flash point, gasoline, 44
float system, carburetor, 49
flow meters, 354–356
fly ladder, 444
flywheel, 26, 98
fog, driving in, 244
foot plates, 441
forward pressure, 400
four-stroke-cycle engine, 3, 9
four-way valve, 404, 405, 406
frame, apparatus, 96
freeway driving, 236
friction horsepower, 35
front-mounted pumps, 293, 294
fuel
diesel, 44
filters, 47
gages, 47
gasoline, 43
injection, 10, 56–58
oil, 73
pumps, 45
systems, 44–58
tanks, 44

gage pressure, 386
gages
    back pressure, 462
    Bourdon-type, 352
    discharge, 175, 184, 351, 352, 353
    engine coolant temperature, 175
    extension, 436
    fuel, 47
    hoisting cylinder, 436
    hydraulic system pressure, 436
    intake, 351, 353, 386
    load indicator, 435
    oil pressure, 175, 318
    pressure, *see* discharge
    vacuum, *see* compound gages
gasoline engines, 2–7
gasoline fuel systems, 44–58
    carburetors, 47–56
    chokes, 54
    filter, 47
    fuel injection, 56
    pump, 45
    tank, 44
gears
    helical, 345
    spur, 345
    timing, 26
gear shifting, 215–220
    after a stop, 220
    double-clutching, 216
    downshifting, 218
    during a turn, 220
    effects of terrain, 219
    upshifting, 218
    using a clutch brake, 216
generator, 83
glow plugs, 67
governors, *see* pressure control governors
governor test, 169, 170
gross vehicle weight, 97
ground jacks, *see* stabilizing jacks

hand rotation control valve, 437
hard intake hose, 180
heat range, spark plugs, 64
heavy streams, *see* master streams
helical gears, 345

hoisting control, 433, 452
hoisting cylinder pressure gage, 436
hoisting operation, 452
hook-up procedures, 374
horsepower, 32
    brake, 34
    friction, 35
    indicated, 34
    rated, 33
    SAE, 33
hose
    hard intake, 180, 371
    intake, 180, 371, 375, 377, 379, 381
hydrants, 371–379
    blocked, 377
    estimating supply from, 380
    hook-ups, 374
    identification, 373
    installations, 372
    selecting for use, 373
hydraulic braking system, 124
hydraulic retarder, 232
hydraulic system pressure gage, 436

idle system, carburetor, 51
ignition systems, 58–72
    battery, 58
    cables, 62, 72
    coil, 59
    condenser, 61
    distributor, 60
    electronic, 71–72
    exhaust oxygen sensor, 69
    glow plugs, 67
    spark plugs, 62–67
    switch, 58
ignition temperature, gasoline, 44
impellers, 266–269
inches of mercury, 174, 321, 351, 387
inclinometer, 447
included angle, 137
indicated horsepower, 34
intake gages, 351, 353, 386
intake hose, 180, 371, 375, 377, 379, 381
intake manifold, 30
intake stroke, 4
intersections, 240, 249

# Index

jackknife, 422, 439, 447, 448, 449

Kimball, Warren, 380
knuckle pin inclination, 137, 139
knuckle pins, 136

ladder cables, 444–446
ladder locks, 446–447
ladder pipe, 460
ladder rungs, 427
leading, 138
lift, see vertical lift
lifting water, 388
live axles, 117
load indicator gage, 435
load limitations, 454
load test, 199
lock pawls, 434
lock valve, 434, 454
lock valve control, 437
lowering operations, 454, 456
lubrication systems, 73–75
    engine oils, 73
    oil changes, 74
    oil coolers, 75
    oil filters, 75
    oil pumps, 75
lugging an engine, 221

magnetic particle inspections, 197
main bearings, 25
main metering system, carburetor, 49
main pumps, 257, 259
main pump transmissions, 344–347
manifold
    exhaust, 30
    intake, 30
    water, 78
manual transmissions, 104
Manufacturer's Record of Pumper Construction Details, 172
master control console, 488
master cylinder, brakes, 125
master streams, 412–416
mean effective pressure, 34
mechanical advance, 61
mechanical efficiency, 37

mercury, inches of, see inches of mercury
midship-mounted aerial ladders, 422
monitoring devices, 351–358
motor, starting, 81
multi-stage centrifugal pumps, 275–277

National Board of Fire Underwriters, 168
National Fire Protection Association, 168, 174, 195, 196
    Standard 1901, 168, 174
    Standard 1904, 195, 196
net pump pressure, 168, 177, 180, 190, 287, 288
New York Fire Department, 482
night driving, 242
nozzle reaction, 461

octane rating, 43
oil, 73–75
    changes, 74
    coolers, 75
    filters, 75
    fuel, 73
    pan, 31
    properties, 73
    pumps, 75
    viscosities, 73
oil pressure gage, 175, 318
one-way streets, 242
operator's control pedestal, 430
outrigger control station, 489
outriggers, 441, 447, 478
overall efficiency, 38
overload test, 169, 171

parking brakes, 132, 234
penetrant inspections, 197
perception time, 227
performance standards, engine, 31–38
permanent pump shutdowns, 385
Philadelphia Fire Department, 468
piston pumps, 262
piston rings, 15–18
    compression, 15
    oil control, 15, 17

pistons, 12–15
　clearance, 14
　displacement, 12
　pins, 13, 29
　rings, 15–18
　skirts, 13
　slap, 14
Pitman arm, *see* steering gear lever
pitot tube, 183, 184
pivot centers, 136, 140
platforms, 195, 196, 469, 471, 476
positive displacement pumps, 258–266
　oil pumps, 75
　priming pumps, 296
　theoretical displacement, 258
　theory of, 258–261
power enrichment system, carburetor, 53
power stroke, 5, 10
power take-off, 428
power train, 98–114
　axles, 117–123
　clutches, 98–102
　driveline, 114–116
　transmissions, 103–114
pre-glow, 67
preignition, 7
pressure
　absolute, 386
　atmospheric, 386
　back, 400
　forward, 400
　relative, 386
pressure computer, pump, 364
pressure control devices, 327–344
pressure control governors, 335–342
pressure gages, *see* discharge gages
priming
　devices, 320–327
　operations, 325
　pumping, 320–327
　tank, 325
　valve, 322
Prony brake, 34
propeller shafts, 115
pump capacities, 168
pumpers (*see also* pumping operations)
　Class A, 168
　Class B, 168
　rated capacity, 168
　requirements, 168
　testing, 167–195
　　certification test, 169–175
　　delivery test, 177
　　service test, 178–195
　triple-combination, 369
pumper test pits, 179
pumping operations, 382–384
　drafting, 386–396
　permanent shutdowns, 385
　techniques, 409–412
　temporary shutdowns, 384
pumping variables, 190
pump operator panel, 318
pumps
　carburetor accelerating, 53
　fire, *see* fire pumps
　fuel, 45
　oil, 75
　slippage, 261, 274
　troubleshooting, 301–311
　water (engine), 77

quantity of lift, 390
Quinn, Robert J., 467

racing an engine, 221
radiator, 79
radiator fill system, 347
radio-controlled automatic pumper, 359–364
radio-controlled nozzles, 361
radiographic inspections, 198
rain, driving in, 244
rated capacity, pumper, 168
rated horsepower, 33
reaction time, 227
rear axles, 117–123
　dead, 117
　live, 117
　lockout mechanism, 122
　tandem-drive, 120–123
　two-speed, 117
rear-mounted aerial ladders, 422, 423

Index

red lights, use of, 248
regulators
    current, 85, 87
    D.C., 84
    voltage, 85, 87
relay operations, 396–409
relay standards, 402
relief valves, 328–335
    combination, 342
    suction, 335
response routes, use of, 242
retarder, 232
revolution counter, 185, 194
road shock, 138
rocking out, 235
rotary gear pumps, 262–264
rotary lobe pumps, 264
rotary vane pumps, 265
rotation control, 435
rotation lock control, 437
rotation operations, 453
rung alignment indicator, 437
ruptured lines, 406

SAE horsepower, 33
S-cam brake system, 129
screw jacks, 441
selecting a hydrant, 373
self-centering levers, 432, 486
series-parallel centrifugal pumps, 277–283
service tests, 178–195
shifting after a stop, 220
shifting during a turn, 220
shimmy, 138
shutdowns, 385
single-stage centrifugal pumps, 274
siren, use of, 248
skids, 237
slack adjusters, 129
slip joints, 114
slippage, pump, 261, 274
Society of Automotive Engineers (SAE)
    horsepower, 33
    oil viscosity, 73
sodium-cooled valves, 22
spark plugs, 58, 62–67
    conditions, 67

    construction, 63
    electronic system, 72
    heat range, 64
    misfiring, 63, 67
Spicer coaxial damper, 100
spotting of apparatus, 375, 449
spring-load aerial ladders, 421
spur gears, 345
stabilizing devices, 438–444, 477
    axle jacks, 439
    outriggers, 441, 478
    stabilizing jacks, 440
    wheel chocks, 439
stabilizing jacks, 440
Standard 1901, 168, 174
Standard 1904, 195, 196
Stang "Intelligiant" monitor, 347
starter button, 437
starting motors, 81
static balance, 137
steering, 135–141, 153–157
    hydraulic power, troubleshooting, 154–157
    manual, troubleshooting, 153–154
steering control, 222
steering error, 137
steering gear lever, 136
steering gear shaft, 136
steering geometry, 137, 138
steering knuckle, 136
steering knuckle spindle, 136
steering lever, 136
stopping distance, 227
suction gages, see vacuum gages
suction hose, 181, 372
suction relief valve, 335
Suggested Specifications for Motor Fire Apparatus, 168
switch, ignition, 58
systems
    carburetor, 47–54
    cooling, 75–81
    electrical, 81–89
    engine, 43–89
    fuel, 43–56
    fuel injection, 56–58
    ignition, 58–72
    lubrication, 73–75

tachometer, 170, 185
tandem drive axles, 120–123
tanks
    gasoline, 44
    operations, 370
    priming, 325
    water, flow test, 172
telescopic-articulated elevating platforms, 471, 475
telescopic elevating platforms, 471, 474
temporary pump shutdowns, 384
"ten and two" position, 222
tension strength, 426
testing, *see* apparatus testing
theoretical lifts, 389
thermal efficiency, 38
thermo-modulated fan, 80
thermostats, 76, 78
throttle button, 438
throttle sensor, 111
tillerman, 423, 452
timing gears, 26
tire wear, 138, 139
toe-in, 137
toe-out, 140
top dead center (TDC), 4, 5
torque, 35
torque converter, 107
towing, 234
tractor-trailer aerial ladders, 422, 423
tramp, 138
transfer valves, 281, 282
transmissions, 101, 103–114, 160–161, 344–347
    automatic, 106–114
    electronic control, 111
    main pump, 344–347
    manual, 104
    synchronizing devices, 106
    troubleshooting (automatic), 160–161
triple-combination pumper, 369
troubleshooting
    air brakes, 157–159
    automatic transmission, 160–161
    carburetor, 148–151
    clutches, 151–152
    diesel engines, 145–148

    drive train, 152
    fire pumps, 192–194, 301–311
    hydraulic power steering, 154–157
    manual steering, 153–154
truss construction, 425, 426
turning angle, 137
turning radius, 137
turntables, 422, 429, 449, 469
twisting force, 427
two-speed axles, 117, 234
two-stroke-cycle engines, 3, 9, 10, 23

ultrasonic inspections, 197
Underwriters Laboratories, Inc., 169, 174, 176
universal joints, 114, 115, 116
uphill driving, 241
upshifting, 217

vacuum, (see *also* inches of mercury), 321, 386
vacuum advance, 61
vacuum gages, 175
vacuum test, 169, 173
valve lifters, 23
valves, 19–22
    cams, 22
    exhaust, 19, 22
    intake, 19, 22
    lifters, 23
    sodium-cooled, 22
vehicle code, 248
venturi, 49, 51
vertical lift, 387
vibration damper, 26
viscosities, oil, 73
voltage regulator, 85, 87
volumetric efficiency
    of engines, 36
    of pumps, 261
volute, 269

wandering, 138, 139
warning systems, 357
water manifold, 78
water pump, 77
water tank flow test, 169, 172–173
water tower, 448

Index

weather, effect of on drafting, 390
wedge brake system, 129
wet weather, driving in, 244
wheel balance, 137

wheel chocks. 439
wheel fight, 138
winter driving, 245
wrist pins, 13, 29